MR Spectroscopy

CLINICAL APPLICATIONS AND TECHNIQUES

MR Spectroscopy

CLINICAL APPLICATIONS AND TECHNIQUES

Edited by

Ian R Young PhD
Professor, The Robert Steiner Magnetic Resonance Unit
Royal Postgraduate Medical School
Hammersmith Hospital
London, UK

H Cecil Charles PhD
Associate Professor of Radiology
and Chief, Center for Advanced MR Development
Department of Radiology
Duke University Medical Center
Durham NC, USA

MARTIN DUNITZ

© Martin Dunitz Ltd 1996

First published in the United Kingdom in 1996 by Martin Dunitz Ltd, The Livery House, 7–9 Pratt Street, London NW1 0AE

All rights reserved. No part of this publication may be reproduced, stored in a retrieval system, or transmitted, in any form or by any means, electronic, mechanical, photocopying, recording or otherwise, without the prior permission of the publisher or in accordance with the provisions of the Copyright Act 1988, or under the terms of any licence permitting limited copying issued by the Copyright Licensing Agency, 33–34 Alfred Place, London WC1E 7DP.

A CIP catalogue record for this book is available from the British Library

ISBN 1-85317-068-2

For full details of all Martin Dunitz titles, please write to:

Martin Dunitz Ltd
The Livery House
7–9 Pratt Street
London NW1 0AE
UK

Composition by Keyword Typesetting, Wallington, Surrey, United Kingdom
Printed and bound in Great Britain by University Press, Cambridge

For our wives, Sylvia and Mary

Table of Contents

List of contributors viii
Preface x
Abbreviations xii

1 MRS of the brain - prospects for clinical application 1
  James W Prichard

2 The whole body NMR spectroscopy examination 27
  H Cecil Charles

3 Problems in making useful measurement of *in vivo* spectra 41
  Ian R Young, Martyn Paley

4 MRS of muscle 55
  Peter A Martin, Henry Gibson, Richard H T Edwards

5 Human cardiac NMR spectroscopy 75
  Paul A Bottomley

6 Paediatric applications of MRS 93
  David G Gadian, J Helen Cross, Alan Connelly

7 Clinical MRS of the liver and spleen 105
  Chitta R Paul, Barton M Milestone, Robert E Lenkinski

8 Clinical applications of brain MRS 139
  Douglas L Arnold, Paul M Matthews

9 Practical aspects of localized *in vivo* ^{1}H NMR spectroscopy and spectroscopy of imaging of the human brain 161
  Jan A den Hollander, Jan Willem C van der Veen, Peter R Luyten

10 MRS in transplantation   175
*Simon D Taylor-Robinson, Maria L Barnard,
Claude D Marcus*

11 MRS in clinical oncology, with particular reference to applications to monitor therapy   211
Wolfhard Semmler, Peter Bachert, Gerhard van Kaick

Index   247

Contributors

Douglas L Arnold MD, Associate Professor, MR Spectroscopic Unit, Department of Neurology & Neurosurgery, Montréal Neurological Institute, Montréal PQ H3A 2B4, Canada.

Peter Bachert PhD, Leiter der Physikalischen MR-Arbeibgruppe, Abteilung Biophysik und Medizinische Strahlenphysik, Forschungsschwerpunkt Radiologie, Deutsches Krebsforschungszentrum (DKFZ), D-69120 Heidelberg, Germany.

Maria L Barnard BSc MB BS MRCP, The Robert Steiner MR Unit & Department of Medicine, Royal Postgraduate Medical School, Hammersmith Hospital, London W12 0NN, UK.

Paul A Bottomley PhD, Russell H Morgan Professor, Department of Radiology, Johns Hopkins University, Baltimore MD 21287–0843, USA.

Alan Connelly PhD, Radiology and Physics Unit, Institute of Child Health, London WC1N 1EH, UK.

J Helen Cross MB ChB, Neurosciences Unit, Institute of Child Health, London WC1N 1EH, UK.

Jan A den Hollander PhD, Professor of Cardiovascular Disease, Center for NMR Research and Development, University of Alabama, Birmingham AL 35294, USA.

Richard H T Edwards BSc MB BS PhD FRCP, Professor and Head of the Department of Medicine, Director of the Magnetic Resonance Research Centre and of the Muscle Research Centre, The University of Liverpool, Liverpool L69 3BX, UK.

David Gadian DPhil, Professor of Biophysics, The Royal College of Surgeons Unit of Biophysics, Institute of Child Health, London WC1N 1EH, UK.

Henry Gibson BSc MSc PhD, Honorary Non-Clinical Lecturer in Medicine, Department of Medicine, University of Liverpool, Liverpool L69 3BX, UK.

Robert Lenkinski PhD, Departments of Biochemistry and Biophysics of Radiology, University of Pennsylvania, Philadelphia, USA.

Peter R Luyten PhD, MR Clinical Science, Philips Medical Systems, 5680 DA Best, The Netherlands.

Claude D Marcus MD, The Robert Steiner MR Unit, Royal Postgraduate Medical School, Hammersmith Hospital, and Service de Radiologie Pr Menanteau, Hôpital Robert-Debré, Reims, France.

Peter A Martin BSc PhD, Luminary Programs, GE Medical Systems, 78533 Buc, France.

Paul M Matthews MD DPhil, Assistant Professor, MR Spectroscopic Unit and Neurometabolic Disease Laboratory, Department of Neurology & Neurosurgery, Montréal Neurological Institute, and Department of Human Genetics, McGill University, Montréal PQ H3A 2B4, Canada.

Barton M Milestone MD, Associate Professor, Department of Radiology, Temple University, Philadelphia PA 19140, USA.

Martyn N J Paley PhD, Senior Lecturer, Department of Neurology, University College, London Hospitals NHS Trust, London W1N 8AA, UK.

Chitta R Paul PhD, Research Associate, Department of Radiology, University of Pennsylvania, Philadelphia PA 19140, USA.

James W Prichard MD, Professor of Neurology, Department of Neurology, Yale University School of Medicine, New Haven CT 06510, USA.

Wolfhard Semmler MD PhD Dipl Phys, Privatdozent, Direktor und Geschäftsführer, Institut für Diagnostikforschung GmbH an der Freien Universität Berlin (IDF), D-14050 Berlin-Charlottenburg, Germany.

Simon D Taylor-Robinson MB BS MRCP, The Robert Steiner MR Unit and Department of Gastroenterology, Royal Postgraduate Medical School, Hammersmith Hospital and The University Department of Medicine, Royal Free Hospital, London NW3 2QG, UK.

Jan Willem C van der Veen PhD, Center for NMR Research and Development, University of Alabama at Birmingham, Birmingham AL 35294, USA.

Gerhard van Kaick MD, Professor und Direktor der Abteilung Onkologische Diagnostik und Therapie, Forschungsschwerpunkt Radiologie, Deutsches Krebsforschungszentrum (DKFZ), D-69120 Heidelberg, Germany.

Preface

Whole body magnetic resonance spectroscopy (MRS) is a unique, non-invasive tool with which to probe tissue metabolism in a clinical setting. In contrast with morphological imaging which shows macroscopic tissue structure, it allows investigation to look inside cell populations at 'steady state' metabolic processes. Scientists and clinicians are able to sample this metabolic information by using largely standard clinical MR machines, with spatial localization assisted by magnetic resonance imaging (MRI). Because of the non-invasive nature of the method, it is easy to monitor a variety of very different interventional procedures in patients, yielding 'on-line' information about the metabolic status of the pathologies being treated, even while therapy is under way. Access to such metabolic information has been utilized by a number of interdisciplinary groups to study a large variety of disease processes.

As of today MRS provides a tool to monitor tissue energy utilization as well as providing various other markers of its functional behaviour. In contrast to other approaches to metabolic imaging, MRS can be performed during the same study as imaging so that there is intrinsic spatial registration between its data and that of high resolution MRI, eliminating the practical problems of inter-technology, or study, image registration. The non-invasive nature of the method allows serial sampling of a region following an intervention, as it were allowing multiple 'biopsies' of the same tissue.

Over the past few years a very significant body of work has been published concerning *in vivo* MRS. This book is conceived as an update on that part of the work which is related to the aim of achieving clinical utilization of MRS. It is aimed primarily at the clinicians, scientists and technologists who have an interest in the current status of MRS, both to become better informed, and because they are contemplating working on it, as well as those currently practising in the field. The reader is not assumed to be an expert in MRI or MRS or even a practitioner of these arts. Background information on the basis of MRS is provided from a biological as well as a technical perspective.

This book is intended to provide a background and firm starting point for the clinicians with interests in clinical spectroscopy, particularly as they are approaching issues such as whether they should actually start work, and, if so, in what area, and with what nuclei and techniques. Nevertheless, it is not designed to be a 'how-to' handbook or an inclusive physics or biochemistry reference book for MRS. It does, however, address relevant and practical questions such as:

What does MRS have to offer the clinician? What are the practical constraints of clinical MRS?

The general format is organized to flow from technology to clinical study and is arranged by anatomic regions (much as one might do in an atlas) to facilitate reading by those with specific interests (eg. in cardiac spectroscopy). While the contributors to this volume are primarily spectroscopists rather than imaging scientists, the authors are well aware of the need for image guidance of MRS and have focused primarily on such studies and the relevance they bear to clinical practice.

We are very grateful to our contributors for their efforts. Putting together substantial chapters is no small task, and they have responded magnificently to our requests. We are very grateful too to Alan Burgess at Martin Dunitz for his tenacity in shepherding the flock of devious and somewhat dilatory sheep in the right direction.

We should like to record the debt of all *in vivo* spectroscopists to George Radda CBE FRS and his Oxford group who pioneered so much of this area of science.

Finally we are grateful for the support of our colleagues and families in allowing us to get on with what, to many of them, must have seemed an entirely superfluous activity.

Ian R Young
London, UK

H Cecil Charles
Durham NC, USA

Abbreviations

5-FU	5-fluorouracil	EM	electron microscopy
5-FUranuc	5-FU nucleosides and nucleotides	EPI	echoplanar imaging
ADP	adenosine diphosphate		
AFP	adiabatic fast passage pulse	FBAL	α-fluoro-β-alanine
Asp	aspartate	FDG	2-fluoro-2-deoxy-glucose
ATP	adenosine triphosphate	FID	free induction decay
		FOV	field of view
bs	body surface	FROGS	fast rotating gradient spectroscopy
BCNU	1,3-bis-(2-chloroethyl)-1-nitrosourea	FUPA	α-fluoro-β-ureido propanoic acid
BPH	benign prostatic hyperplasia		
		GABA	γ-aminobutyric acid
CAD	coronary artery disease	Gln	glutamine
CEB	calcium entry blockers	Glu	glutamate
CHESS	chemical shift selective	GPC	glycerophosphocholine
Cho	choline-containing compounds	GPE	glycerophosphoethanolamine
CK	creatine kinase	GVHD	graft-versus-host disease
CMAP	compound muscle action potential	HCM	hypertrophic cardiomyopathy
COPE	centrally ordered phase encode	HEP	high energy phosphate
Cr	creatine	HFF	high frequency fatigue
CsA	cyclosporin A	HPLC	high pressure liquid chromatography
CSF	cerebrospinal fluid		
CSI	chemical shift imaging	HRBP	heart-rate-blood-pressure
CT	computed tomography	HSVD	Hankel singular
Ct	creatinine		
		Ins	inositols
DCM	dilated cardiomyopathy	ISIS	image selective *in vivo* spectroscopy
DHFU	5-fluoro-5,6-dihydrouracil		
DRESS	depth resolved selective spectroscopy	INR	international normalized ratio
		ITU	intensive therapy unit
DWI	diffusion-weighted imaging		
		Lac	lactate
EC	EuroCollins	LFF	low frequency fatigue
ECT	electro-convulsive therapy	LOCUS	localization of unaffected spins

LVH	left ventricular hypertrophy	PME	phosphomonoester
MAST	motion artifact suppression technique	PNF	primary non-function
		POCE	proton observe-carbon edited
MDP	methylene diphosphate	PRESS	pixel resolved spectroscopy
MESA	multiple echo selective acquisition		
MRA	magnetic resonance angiography	rf	radiofrequency
MR	magnetic resonance	RFZ	rotating frame zeugmatography
MRI	magnetic resonance imaging	ROPE	respiratory ordered phase encode
MRS	magnetic resonance spectroscopy		
MS	multiple sclerosis	SDAT	senile dementia of the Alzheimer's type
NA	N-acetyl	SE	status epilepticus
NAA	N-acetylaspartate	SHR	signal-height ratio
NAAG	N-acetyl-aspartyl-glutamate	SI	spectroscopic imaging
NAD	nicotinamide adenine dinucleotide	STEAM	stimulated acquisition mode
NADH	reduced NAD	STIR	short Tau inversion recovery
NDP	nucleotide diphosphate		
NTP	nucleoside triphosphate	T	tesla
NHL	non-Hodgkin's lymphoma	TAG	triacyl glyceride
NMR	nuclear magnetic resonance	TAN	total adeninine nucleotides
nOe	nuclear Overhauser enhancement	tCho	total choline (free choline, PC and other choline-containing metabolites) in ^{1}H MRS
NTP	nucleotide triphosphates		
NYHA	New York Heart Association		
		TE	echo time
OCT	ornithine carbamoyl transferase	TMR	topical magnetic resonance
		TNR	tumour necrosis factor
PC	phosphocholine	TR	repeat time
PCr	phosphocreatine		
PDE	phosphodiester	UKTSSA	United Kingdom Transplant Support Services Authority
PE	phosphoethanolamine		
PET	positron emission tomography	UNOS	United Network for Organ Sharing
PFK	phosphofructokinase		
Pi	inorganic phosphate		
PIQABLE	peak identification, quantitation, and automatic baseline estimation	VOI	volume of interest
		voxels	volume elements
PL	phospholipids		

MRS of the brain – prospects for clinical application

James W Prichard

INTRODUCTION

Magnetic resonance spectroscopy (MRS) is a term in general, if imprecise, use in the biomedical world to designate NMR observations of any nuclei except the water protons that are the basis of magnetic resonance imaging (MRI). Used in a strictly logical way, MRS would include MRI, which is the spectroscopy of water. However, the convenience of different acronyms for observations based on a single very strong NMR signal (MRI) and those based on a wide range of much smaller signals (MRS) will not be sacrificed here in a lonely campaign for terminological precision. The relationship between MRI and MRS is worth considering for an entirely different reason, which is quite pertinent to this chapter: NMR methods as a whole are becoming so valuable in medicine that the time has arrived to think carefully about the combinations in which they can be used most efficiently.

Every possible NMR measurement cannot be made on every patient every time. 'Conventional' MRI (T_1/T_2/proton density) already presents the medical imaging expert with a daunting array of choices for best demonstration of lesions that have known contrast properties. Congenital, ischemic, neoplastic, and inflammatory lesions and their subdivisions all have slightly different properties that can be exploited to differentiate them from each other, if imaging time is available within the limits imposed by patient tolerance and administrative fiat.

Now – as if with no regard for the plight of the harried expert – come the additional demands of magnetic resonance angiography (MRA), diffusion-weighted imaging (DWI), and imaging of task-activated brain areas detected through local changes in deoxyhemoglobin concentration ('functional neuroimaging' in the absence of a sufficiently compelling acronymn). Echoplanar imaging (EPI) and other fast imaging methods appear likely to reduce motion artifact to the point that *all* of these ways of using the water signal can be applied to heart, bowel, liver, kidneys, and joints, with time resolution in the sub-second range. All this just with the water signal.

And on top of all that, MRS. Out of its infancy but still in its toddler years in 1994, MRS already offers signals from more than two dozen compounds in the brain and invites regard as a 21st-century kind of neuropathology. Years will be needed to reach so grand a goal, because small signals are hard to get. Reasons for their routine use in a clinical setting have to be strong enough to justify the time taken to get them, which is inversely proportional to the square of signal intensity and therefore increases rapidly as they become small. The range of acquisition times involved is quite large: a single two-dimensional slice through the brain with good anatomical resolution can be obtained by EPI in less than 100 milliseconds, while a two-dimensional lactate image adequate to show elevated lactate in a brain tumor took about 30 minutes in 1993. Nevertheless, metabolite maps have such strong appeal to the imagination and so much potential for systematic quantitative analysis that methods for producing them – known collectively as 'spectroscopic imaging' (SI) – are being improved rapidly. Examples appear later in this chapter.

In the mid-1990s, a major problem for everyone working with medical application of NMR methods is to decide which of the now wide array of possible measurements should be made in each clinical situation. Some choices are already obvious – all patients presenting with stroke syndrome should have DWI. But even in stroke, the optimum combination of measurements depends on the stage of the lesion and strategic questions such as whether the test is being done to estimate rehabilitation potential or to rule out hemorrhage prior to anticoagulation or reperfusion therapy. Only one thing is certain: the optimum combination will vary with each kind of lesion and often among individual patients with the same lesion. Finding the optimum will require an unprecedented degree of case-by-case co-operation between clinical imaging experts (now usually called diagnostic radiologists) and other clinical experts identified by the kind of disease they treat or the way they treat it (now usually called neurologists, internists, psychiatrists, oncologists, surgeons, therapeutic radiologists, etc). Much conceptual reorganization of medicine will come from this process. It will be precipitated by the unique versatility of non-invasive NMR methods, which in the perspective of medical history are likely to be comparable in effect to antibiotics and molecular biology.

The job of this chapter is to present data that will help the imaging expert to judge how near various kinds of MRS measurements are to routine clinical application. It deals exclusively with MRS observations on diseases of the brain, both because that is where the author's expertise lies and because technical factors have favored MRS work on that organ relative to others: motion is a relatively smaller problem, shimming is less disturbed by the mass of nearby tissue as well as by motion, and lipids interfere less with ^{1}H MRS because most of them in their normal state are parts of myelin or other large molecules that are not mobile enough to give narrow signals.

Published MRS data on brain diseases through 1993 are abundant enough to demonstrate beyond doubt that the chemical specificity and non-invasiveness of the technique will allow metabolic fingerprinting of pathophysiological processes throughout the natural history of a wide variety of diseases. Together with MRI, MRS is certain to improve understanding of many diseases and become central to guiding management of some of them.

However, because of the price that must be paid in time taken from other diagnostic procedures – or from therapeutic ones – when small MRS signals are acquired in sufficient detail to be useful, no MRS measurement on the brain can be considered essential for routine diagnosis as of early 1994. That is likely to change as techniques improve and research workers identify clinical situations in which MRS information is of critical importance for guiding care of individual patients.

An extensive clinical literature of ^{31}P MRS of muscle exists and can be studied with profit by anyone interested either in muscle physiology and disease or in exploring the full reach of current clinical MRS methods. Access to this work is available through reviews.[1,2]

STROKE

Stroke accounts for about half of all brain disease and extracts a heavy toll from society in lost productivity and in consumption of resources for management of disability that often lasts for years. It is a natural place to look for benefits from new biomedical technology. X-ray computed tomography (CT scanning) greatly advanced understanding of stroke by its unequivocal distinction between hemorrhage and infarction, which is far less clear on clinical grounds than clinicians had thought prior to introduction of CT in the early 1970s. As recounted in other chapters of this volume, MRI quickly added new insights and clinical utility to imaging science from the late 1970s on, as it brought posterior fossa structures under routine scrutiny, detected large volumes of clinically silent neuropathology in multiple sclerosis, delineated ischemic pathology in more detail than CT scanning can, and provided in DWI the most definitive marker of fresh brain ischemia available from any technique.

By late 1993, MRS observations on human stroke and related experimental work on animal models had outlined well enough what measurements are possible to allow initial assessment of the role MRS may come to play in patient management. The lactate and N-acetyl (NA) resonances in the ^{1}H spectrum and intracellular pH (pHi) derived from the ^{31}P spectrum appear to have the best prospects for clinical utility.

^{1}H MRS and human stroke

Figure 1.1 shows ^{1}H spectra acquired from a patient who had suffered multiple cerebral infarctions two weeks before MRS examination.[3] In the infarcted areas, loss of NA (NAX in the figure) and elevation of lactate are clearly evident. Loss of the NA signal is expected when brain is infarcted and loses neurons, which are thought to be the only cells in mature brain that contain N-acetylaspartate (NAA), the principal source of the NA resonance. Animal experimentation shows that the NA signal can begin to fall as early as two hours after onset of focal infarction.[4] It was depressed in both the acute and chronic periods in 6 patients studied at least twice,[5] and it disappeared almost completely between 1 and 9 days after infarction in another patient studied serially.[6] In 10 patients studied by ^{1}H MRS within the first 60 hours after stroke, the NA signal in the lesion was reduced relative to the contralateral hemisphere in all but two; repeat examination of 7 patients 8–17 days after stroke onset showed additional decline at an overall rate calculated at $-29 \pm 9\%$ per week.[7] Reports from other groups are consistent with these observations.[8,9]

To the extent that continued NA decline over the first two weeks after stroke does not merely reflect clearing of debris from neurons killed at the outset, it is evidence of continuing neuronal loss after the acute period, possibly in the ischemic penumbra.[10] Therapy effective at preventing such delayed neuronal loss would presumably also reduce eventual fixed deficit. Monitoring of the NA signal in the subacute poststroke period could serve as a surrogate end point for therapeutic trials aimed at achieving such a result.

Lactate elevation by stroke is a large signal change that is readily detectable by ^{1}H MRS (Fig. 1.1). The earliest report of it in human

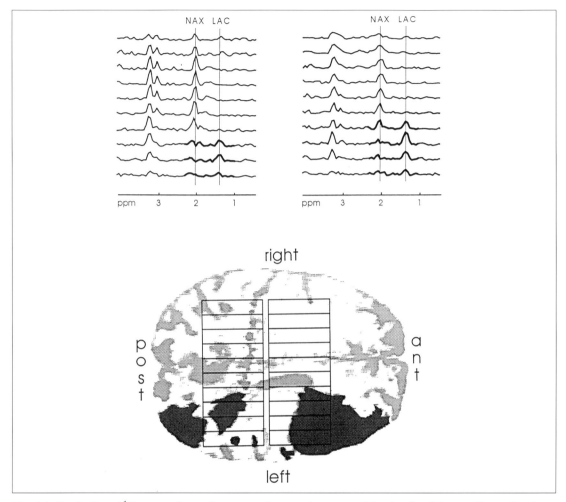

Fig. 1.1. Stack-plots of ^{1}H spectra from adjacent voxels of two volumes of interest in the brain of a patient with infarcts of the left frontal and parietal lobes. The brain image cartoon was made from a single axial view from a set of eight multislice, T_2-weighted MRIs (TR 2.05 sec; TE 94.8 msec; slice thickness 0.4 cm). It shows the location of the infarcts (dark shaded areas) and the 20 voxels from which spectra were obtained. Voxels in each stack correspond to spectra in the same relative vertical position in the spectral stack-plots. The stack of ^{1}H spectra on the left was obtained from the posterior (post) volume of interest, the right one from the anterior (ant) volume. A chemical shift axis in parts per million (ppm) appears below each stack. Lactate (LAC) is the main source of the signal at 1.3 ppm. N-acetylated compounds (NAX) including N-acetylaspartate are the main compounds contributing to the signal at 2.0 ppm. Creatine and trimethylamines are the sources of the signals at 3.0 and 3.2 ppm, respectively. The spectra are heavily T_2-weighted (STEAM sequence: TM 76 msec, TE 270 msec, TR 4 sec, operating frequency 89 MHz) to attenuate lipid signals with short T2s. The echo time TE (270 msec) was chosen to optimize the lactate doublet, which has a 7.4 Hz J-coupling constant. Each spectrum was obtained from a 4.8 ml voxel measuring 3 cm in the anterior–posterior, 0.8 cm in the transverse, and 2 cm in the superior–inferior directions. (Reprinted with permission from Petroff et al.[3]

stroke[11] implied that stroke-associated lactate could remain elevated for months after stroke onset. Despite initial skepticism by many observers, including the present writer, this proved to be correct. Serial study of the same patients showed that lactate elevation usually persists for weeks after stroke and may last for months,[5] and observations of other groups are consistent (Fig. 1.2).[8,9]

More than one mechanism must operate during so long a period of lactate elevation. Within minutes of losing its blood supply, an infarcted region of brain accumulates lactate to a concentration of 15–30 mM, as glucose and glycogen in the unperfused tissue are metabolized in the absence of oxygen, which is exhausted in the first few seconds. If the region is never reperfused and dies, the lactate in it can dissipate only by diffusion, but the process would not take weeks to months. Other sources of elevated lactate that might be associated with infarction include:
- adjacent regions of surviving but impaired tissue (the ischemic penumbra);[10]
- infiltrating cells involved in the tissue's reaction to injury; and
- altered metabolism of surviving brain.

MRS has provided circumstantial evidence for the second of these:[3] a patient was studied two weeks after a stroke and died a week later, at which time autopsy verified the presence of large numbers of macrophages in the regions where ^{1}H MRS had shown elevated lactate. The role of altered metabolism remains conjectural; the long-term metabolic response of surviving brain tissue is not well understood.

Thus, at various times after infarct onset, lactate concentrations above those of normal brain might be present in any of at least four compartments:
- killed tissue;
- an ischemic penumbra;
- infiltrating cells; or
- metabolically altered brain tissue.

In human stroke, however, only the first two of these sources would be present in the first 24–48 hours after onset, the time when initial diagnostic procedures are done and treatment started.

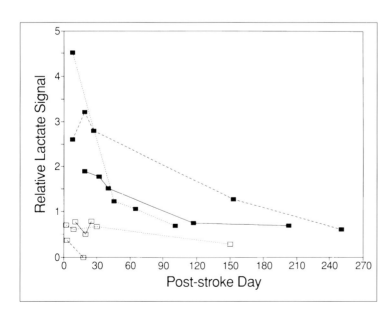

Fig. 1.2. Time course of lactate associated with infarct in six patients studied by ^{1}H MRS more than once. Each point represents a separate study; lines connect points from the same patient. In each study after the first, the lactate signal was taken from the region in which it was largest in the first study; it is plotted as its ratio to the NA signal in an unifarcted region of the same brain on the same day. (Reprinted with permission from Graham et al.[5])

Infiltrating cells are not numerous until around 72 hours, and changes in relative populations of native brain cells – the likeliest mechanism of intrinsic metabolic alteration – would require days or longer to develop. Measurement of the lactate signal in the first 24–48 hours can in principle provide important information about subsequent evolution of the infarct.

Lactate in killed brain is metabolically inert; lactate in an ischemic penumbra turns over, although perhaps at a different rate than in normal brain tissue. The anatomical boundary between pools of lactate in these two states need not be sharp, but their relative sizes reflect the proportion of killed to surviving tissue at a time when no other currently available measurement can do so. Metabolically active and metabolically inert lactate can be distinguished from each other by labeling with ^{13}C from blood glucose, as described in the last section of this chapter.

If that were done as soon as a patient with stroke syndrome reached a hospital, the treatment team would learn in less than two hours whether they were dealing with a large mass of killed tissue likely to soften and become necrotic over the next few days, a mass of impaired tissue still able to recover, or some combination of these. The knowledge could well be critical for decisions between aggressive therapies that increase the risk of hemorrhagic transformation and more conservative ones.

However, labeling of stroke-elevated brain lactate from 1-^{13}C-glucose would cost several hundred dollars per patient and entail a small risk of worsening the situation by elevating blood glucose at a time when the patient might have a second stroke. The panoply of modern NMR techniques includes some simpler ones that may provide similar prognostic information: DWI is the most promising of these.[12] Until the simpler procedures have been more completely evaluated in clinical trials, labeling of stroke-elevated lactate with ^{13}C is unlikely to become routine in diagnostic evaluation of stroke.

Other characteristics of stroke-elevated lactate apart from its metabolic turnover rate may be useful for understanding pathophysiology and for guiding therapy. There has appeared a preliminary report of retrospective correlation between intensity of the lactate signal in the voxel where it was most intense in the first days after stroke onset and clinical outcome as judged by the Barthel index at discharge several weeks later.[13] The correlation is the first of its kind, and it is encouraging because the lactate measurement can be refined considerably if the information it provides justifies the effort. A step in that direction has been taken in studies of relaxation times of the NA, choline, and total creatine signals in stroke patients, which were not different from age-matched controls.[14,15] The implication of the finding is that the increase in lactate signal intensity associated with stroke is due to increased concentration of lactate rather than to a change in its relaxation properties.

The comparison is hard to make for lactate itself because of its low concentration in normal brain. One of the studies cited above also reported relaxation times of stroke-associated lactate: at 2250 ± 610 (sem) msec, T1 was in the same range as the other metabolites, while T2 was 780 ± 257 msec, 2–3 times longer.[15] Until lactate relaxation properties have been measured in normal brain, the possibility that they influence the appearance of ^{1}H spectra in stroke must be held open. One should also note that the error ranges of existing measurements are rather large; improved techniques may yet lead to discovery of relaxation property changes that figure in the pathophysiology of metabolites observable by ^{1}H MRS in stroke.

^{31}P MRS and human stroke

The phosphorus nucleus is considerably less sensitive than the proton. In consequence, ^{31}P spectra take longer to acquire and have coarser anatomical resolution than ^{1}H spectra. Nevertheless, ^{31}P MRS has been done successfully in human stroke. The first report was of normal metabolite ratios with reduced total phosphorus signal in chronic stroke, consistent with replacement of infarcted tissue by cerebrospinal and interstitial fluid.[16] A group at Henry Ford Hospital in Detroit developed the capability to study acutely ill patients – itself a nontrivial achievement – and used it to monitor ^{31}P changes from the acute to the subacute period.[17] Acutely, ATP was reduced, inorganic phosphate was elevated, and intracellular pH calculated from the phosphate resonant frequency was acidotic, all changes consistent with traditional understanding of stroke pathophysiology. The ^{31}P MRS changes did not correlate with clinical severity of the stroke. Over ensuing days, alkalosis supervened. The authors suggested that acute therapy might have to be instituted within the period of acidosis in order to be effective.

From information available in ^{31}P spectra and with certain assumptions, an estimate of tissue Mg^{2+} concentration can be made.[18] The Detroit group exploited this capability to calculate Mg^{2+} changes associated with stroke; Mg^{2+} was elevated in the acidotic period and might be a pathophysiological factor or a marker of cellular injury.[19]

Both ^{1}H and ^{31}P spectra can be obtained from the same patient in the same session. Such combined observations are a powerful way of analyzing relationships among a wide range of brain metabolites, as was first demonstrated in animal work on hypoglycemia a decade ago.[20] Two groups have overcome the rather considerable technical obstacles to doing this in human stroke patients, with illuminating results.[21,22] Both groups found that stroke-associated lactate and pH were usually not inversely correlated, as casual thinking suggests they ought to be. In samples ranging collectively from the acute to the chronic period, the more common association was of elevated lactate with alkalosis, rather than acidosis or normal pH. Several processes involved in ischemic tissue damage and repair could account for this, because ^{31}P and ^{1}H spectra come from tissue volumes that are not the same size and are both large compared to the dimensions of any of the several metabolic compartments they contain. One should not expect metabolites in such tissue samples to behave as though they were in a well-stirred test tube. In experimental animals, dissociation of lactate from pH has been observed in status epilepticus by NMR[23] (see Epilepsy below) and, by biochemical techniques, in tumors and 24 hours after global ischemia.[24] The last observation is a direct predecessor of the combined ^{1}H–^{31}P findings in human stroke, which with further analysis are sure to improve the understanding of the pathophysiology of stroke. The work is a good example of the way in which powerful new NMR methods can be applied to such a problem directly in the species of interest.

Other nuclei

Use of ^{13}C labeling of stroke-elevated lactate was mentioned above and is discussed further in the last section of this chapter. Brain oxygen consumption and blood flow have been measured simultaneously in cats by combined ^{17}O and ^{19}F NMR during ventilation with $^{17}O_2$ and CHF_3.[25] Like ^{13}C, ^{17}O is expensive, and the experiments were done at 4.7 T. Nevertheless, the prospect of measuring these important quantities in humans for various purposes – including analysis of stroke pathophysiology and perhaps, one day, guidance of stroke management – is sufficiently attractive to ensure that it will be explored further.

Sodium NMR was applied to brain research a decade ago,[26] and it was a component of a still unique experiment in which three nuclei of biological importance were observed during status epilepticus in cats[27] (see Epilepsy below). Recently, it was used to investigate the role of glucose availability in ion homeostasis during experimental cerebral ischemia.[28]

A recent demonstration of the feasibility of ^{133}Cs NMR to distinguish between the extracellular and intracellular compartments in living tissue shows that the inventiveness of NMR scientists continues undiminished.[29] In normal rats, the distinction was better in kidney than in brain, but it might well be useful in vasogenic and interstitial edema, in which brain extracellular space is increased.

Summing up: stroke and MRS in the mid-1990s

Must a state-of-the-art clinical NMR facility in a major medical center provide MRS for stroke diagnosis in 1994? No. 1996? Maybe. 1998? Yes. In the present writer's opinion, lactate in the ^{1}H spectrum, together with diffusion-weighted imaging, is likely to become a standard part of stroke syndrome evaluation, because both measurements contain information about the lesion that helps predict its future course. In principle, much additional information of similar value could come from ^{31}P MRS, but the relatively low sensitivity of the ^{31}P nucleus imposes a burden of additional acquisition time on an acute clinical situation already crowded with other kinds of medical imperatives. The cost of ^{13}C and ^{17}O in the quantities that would be needed for human studies is not out of the range of diagnostic procedures in use today, but it is high enough to require special justification by demonstrations of efficacy that are not yet in the literature.

EPILEPSY

Animal studies by ^{31}P MRS

Changes in brain energy stores and intracellular pH (pHi) observed by ^{31}P spectroscopy during status epilepticus (SE) in rabbit[30] was the first NMR study of seizure phenomena *in vivo*. SE induced by the gamma-aminobutyric acid (GABA) blocker bicuculline caused a fall in phosphocreatine (PCr) accompanied by a rise in inorganic phosphate (Pi), and, somewhat later, a fall in pHi measured from the chemical shift of Pi. ATP remained near its control level. These changes are all well-known consequences of SE that had been studied by earlier techniques requiring removal of tissue. The significance of the MRS study was its demonstration of them *in vivo*. Several more studies in rabbit[31] and neonatal dog[32,33] refined and extended the initial observations.

Animal studies by ^{1}H MRS

Brain lactate is elevated by seizures, as had been known for many years from conventional biochemical studies of excised tissue. *In vivo* observations by ^{1}H MRS led to a new appreciation of how persistent the lactate elevation is after even brief seizure discharge. Lactate remained elevated in SE after other metabolic variables returned to normal in rabbits[23] and neonatal dogs;[34] in the rabbit, it required more than an hour to reach control levels after shock-induced seizure discharge lasting less than 5 minutes.[35]

The extended persistence of lactate elevation in these experiments could have come about either because lactate was trapped in killed cells, the extracellular space or some other non-metabolizing compartment, or because the lactate rise was caused by stimulus-induced elevation of glycolysis not matched by increased respiration during the period of observation. The second mechanism could operate in the absence of brain damage, but in that case, failure of intact Krebs cycle machinery to metabolize the excess lactate presents a new problem, one not appreciated until *in vivo* MRS experiments made it apparent. As described in the last section of this chapter, advanced MRS experiments in which shock-elevated brain lactate was labeled with ^{13}C from blood glucose showed that the second explanation is correct; brain damage is unlikely to be the reason for the long persistence of brain lactate after brief brain stimulation.

The persistence might have adaptive utility. By reason of its metabolic role at a key interface between biochemistry and brain function, lactate is well suited to act as an inhibitory neuromodulator. One experimental study is consistent with this idea: in concentrations easily produced by epileptic discharge in the mammalian brain, lactate inhibited release of acetylcholine from *Torpedo* electroplax.[36] Another possible adaptive function is priming of neurons for maximum firing rates by storage of potential energy near mitochondria on receipt of a warning that heavy energy demand is imminent. Maximum firing could be sustained longer that way than if it were limited by the several seconds required to increase rates of glycolysis and glycogenolysis. Elevation of lactate in the human visual system by physiological photic stimulation might be another expression of the same adaptive mechanism.[37] No direct evidence for this kind of adaptive priming by lactate exists, but it would be hard to get, because it would ordinarily occur in tissue compartments far smaller than the anatomical resolution of current measurement techniques. It would become detectable by them only during highly synchronized neuronal activity, like the seizure discharge and evoked potential generation that occurred in the experiments mentioned above.

Animal studies by combined ^1H–^{31}P MRS

Spectra from ^1H and ^{31}P can be obtained from the same subject in the same experiment, a capability first demonstrated in an experimental study of hypoglycemic encephalopathy.[20] Combined ^1H–^{31}P observations on bicuculline-induced SE in the rabbit showed that lactate and pHi became dissociated, lactate concentration remaining elevated while pHi returned to normal.[23] The phenomenon occurred only when seizure discharge was of low to medium intensity; during the most intense discharge, pHi remained acidotic.[23,27]

Although such dissociation had not been observed before in SE, it has been observed by biochemical techniques in experimental tumors and 24 hours after prolonged cerebral ischemia.[25] Glycolysis produces lactate and hydrogen ion in equal quantities, but their biochemical roles in the brain are quite different; the fact that they move into separate compartments and consequently develop non-equivalent concentrations should not be surprising. Because both substances are involved in rapid biochemical reactions, their concentrations change quickly during sampling of tissue for conventional biochemical analyses. Non-invasive MRS is an especially powerful tool for investigation of such labile compounds. Much is surely still to be learned about the parts they play in normal and disordered brain function.

Human studies by ^{31}P MRS

The first MRS observations on epileptic human subjects were ^{31}P spectra from infants.[38] As in animal experiments, the PCr:Pi ratio was decreased about 50% during seizures, and it returned to normal afterwards. The lowest ratios during seizures were in infants who developed long-term neurological sequelae.

Adults with chronic temporal lobe epilepsy have been studied by two groups, who obtained different results. One group reported alkaline pHi on the side of the seizure focus in eight patients, and, in seven of them, increased Pi and decreased phosphomonoesters (PME) as well.[39,40] The other group found reduced PCr:Pi ratios without other changes.[41] Whether the discrepancy reflects actual differences between patient populations or merely differences in technique is not yet known.

Use of ^{31}P MRS for detection of unilateral interictal hypometabolism in patients with chronic focal epilepsy has been reported in preliminary form.[42]

Human studies by ^{1}H MRS

An early use of ^{1}H MRS related to human epilepsy was analysis of extracts of histologically normal samples of temporal lobe resected for relief of intractable complex partial seizures.[43] The data are valuable because agonal artefacts were minimized by freezing the tissue less than an hour after its blood supply was interrupted. As human brain samples cannot be frozen *in situ*, such material is close to the freshest that can be obtained. The study is a useful source of information about concentrations of several human brain metabolites that are observable by *in vivo* MRS.

Free fatty acids contribute to the ^{1}H spectrum, but the contribution is small in normal brain, because most brain lipids exist as myelin and other large molecules that do not tumble freely and therefore do not produce sharp resonance lines. Appearance of large lipid signals in ^{1}H spectra from brain has usually been due to contamination of the spectra by free lipids in bone marrow and scalp. As increasingly effective localization methods have been implemented, this problem has become manageable, so that free fatty acids in the brain itself can be reliably detected. One might expect tissue disruption of any origin to increase brain free fatty acid content, thereby providing an additional metabolic signature of the pathophysiological process detectable non-invasively by ^{1}H MRS. Systematic efforts to identify such signatures are warranted in stroke, multiple sclerosis, and other diseases known to include membrane disruption in their pathophysiology. A possible instance of lipid mobilization by disease has been reported in patients who had received electroconvulsive therapy or suffered from untreated temporal lobe epilepsy.[44] The finding implies that epileptic discharge in the human brain can mobilize small lipid molecules so as to produce a much larger signal than is recorded from brain lipids in their normal physical state. If the observation is confirmed and other similar ones made, the relatively simple spectroscopy involved will become a valuable source of information about the time course of brain damage and repair in epilepsy and a number of other disorders.

Rasmussen's syndrome is a form of chronic localized epileptogenic encephalitis. Observation by ^{1}H MRS of two patients with this condition produced the first report of elevated lactate associated with seizure discharge in human brain.[45] Similar observations by another group have appeared in preliminary form.[46]

The NA signal in the ^{1}H spectrum includes a major component from the NAA, which is localized principally if not exclusively to the neurons. Reduction of it appears to be a common feature of chronically epileptogenic brain tissue, having been documented in full papers by two groups,[45,47] and, furthermore, in no fewer than six additional preliminary reports by these groups and others.[46,48–52]

The usual interpretation of reduced NA is that it reflects loss of neurons. If that is so, loss of NA in temporal regions of patients with complex partial epilepsy might be expected to be associated with atrophy, since glial replacement of neurons would be unlikely to match the original volume exactly. In accordance with this idea, one of the above studies found that the NA:creatine ratio correlated with volume loss on MRI in nine of 10 patients. The one patient with a normal MRI nevertheless had a reduced NA:creatine ratio in the temporal region, and pathological examination of the resected tissue revealed mild mesial temporal sclerosis. These data imply that chemically specific ^{1}H MRS abnormalities may be detectable *in vivo* before structural changes in some patients. Even if MRI techniques that are especially sensitive to mesial temporal sclerosis come into routine use,[53] preoperative detection of chemical abnormality may contribute to accurate selection of the tissue to be resected for relief of complex partial epilepsy. One can reasonably expect to see this selection made entirely by non-invasive electroencephalographic and NMR techniques in the near future, obviating in most if not all cases the need to implant intracranial electrodes.

GABA can be detected in the human brain by ^{1}H MRS.[54] Monitoring of GABA and glutamine signals has been used for direct observation of the effects of vigabatrin, a new antiepileptic drug.[55] Capability of this kind opens the way to a new level of refinement in neuropharmacology comparable to the one made possible by measurement of antiepileptic drug levels in the blood.

A note on DWI in status epilepticus

Although in general MRI subjects are outside the purview of this chapter, mention of DWI is warranted because of its recently reported sensitivity to epileptic activation of the brain.[56] DWI is thought to measure the mean diffusion path length of water molecules in tissue.[57] Its great clinical importance in stroke – noted at the beginning of the chapter – rests on the fact that it detects ischemic brain earlier than CT scanning or conventional MRI.[12] Increased signal is seen in the ischemic tissue, presumably because of a 30–40% shortening of the mean path length of water diffusion. The mechanism is uncertain. The leading possibility is net movement of water from extracellular to intracellular spaces as cytotoxic edema develops. Theory suggested that DWI might show a similar change in SE, and appropriate observations during bicuculline-induced SE in rat showed that it did.[56] The change was about half that seen in ischemia. Its presence in the brain at the same time as the MRS-detectable chemical changes described above suggests that the two classes of phenomenon may shed light on each other as they are investigated further. Of special interest for clinical application is the possibility that DWI changes may preserve a record of recent seizure activity in epileptic regions of human brain. If that is true, DWI will become a powerful tool for non-invasive detection and analysis of seizure activity in the depths of the human brain.

Summing up: epilepsy and MRS in the mid-1990s

The data reviewed above establish an unequivocal place for MRS in animal and human research related to epilepsy. A role for MRS in routine diagnosis of clinical seizure states has not yet been defined. In the opinion of this writer, the likeliest prospects are detection of reduced NA signals in the ^{1}H spectrum associated with mesial temporal sclerosis, and detection by ^{31}P MRS of depressed temporal lobe metabolism that has been shown by positron emission tomography to be present in many patients with complex partial epilepsy. Both observations are completely non-invasive and can be made on properly equipped clinical imaging machines. Their cost is tens of minutes of acquisition time, which requires that the information obtained be more valuable than anything else that could be done in the same time. NMR measurements based on the water signal are developing rapidly, and they may solve some of the same clinical problems for which MRS looks promising, at a lower cost in acquisition time. Until research makes clear some unique advantage of MRS in routine epilepsy diagnosis, the clinical imaging expert should concentrate on water-based MRI.

DEMENTIA, TUMORS, AND MULTIPLE SCLEROSIS

Dementia

Alzheimer's disease is the major cause of dementia in the industrial world and a major source of strain on facilities for dealing with chronic disability. In public health terms, dementia caused by multiple strokes is a distant second, and several specifically treatable dementias are even less common. The etiology of Alzheimer's disease remains unknown. New insight concerning the disease is a reasonable goal for any new biomedical technology, and increasing numbers of MRS workers are seeking it. So far, the only unequivocal MRS finding is loss of NA signal, which has been reported recently by several groups (see below).

Alzheimer studies by ^{31}P MRS

Consensus has not been reached among workers who have done ^{31}P studies of Alzheimer patients. Conflicting claims based on apparently conflicting observations have persisted for several years. One series of studies that has been summarized recently[58] holds that phosphomonoester (PME) and Pi are increased in Alzheimer patients compared to controls.[59] Another group finds no clear ^{31}P changes associated with the disease.[60,61] The different conclusions are not explained by any obvious difference in patient selection or characterization; ^{31}P MRS acquisition methods could be a factor, as they were not the same in each study.

Alzheimer studies by ^{1}H MRS

In contrast to the ^{31}P data, ^{1}H observations by several groups have produced general agreement that a reduced NA signal is characteristic of Alzheimer's disease. An *in vitro* study of extracts from Alzheimer brains showed that NAA was selectively reduced compared to age-matched controls.[62] Preliminary reports of no fewer than nine studies of living Alzheimer patients by ^{1}H MRS all describe decreased NA signals in one way or another. NA was found to be reduced in the frontal lobes of 12 patients, and its T_2 was significantly prolonged; neither intensities nor T_2s of other resonances were different from controls.[63] Other studies describe reduced NA in the parietal lobes[64] and in the frontal and parietotemporal lobes.[65] Two studies using different forms of ^{1}H SI both observed decreased NA; one also reported an increased choline signal,[66] and the other found NA loss predominantly in the temporal lobes early in the disease, whereas it was reduced in almost all brain voxels in advanced cases.[67] A result similar to the second of these was obtained by workers using single volume localized ^{1}H MRS; NA was reduced in the hippocampus–amygdala region, but was not different from controls in a parieto-occipital volume.[68] A study that employed special methods to isolate signal from parietooccipital grey matter did find reduced NA there, as well as increased myoinositol.[69] The most convincing single report concerning NA is from a cooperative study involving 35 institutions, in which 30 Alzheimer patients studied by ^{1}H MRS according to a standard protocol showed 30% reduction of the NA signal compared to controls.[70] In a study that highlights the synergism between MRS and MRI in a very specific way, NA reduction in Alzheimer patients, although significant, included considerable overlap between patient and controls; however, combination of the ^{1}H MRS data with morphometric measures of hippocampal and total intracranial volume separated the groups completely.[71]

This impressive body of data demonstrates two things:

1. NA is reduced in Alzheimer's disease, and although biological significance and clinical utility remain to be determined, a definite MRS abnormality is available for use in Alzheimer research.
2. ^{1}H MRS suitable for clinical research is spreading rapidly. This fact is of general interest, as the technique can be applied to many other disease problems. It is likely to be most effective when combined with various kinds of MRI in the manner demonstrated by the last study cited above.

Brain tumors

Progress in understanding neoplasia in the nervous system by any *in vivo* technique is hampered by the cellular heterogeneity of brain tumors. The problem is a familiar one to workers who have studied them with positron emission tomography (PET). The heterogeneity is below the resolution of both current and anticipated MRS techniques. It was evident in an early ^{31}P MRS study, which commented on the 'striking diversity' of metabolic patterns observed.[72] Later studies associated increased PME, decreased PCr, alkaline pH, and altered PDE signals with heightened aggression of gliomas.[73,74]

The finer anatomical resolution of ^{1}H MRS, particularly in the form of SI, allows more detailed observation of metabolic heterogeneity in brain tumors. An initial study that combined ^{1}H SI with PET suggested that regions of high lactate coincided with regions of high glucose uptake.[75] In later work with improved techniques, the same group found a more complicated situation: high lactate was also associated with loculations of extracellular fluid.[76] Another group able to study patients with both ^{1}H SI and PET reported similar metabolic findings and variability.[77,78] Both groups observed variable reductions in NA and increased choline signals.

The SI methods used in these studies have importance beyond their application to brain tumors. Rapid development in recent years has brought them to the verge of feasibility for routine practice. Figure 1.3 shows distribution of four metabolites in two-dimensional spectroscopic images from a patient treated by radiation three years earlier for a grade 2–3 astrocytoma of the left temporal lobe. Dr J R Alger's description follows:

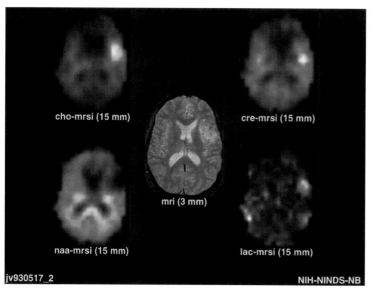

Fig. 1.3. ^1H *spectroscopic images from a 43-year-old patient irradiated 3 years earlier for a grade 2–3 astrocytoma of the left temporal lobe. (Patient left is viewer right in the images.) mrsi – magnetic resonance spectroscopic image; cho – choline-containing compounds; cre – total creatine (creatine + phosphocreatine); naa – N-acetylaspartate; lac – lactate. (Original print supplied by Dr J R Alger.)*

The ^{1}H-MRSI study was performed using a 4-slice spin echo (TE = 272 ms) axial-oblique acquisition procedure (only one section is shown) on a conventional 1.5 T GE Signa. The MRSI slice thickness was 15 mm. Two-dimensional phase-encoding was performed with a matrix size of 32×32 over a field-of-view of 240 mm. A gradient-echo MRI scan (3 mm thick) taken from the center of the slice that provided the MRSI is also shown for orientation purposes. The smoothed MRS images show the variation in the signal intensities of the choline (cho), creatine (cre), NAA and lactate (lac) signals. The images have been "windowed and leveled" until they are pleasing to the eye, a procedure used extensively in MRI. The patterns are characteristic of many tumor patients: the lesion shows a prominent elevation of the cho signal, a loss of the NAA signal, and an area of increased cre signal at the periphery of the area showing increased cho. This one is unusual in that it shows an elevated lac signal (confirmed by viewing the spectra) at the lesion site; other areas of apparently increased lac are either noise or lipid bleed from extracerebral tissues. In general, the choline signal elevation serves as a better tumor marker than the lactate signal.[77]

The amount of information in Fig. 1.3 is very large. As capability to make such metabolic maps spreads from research laboratories to the clinical venue, SI will become the standard way of presenting MRS data, and it will change the way physicians think about disease.

Metabolic maps can be made from ^{31}P spectra as well as ^{1}H spectra. Owing to the lower sensitivity of ^{31}P, their anatomical resolution is not so fine, but it is adequate to show metabolite distributions across major brain structures and defects caused by large lesions.[79,80] The information about energy state, pH, and phospholipids that can be obtained from the ^{31}P spectrum provides strong motivation for continued refinement of ^{31}P SI.

Multiple sclerosis

One of the earliest biomedical advances produced by MRI was appreciation of extensive, clinically silent pathology of cerebral white matter in multiple sclerosis. For the usual reason of much lower signal intensity, MRS of the disease was not practical until several years later. In the 1990s, MRS technology had advanced enough to support increasing numbers of studies; the 1993 *Proceedings of the Society of Magnetic Resonance in Medicine* contains nine preliminary reports on ^{1}H MRS observations in patients with multiple sclerosis. The commonest findings are reduced NA and increased choline signals associated with plaques. Elevated lactate has also been observed; it may reflect some aspect of active lesion evolution, possibly the presence of highly glycolytic white cells, as in subacute cerebral infarction.[3]

Figure 1.4 illustrates all of these metabolic abnormalities in a multiple sclerosis patient with an unusually large cerebral plaque. It also demonstrates the complementary advantages of metabolic mapping by ^{1}H SI and examination of individual spectra.[81]

Figure 1.5 shows the time course of the plaque-related metabolic changes in the same patient over a period of months. Metabolic aspects of plaque evolution were not previously accessible for study in human patients. New understanding of the pathophysiology of multiple sclerosis will certainly be achieved as MRI and the non-invasiveness and chemical specificity of MRS are used together to obtain new information on the natural history and therapeutic responsiveness of the disease in individual patients. All of the data in Figs 1.4 and 1.5 are new information bearing on the underlying cellular and molecular biology. An additional example is a preliminary report of slight recovery of the NA signal in patients with the relapsing–remitting form of multiple sclerosis, an observation that may reflect neuronal recovery not previously appreciated as part of the disease process.[82]

Summing up: MRS in dementia, tumors, and multiple sclerosis in the mid-1990s

MRS studies have not yet led to fundamental new understanding of the dementias or of tumors in the brain, nor is MRS necessary for routine diagnosis of these conditions. However, research motivated by them has played an important part in development of powerful new NMR techniques, notably SI. In multiple sclerosis, the situation is different: ^{1}H MRS combined with various kinds of MRI is likely to become an important adjunct to clinical research, and a role for it in management of individual therapy may emerge.

PROSPECTS FOR ^{13}C MRS IN CLINICAL NEUROLOGY

Most practical applications of ^{13}C spectra must await some improvement in the signal-to-noise ratio. One application that might be made immediately is in the use of ^{13}C as an isotopic tracer in reactions. Various chemical forms of carbon could be identified by their chemical shifts and fine structures, even in complex mixtures, after the introduction of a compound enriched in ^{13}C. Even some biological systems might be studied by this technique.

P C Lauterbur, 1958[83]

The history of biomedical ^{13}C MRS began with the above passage.[82] The previous year, its author

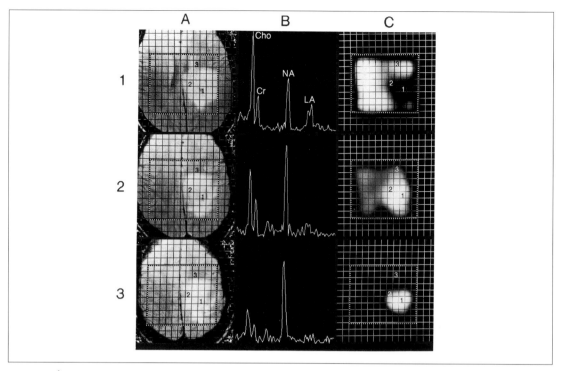

Fig. 1.4. ^{1}H spectroscopic imaging (SI) in a multiple sclerosis patient with a large cerebral plaque. (a) Axial conventional water-based MRIs (TR 2100, TE 30) and (c) metabolite-based SIs (TR 2000, TE 272) of the lesion at 19 days after onset of symptoms. Each of the SIs is based on the signal from a different metabolite: C1 is based on signals from N-acetylaspartate (NA); C2 is based on signals from choline (Cho); C3 is based on signals from lactate (LA). The three slices of the conventional MRI span the thickness of each of the metabolite images. Note that there is abnormal hypointense signal in the NA image and abnormal hyperintense signals on the Cho and LA images centred on the same volume that gives an abnormal hyperintense signal on the water-based MRI. (b) Representative ^{1}H spectra from the voxels labeled 1, 2, and 3 in the images. B1 corresponds to voxel 1, B2 to voxel 2, and B3 to voxel 3. Line width and absolute peak heights vary between voxels because of differences in magnetic field and radio frequency homogeneity. However, relative resonance intensity ratios for individual voxels are not affected by this. Note the relatively high Cho:creatine (Cr), low NA:Cr, and high LA:Cr ratios in and around the lesion. (Reprinted with permission from Arnold et al.[81] Original print supplied by Dr D L Arnold.)

had published the first report of ^{13}C chemical shifts,[84] which was followed shortly by another.[85]

Interest in ^{13}C spectroscopy centered at first on its usefulness in development of chemical shift theory, because of the large number of readily available compounds spread over a wide chemical shift range. Although Lauterbur's words from 1958[83] show that the biological potential of ^{13}C MRS was appreciated from the beginning, years became decades while NMR technology evolved to allow the first ^{13}C labeling studies on enzyme

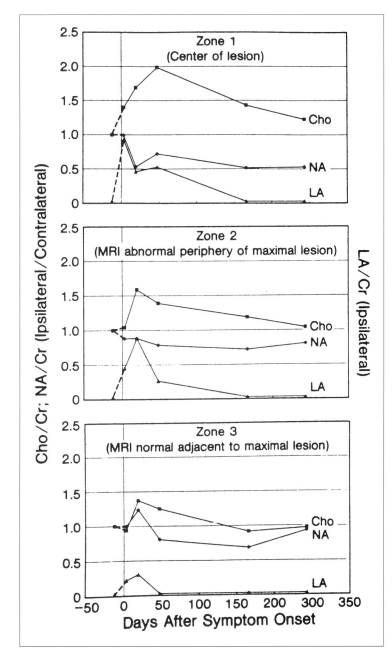

Fig. 1.5. Graphic representation of changes in relative metabolite ratios over time in the patient whose data are shown in Figure 1.4. Choline (Cho):creatine (Cr) (black squares) and N-acetylaspartate (NA):Cr (black diamonds) ratios are expressed relative to voxels in the same relative positions in the contralateral hemisphere. The lactate (LA):Cr ratio (black triangles) is expressed as the ipsilateral ratio of resonance intensities because LA cannot be reliably measured in the normal hemisphere with the signal-to-noise ratio available for individual voxels. The initial and final values for LA in the lesion are shown as near zero, again because they cannot be measured in individual voxels with the available signal-to-noise. Each ratio is extrapolated back in time (dashed lines) to before onset of symptoms, to illustrate graphically the change from normal ratios that are assumed to have been present before the lesion formed. (Reprinted with permission from Arnold et al.[81])

systems and functioning cells,[86] intact animals,[87] animal brain,[88] and human brain.[89] The potential is so great that the most advanced of these studies belong to ^{13}C's early biomedical history.

Special characteristics of ^{13}C MRS

Natural carbon is nearly 99% non-magnetic ^{12}C, which gives no NMR signal. The 1.1% naturally abundant ^{13}C in organic molecules gives signals that can be detected in sufficiently concentrated solutions, and their wide chemical shift range makes observation of them useful enough for studies of structure and conformation that a large literature has accumulated.[90] In living subjects, including humans, natural ^{13}C MRS has far more potential than has been exploited for characterization of normal and pathological variation among tissues, but a more compelling opportunity that has attracted most workers is the possibility of ^{13}C enrichment of molecules observable *in vivo*. By providing the organism with a nutrient enriched in ^{13}C, as demonstrated over 20 years ago in *Candida utilis*,[91] one gains the simultaneous advantages of increased signal-to-noise ratio for observation of molecules that receive the ^{13}C through metabolic processes and a way of measuring metabolic rates *in vivo*.

The ideal ^{13}C source for labeling brain metabolites is glucose, from which the adult organ normally derives nearly all of its energy. Most glucose molecules metabolized in the brain are converted to two molecules of lactate, one of which has the C1 from glucose in its methyl position. Further metabolism via the Krebs cycle creates several other ^{13}C-labeled molecules that are observable *in vivo*, notably 4-^{13}C-glutamate. As was first shown in rabbit brain, 1-^{13}C-glucose, 3-^{13}C-lactate, and ^{13}C-labelled amino acids can all be detected *in vivo* by ^{13}C MRS.[86] Recently, similar observations have been made in the human brain.[92]

Proton observe-carbon edited (POCE) MRS

A disadvantage of ^{13}C MRS for *in vivo* work is the long acquisition times that are necessary compared to ^{31}P and ^{1}H. The sensitivity of the proton can be exploited to detect ^{13}C in samples containing enough ^{1}H–^{13}C bonds, because the signal from the proton is split in a characteristic way by the magnetic carbon bonded to it. The principle has been used in studies of molecular structure since the 1960s.[86] Much later, improved technology and diligent effort resulted in its successful adaptation for observation of ^{13}C-labeled compounds in the brains of living animals,[93] and, recently, humans.[89] Among the biologically important measurements made possible by this technique is calculation of cerebral metabolic rates from the time course of ^{13}C accumulation in observable metabolites.[94] POCE MRS and related methods are still in early stages of development; as they mature and are widely implemented, they are likely to become the pre-eminent means of measuring cerebral metabolic rates *in vivo*.

An application of POCE MRS and its implications for stroke

The methyl carbon of lactate is among the chemical groups that can be detected *in vivo* by POCE MRS. As described in an earlier part of this chapter, brief shock-induced seizure discharges in rabbit brain caused lactate elevation more prolonged than expected and raised the question of whether the lactate was trapped in a non-metabolizing compartment or persistent in the presence of competent metabolic machinery.[35] POCE MRS for detection of 3-^{13}C-lactate, combined with measurement of total lactate from the ^{1}H spectrum, showed clearly that the second explanation was the correct one.[95] Infusion of 1-^{13}C-glucose after elevation of brain lactate by electroshock caused the ^{13}C isotopic fraction of brain lactate to reach a steady state at the level predicted from the ^{13}C isotopic fraction of blood glucose. This could have occurred only by metabolic turnover of all observable brain lactate, which therefore cannot have been trapped in a non-metabolizing compartment. The original study[95] can be consulted for the particulars of the experiment. The principle is illustrated in Fig. 1.6.

The same principle can be used to determine the metabolic activity of any metabolite pool that is observable *in vivo* and potentially enrichable in ^{13}C. Such observations are not limited to animals. Their non-invasiveness allows them to be made in humans whenever scientific or clinical justification exists to do so, as labeling of infarct-associated lactate in a stroke patient[97] and labeling of glutamate in the normal human brain[89] have demonstrated.

For instance, the principle illustrated in Fig. 1.6 could be used in initial assessment of stroke in the following way: immediately after a patient presenting to an emergency room with stroke syndrome is evaluated clinically, the airway secured, and the necessary bloods drawn, 10 g of 1-^{13}C-glucose might be administered. Forty minutes later, the patient enters the bore of an NMR machine equipped for conventional MRI, DWI, ^{1}H SI, and irradiation at the resonant frequency of ^{13}C. Old pathology is detected by MRI, acute ischemia by DWI, and two ^{1}H SI images are made through the DWI-defined ischemic lesion – one with

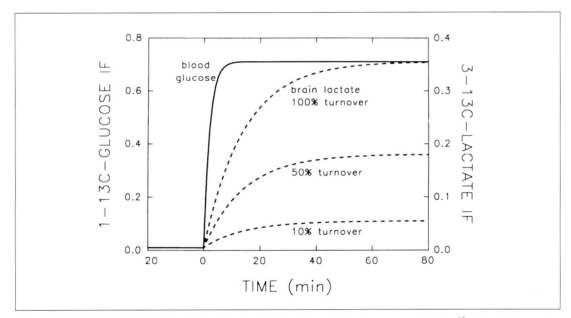

Fig. 1.6. Schematic representation of the time course of an experiment in which brain lactate is labeled with ^{13}C from blood glucose. The isotopic fraction (IF) of 1-^{13}C-glucose in the blood is plotted as a solid line referred to the left ordinate. The IF of 3-^{13}C-lactate in the brain is plotted as dashed lines referred to the right ordinate. At time 0, the ^{13}C IF of blood glucose is raised from its natural abundance level of 0.001 by infusion of 1-^{13}C-glucose and maintained at about 0.7. Three possible responses of brain lactate are depicted by the dashed lines: If all of it receives label, its ^{13}C IF will rise to the level predicted by the ^{13}C IF of blood glucose, as illustrated by the uppermost dashed line. If some of the lactate is trapped in a non-metabolizing compartment, a steady state will be reached at some lower value of 3-^{13}C-lactate IF, illustrated by the middle and lowest dashed lines. The ratio between 3-^{13}C-lactate and 1-^{13}C-glucose in the steady state provides information about the fraction of the lactate pool that is metabolically active. (Reprinted with permission from Prichard.[96])

collapse of ^{13}C-induced splittings by irradiation of ^{13}C, one without. From the two kinds of ^{1}H SI data, distributions of 3-^{12}C- and 3-^{13}C-lactate relative to the DWI lesion are determined, on-line. If the proportion of labeled to unlabeled lactate associated with the lesion is low, most of the lesion is killed tissue, which can be expected to become necrotic over the succeeding few days, presenting a substantial hazard of lethal hemorrhage to aggressive reperfusion therapies. If most lesion-associated lactate has been labeled with ^{13}C, then most of the lesion is still metabolizing and possibly salvageable; aggressive therapies are justified.

This scheme of emergency stroke evaluation is feasible in every technical and logistic particular, but its prospect for routine use depends on other things. It is essentially a way of assessing how much of a fresh stroke is an 'ischemic penumbra' that may be able to recover.[10] Rapid evolution of MRI may soon provide equivalent information from simpler measurements, the best prospects being DWI and magnetization transfer contrast imaging. To paraphrase an axiom of NMR common sense often and properly emphasized by Professor David Gadian: 'Whenever the water signal can be used, it should be, because it is so much bigger than the others.' The point here is that if ^{13}C labeling of stroke-elevated lactate proves to be the best predictor of a fresh lesion's later course, it can be widely implemented for that purpose.

Summing up: clinical prospects for ^{13}C labeling *in vivo*

The clinical prospects are low in the short term, as too much remains unknown about the diagnostic utility of simpler NMR techniques to allow even the most abandoned enthusiast to argue plausibly that ^{13}C labeling will soon be a common clinical tool. Use in stroke as outlined above remains a possibility for the longer term, and the inherent power of the method is so great that important clinical applications may emerge from unexpected quarters. However, for the years that any such developments will take, the clinical imaging expert can safely regard ^{13}C labeling *in vivo* as a research tool.

OVERALL SUMMING UP

The chemical specificity of MRS guarantees its future in clinical neurology. In 1994, no routine clinical application of MRS is standard practice of the kind that every hospital must provide, like x-ray equipment and electrocardiographs, but MRI has not reached that point either. It will. Diagnostic MRI is so much more versatile and efficient than competing technologies that it is certain to become the premier standard medical imaging method of the latter 1990s. The step from MRI to the simpler kinds of MRS is no longer a large one in either technique or money. Widespread availability of MRI machines will facilitate implementation of MRS, first for research, then for routine clinical use.

Chemically specific characterization of disease processes at all of their stages is the most important prospect that MRS offers clinical medicine. In much the same way that microscopic study of human tissue obtained after death and from the occasional biopsy contributed to modern conceptions of disease, non-invasive longitudinal MRS data will provide new understanding of how pathophysiological processes evolve. One can readily imagine biochemical profiles that distinguish ischemic, neoplastic, inflammatory, degenerative, and other categories of lesions from each other and are useful as monitors of therapeutic effectiveness. As this body of knowledge grows, specific kinds of clinical utility in management of individual patients will surely emerge and become routine.

The bias of the present writer is that lactate measurement in early stroke diagnosis is likely to be the first MRS application to become standard in clinical neurology. However, the MRS observations described here are all early ones and are crude compared to what will become available as MRS technology matures. The inherent compatibility of NMR techniques ensures that what are today usually thought of separately as 'MRI' and 'MRS' will merge into a collection of related methods from which a physician can choose the combination most useful for the particular clinical problem at hand. Many patients with diseases inside and outside the nervous system will be managed in the future with the aid of a range of NMR measurements constituting the most important source of diagnostic information after the history and physical examination.

ACKNOWLEDGEMENT

The author's own work was supported by UPSPHS grants NS 27883, DK 34576, and NS 21708.

REFERENCES

1 Argov Z, Bank WJ. Phosphorus magnetic resonance spectroscopy (31P MRS) in neuromuscular disorders. *Ann Neurol* 1991; **30**:90–7.

2 Radda GK. Control, bioenergetics, and adaptation in health and disease: noninvasive biochemistry from nuclear magnetic resonance. FASEB J 1992; **6**:3032–8.

3 Petroff OA, Graham, GD, Blamire AM, et al. Spectroscopic imaging of stroke in humans: histopathology correlates of spectral changes. *Neurology* 1992; **42**:1349–54.

4 Itoh S, Maeda M, Matsuda T, et al. Evaluation of acute brain ischemia with ^{1}H-MRS and diffusion weighted MR image. *Proc Soc Magn Reson Med* 1993; **12**:1488.

5 Graham GD, Blamire AM, Howseman AM, et al. Proton magnetic resonance spectroscopy of cerebral lactate and other metabolites in stroke patients. *Stroke* 1992; **23**:333–40.

6 Barker PB, Gillard JH, Agildere AM, et al. Serial magnetic resonance evaluation of acute stroke: MRI, MRA, perfusion imaging, and multislice proton spectroscopic imaging. *Proc Soc Magn Reson Med* 1993; **12**:1487.

7 Graham GD, Blamire AM, Rothman DL, et al. Early temporal variation of cerebral metabolites after human stroke: A proton magnetic resonance spectroscopy study. *Stroke* 1993; **24**:1891–6.

8 Duijn JH, Matson GB, Maudsley AA, Hugg JW, Weiner MW. Human brain infarction: proton MR spectroscopy. *Radiology* 1992; **183**:711–18.

9 Ford CC, Griffey RH, Matwiyoff NA, Rosenberg GA. Multivoxel 1H-MRS of stroke. *Neurology* 1992; **42**:1408–12.

10 Prichard JW. *The Ischemic Penumbra: Prospects for Analysis by NMR Spectroscopy. Molecular and Cellular Approaches to the Treatment of Brain Disease.* Raven Press: New York, 1993, 153–74.

11 Berkelbach van der Sprenkel JW, Luyten PR, van Rijen PC, Tulleken CA, den Hollander JA. Cerebral lactate detected by regional proton magnetic resonance spectroscopy in a patient with cerebral infarction. *Stoke* 1988; **19**:1556–60.

12 Warach S, Chien D, Li W, Ronthal MB, Edelman RR. Fast magnetic resonance diffusion-weighted imaging of acute human stroke. *Neurology* 1992; **42**:1717–23.

13 Graham GD, Kalvach P, Blamire AM, et al. Clinical correlates of proton magnetic resonance spectroscopy findings after acute cerebral infarction. *Proc Soc Magn Reson Med* 1993; **12**:1483.

14 Gideon P, Henriksen O. In vivo relaxation of N-acetyl-aspartate, creatine plus phosphocreatine, and choline containing compounds during the course of brain infarction: a proton MRS study. *Magn Reson Imaging* 1992; **10**:983–8.

15 Graham GD, Blamire AM, Rothman DL, Petroff OAC, Prichard JW. Proton in vivo relaxation times of lactate and other metabolites in infarcted human brain tissue. *Proc Soc Magn Reson Med* 1993 **12**:1482.

16 Bottomley PA, Drayer BP, Smith LS. Chronic adult cerebral infarction studied by phosphorus NMR spectroscopy. *Radiology* 1986; 160:763–6.

17 Levine SR, Helpern JA, Welch KM, et al. Human focal cerebral ischemia: evaluation of brain pH and energy metabolism with P-31 NMR spectroscopy. *Radiology* 1992; 185:537–44.

18 Halvorson HR, Vande LAM, Helpern JA, Welch KM. Assessment of magnesium concentrations by 31P NMR in vivo. *NMR Biomed* 1992; 5:53–8.

19 Helpern JA, Vande LA, Welch KM, et al. Acute elevation and recovery of intracellular [Mg^{2+}] following human focal cerebral ischemia. *Neurology* 1993; 43:1577–81.

20 Behar KL, den Hollander JA, Petroff OAC, Hetherington H, Prichard JW, Shulman RG. The effect of hypoglycemic encephalopathy upon amino acids, high energy phosphates, and pHi in the rat brain in vivo: detection by sequential ^{1}H and ^{31}P NMR spectroscopy. *J Neurochem* 1985; 44:1045–55.

21 Hugg JW, Duijn JH, Matson GB, et al. Elevated lactate and alkalosis in chronic human brain infarction observed by 1H and 31P MR spectroscopic imaging. *J Cereb Blood Flow Metab* 1992; 12:734–44.

22 Gideon P, Sperling B, Henriksen O, et al. Intracellular pH, lactate content, and regional cerebral blood flow in acute cerebral infarction. *Proc Soc Magn Reson Med* 1993; 121:1484.

23 Petroff OAC, Prichard JW, Ogino T, Avison MJ, Alger JR, Shulman RG. Combined 1H and 31P nuclear magnetic resonance studies of bicuculline-induced seizures in vivo. *Ann Neurol* 1986; 20:185–93.

24 Paschen W, Djuricic B, Mies G, Schmidt-Kastner R, Linn F. Lactate and pH in the brain: Association and dissociation in different pathophysiological states. *J Neurochem* 1987; 48:154–9.

25 Pekar J, Sinnwell TM, Ligeti L, et al. Double-label tracer experiments using multinuclear MRI: Mapping cerebral oxygen consumption and blood flow using ^{17}O and ^{19}F MRI. *Proc Soc Magn Reson Med* 1993; 12:1388.

26 Maudsley AA, Hilal SK. Biological aspects of sodium-23 imaging. *Br Med Bull* 1984; 40:165–6.

27 Schnall MD, Yoshizaki K, Chance B, Leigh JSJ. Triple nuclear NMR studies of cerebral metabolism during generalized seizure. *Magn Reson Med* 1988; 6:15–23.

28 Tyson RL, Peeling J, Sutherland GR. Changes in sodium gradients in short-duration cerebral ischemia studied using in vivo single- and double-quantum ^{23}Na NMR spectroscopy. *Proc Soc Magn Reson Med* 1993; 12:1496.

29 Li Y, Neil JJ, Ackerman JJH. Observation of 133Cs in vivo: A preliminary study. *Proc Soc Magn Reson Med* 1993; 12:1519.

30 Prichard JW, Alger JR, Behar KL, Petroff OAC, Shulman RG. Cerebral metabolic studies in vivo by 31P NMR. *Proc Natl Acad Sci U S A* 1983; 80:2748–51.

31 Petroff OAC, Prichard JW, Behar KL, Alger JR, Shulman RG. In vivo phosphorus nuclear magnetic resonance spectroscopy in status epilepticus. *Ann Neurol* 1984; 16:169–77.

32 Young RS, Osbakken MD, Briggs RW, Yagel SK, Rice DW, Goldberg S. 31P NMR study of cerebral metabolism during prolonged seizures in the neonatal dog. *Ann Neurol* 1985; 18:14–20.

33 Young RS, Cowan B, Briggs RW. Brain metabolism after electroshock seizure in the neonatal dog: a ^{31}P NMR study. *Brain Res Bull* 1987; 18:261–3.

34 Young RS, Chen B, Petroff OA, Gore JC, et al. The effect of diazepam on neonatal seizure: in vivo 31P and ^{1}H NMR study. *Pediatr Res* 1989; 25:27–31.

35 Prichard JW, Petroff OA, Ogino T, Shulman RG. Cerebral lactate elevation by electroshock: a ^{1}H magnetic resonance study. *Ann N Y Acad Sci* 1987; 508:54–63.

36 Gaudry-Talarmain YM. The effect of lactate on acetylcholine release evoked by various stimuli from Torpedo synaptosomes. *Eur J Pharmacol* 1986; 129:235–43.

37 Prichard J, Rothman D, Novotny E, et al. Lactate rise detected by ^{1}H NMR in human visual cortex during physiologic stimulation. *Proc Natl Acad Sci U S A* 1991; 88:5829–31.

38 Younkin DP, Delivoria PM, Maris J, Donlon E, Clancy R, Chance B. Cerebral metabolic effects of neonatal seizures measured with in vivo ^{31}P NMR spectroscopy. *Ann Neurol* 1986; 20:513–19.

39 Hugg JW, Matson GB, Turieg DB, Maudsley AA, Sappey MD, Weiner MW. Phosphorus-31 MR spectroscopic imaging (MRSI) of normal and pathological human brains. *Magn Reson Imaging* 1992; 10:227–43.

40 Laxer KD, Hubesch B, Sappey MD, Weiner MW. Increased pH and inorganic phosphate in temporal seizure foci demonstrated by [31P]MRS. *Epilepsia* 1992; 33:618–23.

41 Kuzniecky R, Elgavish GA, Hetherington HP, Evanochko WT, Pohost GM. In vivo 31P nuclear magnetic resonance spectroscopy of human temporal lobe epilepsy. *Neurology* 1992; 42:1586–90.

42. Hugg JW, Helpern JA, Welch KMA. ^{31}P MRS detects ictal hypermetabolism in migraine and interictal hypometabolism in epilepsy. *Soc Magn Reson Med Abstracts* 1993; **12**:1549.

43. Petroff OA, Spencer DD, Alger JR, Prichard JW. High-field proton magnetic resonance spectroscopy of human cerebrum obtained during surgery for epilepsy. *Neurology* 1989; **39**:1197–202.

44. Woods BT, Chiu TM. Induced and spontaneous seizures in man produce increases in regional brain lipid detected by in vivo proton magnetic resonance spectroscopy. *Adv Exp Med Biol* 1992; **318**:267–74.

45. Matthews PM, Andermann F, Arnold DL. A proton magnetic resonance spectroscopy study of focal epilepsy in humans. *Neurology* 1990; **40**:985–9.

46. Breiter SN, Soher BJ, Lessor RP, Barker PB. Proton spectroscopic imaging and MRI of epilepsy. *Soc Magn Reson Med Abstracts* 1993; **12**:1546.

47. Layer G, Traber F, Muller LU, Bunke J, Elger CE, Reiser M. Spectroscopic imaging. A new MR technique in the diagnosis of epilepsy? *Radiologe* 1993; **33**:178–84.

48. Cendes F, Andermann F, Preul MC, Arnold DL. Proton MR spectroscopic imaging in the investigation of temporal lobe epilepsy. *Soc Magn Reson Med Abstracts* 1993; **12**:431.

49. Connelly A, Jackson GD, Duncan JS, King MD, Gadian DG. The contribution of ^{1}H MRS to the presurgical assessment of temporal lobe pathology in patients with intractable epilepsy. *Soc Magn Reson Med Abstracts* 1993; **12**:1463.

50. Constantinidis I, Epstein CM, Peterman S, et al. Can ^{1}H spectroscopic imaging lateralize temporal lobe seizure foci? *Soc Magn Reson Med Abstracts* 1993; **12**:429.

51. Fujimoto T, Yamada K, Nakano T, Tsuji T, Akimoto H. Proton spectroscopy using MRS with 2T in patients with epileptic psychoses. *Soc Magn Reson Med Abstracts* 1993; **12**:1541.

52. Ng TC, Comair Y, Xue M, et al. Proton chemical shift imaging for the presurgical localization of temporal lobe epilepsy. *Soc Magn Reson Med Abstracts* 1993; **12**:428.

53. Jackson GD, Berkovic SF, Duncan JS, Connelly A. Optimizing the diagnosis of hippocampal sclerosis using MR imaging. *Am J Neurorad* 1993; **14**:753–62.

54. Rothman DL, Petroff OA, Behar KL, Mattson RH. Localized ^{1}H NMR measurements of gamma-aminobutyric acid in human brain *in vivo*. *Proc Natl Acad Sci U S A* 1993; **90**:5662–6.

55. Petroff O, Rothman D, Behar K, Mattson R. The effect of vigabatrin on GABA, glutamate, and glutamine levels in human brain measured in vivo with ^{1}H NMR spectroscopy. *Soc Magn Reson Med Abstracts* 1993; **12**:434.

56. Zhong J, Petroff OAC, Prichard JW, Gore JC. Changes in water diffusion and relaxation properties of rat cerebrum during status epilepticus. *Magn Reson Med* 1993; **30**:241–6.

57. Moseley ME, Wendland MF, Kucharczyk J. Magnetic resonance imaging of diffusion and perfusion. *Top Mag Reson Imaging* 1991; **3**:50–67.

58. Kanfer JN, Pettegrew JW, Moossy J, McCartney DG. Alterations of selected enzymes of phospholipid metabolism in Alzheimer's disease brain tissue as compared to non-Alzheimer's demented controls. *Neurochem Res* 1993; **18**:331–4.

59. Brown GG, Levine SR, Gorell JM, et al. In vivo 31P NMR profiles of Alzheimer's disease and multiple subcortical infarct dementia. *Neurology* 1989; **39**:1423–7.

60. Bottomley PA, Cousins JP, Pendrey DL, et al. Alzheimer dementia: quantification of energy metabolism and mobile phosphoesters with P-31 NMR spectroscopy. *Radiology* 1992; **183**:695–9.

61. Murphy DG, Bottomley PA, Salerno JA, et al. An in vivo study of phosphorus and glucose metabolism in Alzheimer's disease using magnetic resonance spectroscopy and PET. *Arch Gen Psychiatry* 1993; **50**:341–9.

62. Klunk WE, Panchalingam K, Moossy J, McClure RJ, Pettegrew JW. N-acetyl-L-aspartate and other amino acid metabolites in Alzheimer's disease brain: a preliminary proton nuclear magnetic resonance study. *Neurology* 1992; **42**:1578–85.

63. Christiansen P, Schlosser A, Henriksen O. Reduced N-acetyl aspartate content in frontal brain of patients with Alzheimer's disease. *Proc Soc Magn Reson Med* 1993; **12**:226.

64. Renshaw PF, Satlin A, Johnson KA. Parietal lobe proton MRS in patients with Alzheimer's disease. *Proc Soc Magn Reson Med* 1993; **12**:232.

65. Itoh S, Kimura H, Matsuda T, et al. Evaluation of neuronal changes in brain with Alzheimer's disease using localized proton MR spectroscopy. *Proc Soc Magn Reson Med* 1993; **12**:229.

66. Charles HC, Boyko OB, Lazeyras FS, Tupler L, Krishman R. Metabolic brain mapping in Alzheimer's disease using high resolution proton spectroscopy. *Proc Soc Magn Reson Med* 1993; **12**:228.

67. Ide M, Naruse S, Furuya S, et al. ^{1}H-CSI study of the Alzheimer's disease. *Proc Soc Magn Reson Med* 1993; **12**:231.
68. Jungling FD, Wakhloo AK, Stadtmuller G, Hennig J. Localized ^{1}H-spectroscopy in the hippocampus of normals and patients with Alzheimer's disease. *Proc Soc Magn Reson Med* 1993; **12**:1555.
69. Moats RA, Shonk T, Ernest T, Miller BL, Ross BD. In vivo quantitative ^{1}H difference spectroscopy in patients with probably Alzheimer's disease. *Proc Soc Magn Reson Med* 1993; **12**:230.
70. Kesslak JP, Drost DJ, Naruse S, et al. Single volume proton spectroscopy in Alzheimer's disease patients: A multicenter pilot study. *Proc Soc Magn Reson Med* 1993; **12**:227.
71. Carr CA, Guimaraes AR, Growdon JH, Gonzales RG. Combining proton MRS and MRI morphometry increases accuracy in the diagnosis of Alzheimer's disease. *Proc Soc Magn Reson Med* 1993; **12**:233.
72. Oberhaensli RD, Hilton JD, Bore PJ, Hands LJ, Rampling RP, Radda GK. Biochemical investigation of human tumours in vivo with phosphorus-31 magnetic resonance spectroscopy. *Lancet* 1986; **2**:8–11.
73. Jeske J, Herholz K, Heindel W, Heiss WD. Metabolic studies of gliomas with positron emission tomography and phosphorus 31 MR spectroscopy in diagnosis and treatment planning. [German]. *Onkologie* 1989; **1**:42–5.
74. Heiss WD, Heindel W, Herholz K, et al. Positron emission tomography of fluorine-18-deoxyglucose and image-guided phosphorus-31 magnetic resonance spectroscopy in brain tumors. *J Nucl Med* 1990; **31**:302–10.
75. Luyten PR, Marien AJ, Heindel W, et al. Metabolic imaging of patients with intracranial tumors: H-1 MR spectroscopic imaging and PET. *Radiology* 1990; **176**:791–9.
76. Herholz K, Heindel W, Luyten PR, et al. In vivo imaging of glucose consumption and lactate concentration in human gliomas. *Ann Neurol* 1992; **31**:319–27.
77. Alger JR, Sillerud LO, Behar KL, et al. In vivo carbon-13 nuclear magnetic resonance studies of mammals. *Science* 1981; **214**:660–2.
78. Fulham MJ, Bizzi A, Dietz MJ, et al. Mapping of brain tumor metabolites with proton MR spectroscopic imaging: clinical relevance. *Radiology* 1992; **185**:675–86.
79. Brown TR. Practical applications of chemical shift imaging. *NMR Biomed* 1992; **5**:238–43 [review].
80. Murphy BJ, Stoyanova R, Srinivasan R, et al. Proton-decoupled 31P chemical shift imaging of the human brain in normal volunteers. *NMR Biomed* 1993; **6**:173–80.
81. Arnold DL, Matthews PM, Francis GS, O'Connor J, Antel JP. Proton magnetic resonance spectroscopic imaging for metabolic characterization of demyelinating plaques. *Ann Neurol* 1992; **31**:235–41.
82. De Stefano N, Francis G, Antel JP, Arnold DL. Reversible decreases of N-acetylaspartate in the brain of patients with relapsing remitting multiple sclerosis. *Proc Soc Magn Reson Med* 1993; **12**:280.
83. Lauterbur PC. Some applications of C^{13} nuclear magnetic resonance spectra to organic chemistry. *Ann N Y Acad Sci* 1958; **70**:841–57.
84. Lauterbur PC. C^{13} nuclear magnetic resonance spectra. *J Chem Phys* 1957; **26**:217–18.
85. Holm CH. Observation of chemical shielding and spin coupling of C^{13} nuclei in various chemical compounds by nuclear magnetic resonance. *J Chem Phys* 1957; **26**:707–8.
86. Matwiyoff NA, Ott DG. Stable isotope tracers in the life sciences and medicine. *Science* 1973; **181**:1125–33.
87. Alger JR, Frank JA, Bizzi A. et al. Metabolism of human gliomas: assessment with H-1 MR spectroscopy and F-18 fluorodeoxyglucose PET. *Radiology* 1990; **177**:633–41.
88. Behar KL, Petroff OAC, Prichard JW, Alger JR, Shulman RG. Detection of metabolites in rabbit brain by ^{13}C-NMR spectroscopy following administration of [1-^{13}C] glucose. *Magn Res Med* 1986; **3**:911–20.
89. Rothman DL, Novotny EJ, Shulman GI, et al. ^{1}H-[^{13}C] NMR measurements of [4-^{13}C] glutamate turnover in human brain. *Proc Natl Acad Sci U S A* 1992; **89**:9603–6.
90. Kalinowski HO, Berger S, Braun S. *Carbon-13 NMR Spectroscopy*. John Wiley and Sons: Chichester, 1988.
91. Eakin RT, Morgan LO, Gregg CT, Matwiyoff NA. Carbon-13 nuclear magnetic resonance spectroscopy of living cells and their metabolism of a specifically labeled ^{13}C substrate. *FEBS Lett* 1992; **28**:259–64.
92. Gruetter R, Novotny EJ, Boulware SD, et al. Direct measurement of brain glucose concentrations in humans by 13C NMR spectroscopy. *Proc Nat Acad Sci U S A* 1992; **89**:1109–12 [published erratum appears in *Proc Natl Acad Sci U S A* 1992; **89**:12208].
93. Rothman DL, Behar KL, Hetherington HP, et al. ^{1}H observed ^{13}C decoupled spectroscopic measurements of

lactate and glutamate in the rat brain in vivo. *Proc Natl Acad Sci U S A* 1985; **82**:1633–7.
94 **Mason GF, Rothman DL, Behar KL, Shulman RG.** NMR determination of the TCA cycle rate and alpha-ketoglutarate exchange rate in rat brain. *J Cereb Blood Flow Metab* 1992; **12**:434–47.
95 **Petroff OA, Novotny EJ, Avison M, et al.** Cerebral lactate turnover after electroshock: in vivo measurements by 1H/13C magnetic resonance spectroscopy. *J Cereb Blood Flow Metab* 1992; **12**:1022–9.
96 **Prichard JW.** Magnetic resonance spectroscopy of cerebral metabolism in vivo. In: Asbury AK, McKhann GM, MacDonald WI. *Diseases of the Nervous System.* Saunders: Philadelphia, 1992;1589–1605.
97 **Rothman DL, Howseman AM, Graham GD, et al.** Localized proton NMR observation [3-13C]lactate in stroke after [1-13C]glucose infusion. *Magn Reson Med* 1991; **21**:302–7.

2 The whole body NMR spectroscopy examination

H Cecil Charles

INTRODUCTION

Early studies of NMR spectroscopy of tissue samples suggested the possibility of significant clinical impact if the acquisition of localized spectroscopic information from human subjects should prove feasible. The first human spectroscopic studies were obtained without the benefit of image guidance and were often limited to diffuse processes such as metabolic diseases (e.g. myopathies). With the advent of MRI technology, the basic tools for MRS acquisition became more widely available in the academic medical community, and the problem of evaluation of focal lesions was addressed by the development of a variety of image-guided localization schemes. This chapter focuses on the underlying principles of the NMR experiment as well as on the practical requirements for image-guided NMR spectroscopy data acquisition in the clinical environment.

MR PHYSICS: A VERY SHORT COURSE

As generally practised, three different magnetic fields are utilized in image-guided MRS studies. The static magnetic field (B_0) establishes longitudinal magnetization in the sample and defines the chemical shift dispersion. When a group of magnetically active spin-1/2 nuclei (e.g. ^1H, ^{31}P, ^{19}F, ^{13}C) are placed in a homogeneous static magnetic field the nuclei populate two energy states – a lower energy state, said to be parallel to the magnetic field, and an upper energy state, said to be antiparallel to the magnetic field with a very slight excess in the parallel state. These magnetic moments are oriented at an angle to the field and precess about the field at the Larmor frequency (Equation 1),

$$f = \frac{\gamma}{2\pi} B_0 \qquad (1)$$

where γ is the gyromagnetic ratio, a constant for any specific nucleus, as represented by the vector diagram in Fig. 2.1. The resultant longitudinal magnetization can be represented as the vector sum (M_z) of the ensemble of nuclei in the magnetic field (Fig. 2.1). The energy difference between these states is given by equation 2,

$$\Delta E = h \times f \qquad (2)$$

where f is the frequency of the energy to be supplied by the second magnetic field, a time-varying radiofrequency magnetic field.

The application of the second radiofrequency (rf) magnetic field orthogonal to the static field produces a net torque on the magnetization and flips the resultant vector, M_z, through an angle, θ

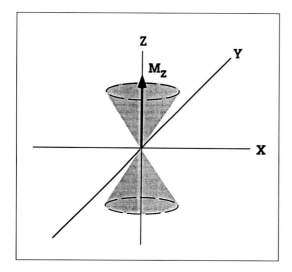

Fig. 2.1 The slight excess magnetization aligned 'parallel' to the static field (Z) generates a resultant bulk longitudinal magnetization (M_z).

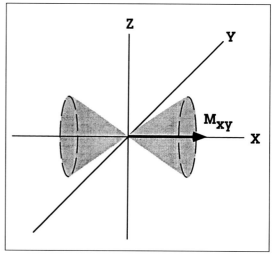

Fig. 2.2 Following a '90°' radiofrequency pulse (the second magnetic field) longitudinal magnetization nutates to yield transverse magnetization (M_{xy}).

(Fig. 2.2). When $\theta = 90°$, the resultant vector will lie in the x–y plane and has become transverse magnetization (M_{xy}). If the rf field is discontinued, a signal can be detected from this precessing transverse magnetization and digitized in a computer for further analysis (Fig. 2.3). This signal, a free induction decay (FID), when subjected to a fast Fourier transformation (FFT) yields a spectrum such as that in Fig. 2.4.

The fact that different signals are seen in the spectrum is due to the chemical shift. The same nuclei in different parts of a molecule (e.g. water or a triglyceride) are exposed to the same static magnetic field (B_0) but are shielded from that field to varying degrees by the differing electron density about each nucleus. Thus a nucleus that is shielded from the static field has a lower precession frequency whereas a nucleus where the local field is enhanced has a higher frequency. Because of practical considerations, the chemical shift is generally stated in parts per million Hz (ppm) (Equation 3) and is usually referenced to a specific material. For instance, *in vivo* references include water at 4.7 ppm for ^{1}H spectroscopy or creatine phosphate at 0 ppm for ^{31}P spectroscopy (Fig. 2.5).

$$\text{ppm} = \frac{\text{observed frequency (Hz)} - \text{reference frequency (Hz)}}{\text{Larmor frequency (MHz)}}$$

(3)

The need to obtain localized spectra from specific regions in the body demands the use of a third magnetic field in the form of a field gradient in order to provide the link between frequency and space (and thus localization). While the basic concept of changing the local magnetic field to alter the resonant frequency appears obvious today, Lauterbur's observation that the alterations in the local field by application of a (linear) magnetic field gradient created a 'yoke' between space and frequency was a critical step in the evolution of MRI and MRS.

The whole body NMR spectroscopy examination

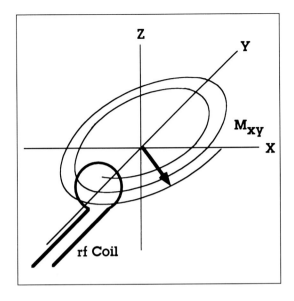

Fig. 2.3 *After the radiofrequency (rf) pulse has been turned off the precessing transverse magnetization induces a voltage in the rf coil to yield a free induction decay (FID).*

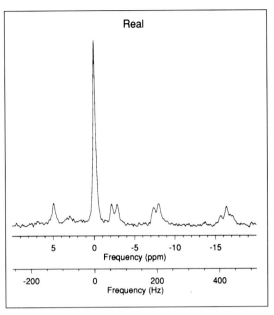

Fig. 2.5 *The phosphorus spectrum from resting human quadraceps is shown with frequency scales in Hz and ppm. In both cases, the spectral components are referenced to phosphocreatine set as 0 ppm. Since ppm is normalized to the Larmor frequency, it is independent of magnetic field strength and makes comparisons between different systems easier that simply reporting the frequency scale in Hz.*

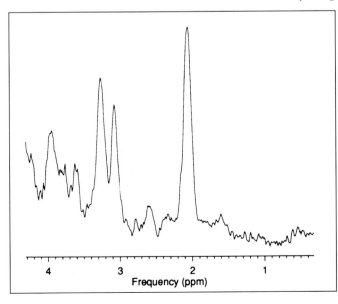

Fig. 2.4 *A localized proton spectrum (STEAM, TE 270 ms) obtained from normal human brain. The signals at 2, 3 and 3.2 ppm represent N-acetylaspartate, creatine + phosphocreatine and the trimethylamine moiety of choline containing metabolites (e.g. choline, phosphocholine and glycerophosphocholine), respectively.*

29

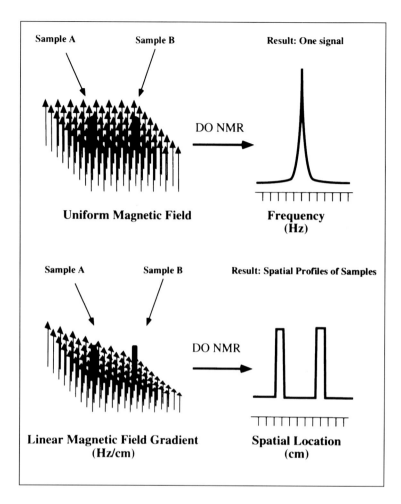

Fig. 2.6 In the presence of a uniform magnetic field (upper diagram) two water samples (A and B) would simply yield a single spectral component at the resonant frequency for the water molecule. The imposition of an inhomogeneous field, in this case a linear gradient, would cause the spatial distribution (profiles) of the two tubes. The chemical shift information has been sacrificed to obtain spatial information. Localized spectroscopy uses a combination of these two tactics to yield spatially resolved spectra.

A graphic representation of this phenomenon is depicted in Fig. 2.6, in which two discrete samples yield a single signal in a homogeneous field, in contrast to a projection image in a linear field gradient imposed along one axis, which encodes frequency as spatial distribution. When this tactic is applied in a spatially systematic fashion, a two- or three-dimensional image can be reconstructed from the spatially encoded frequency data. This general encoding tactic is utilized in various localization schemes for clinical spectroscopy (see below).

THE CLINICAL NMR SCANNER

A general high field (greater than 1 Tesla) clinical NMR scanner possesses most of the hardware components required for hydrogen spectroscopic acquisitions as it is delivered from its manufacturer. However, the instrumental criteria necessary for localized NMR spectroscopy, though similar to those required for imaging, are somewhat more stringent in the areas of magnetic field homogeneity and stability. The components of the scanner

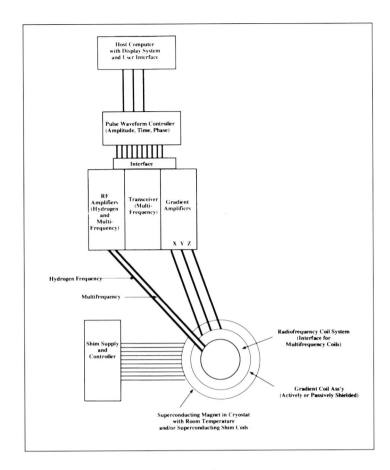

Fig. 2.7 A general block diagram of a magnetic resonance scanner is shown. The major hardware enhancements required for spectroscopy are a multifrequency (or broadband) transceiver for accessing nuclei other than hydrogen, radiofrequency amplifiers capable of multifrequency (or broadband) operation ideally with a separate cabling and interface to the rf coil system in the magnet. A method of automatic shim control, either at the shim supply or using the gradient amplifiers is required. The software and display for data acquisition should be integrated with the imaging software for image guided spectroscopic acquisition.

include the magnet subsystem, gradient subsystem, radiofrequency subsystem, data acquisition subsystem, and display and analysis subsystems (Fig. 2.7). The three magnetic field components will be addressed separately (and briefly) as to relevant requirements for the study of NMR spectroscopy. The additional components required for studies of nuclei other than hydrogen are also described.

The magnet subsystem (B_0)

The static magnetic field strength must be strong enough to provide adequate dispersion for differentiation of the desired spectral components. Most clinical spectroscopy has been accomplished at the commercially available strength of 1.5 Tesla, although work has been reported at higher field strengths (up to 4 Tesla) and at lower field strengths (0.5 Tesla and 1.0 Tesla). The stability of the field is critical, since the frequency resolution we seek is often on the order of 1 Hz (1 part in 63.8 million at 1.5 Tesla, 1 part in 170 million at 4 Tesla). Consequently, magnet field drift must be better than 1 Hz over the period of a clinical spectroscopy examination, a specification met by most superconducting magnet designs.

The volume of the region of high quality magnetic field homogeneity is just as important as field strength once adequate dispersion has been

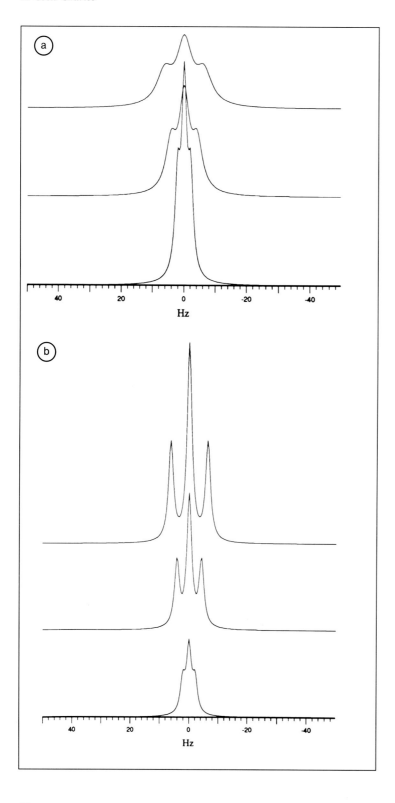

Fig. 2.8 *The need for excellent magnet homogeneity is shown in the computed spectra. The parameters in Table 2.1 have been modeled in these spectra showing the impact of increasing the magnetic field from 0.5 T to 1.5 T with homogeneity constant in ppm (2.8a) and Hz (2.8b). It is clear that the magnet homogeneity in ppm must be improved at higher magnetic fields to gain the advantages of signal to noise and chemical shift dispersion.*

Table 2.1. Model parameters for Fig. 2.8.

Field strength	0.5 Tesla	1.0 Tesla	1.5 Tesla
Area	100:200:100	200:400:200	300:600:300
Width (constant ppm)	2 Hz	4 Hz	6 Hz
Intensity (constant ppm)	50:100:50	50:100:50	50:100:50
Width (constant Hz)	2 Hz	2 Hz	2 Hz
Intensity (constant Hz)	50:100:50	100:200:100	150:300:150

obtained. Increases in the volume of high homogeneity affects the signal to noise ratio (S/N) and can result in smaller volume elements as well as allowing access to regions of the patient physically further away from the isocenter of the magnet. Unfortunately, specifications of magnetic field homogeneity are often stated in parts per million (ppm) maximum deviation from the nominal value in a prescribed volume. It is important to consider not only the 'best' homogeneity at the center of the magnet, typically in a 10 cm diameter spherical volume (DSV), but also performance in larger volumes and even alternative shapes of high quality field (e.g. cylindrical volumes).

If the magnetic field homogeneity is maintained with increasing field strength, two important criteria improve – signal to noise ratio and chemical shift dispersion. Improvements in signal to noise ratio with field strength have been exploited in MRI and this effect is well described in the refereed literature as well as in manufacturers' literature. The increase in chemical shift dispersion occurs because the observed resonant frequency is proportional to B_0 (the Larmor relationship), so that differences in resonant frequency of individual spectral lines (or chemical shift) are linearly proportional to magnetic field strength. The degree to which this enhancement in frequency resolution can be exploited is a function of magnetic field homogeneity. Figure 2.8 shows the results of a model of three spectroscopic signals with 1:2:1 concentration, separated by 0.1 ppm. In this model, the signal was assumed to increase linearly with B_0, and the homogeneity was kept constant either in ppm (Fig. 2.8a) or Hz (Fig. 2.8b). The parameters for each spectrum are summarized in Table 2.1. Notice that the improvements in chemical shift dispersion are evident only with improved homogeneity of the magnetic field. The adjustment of magnetic field homogeneity (known as 'shimming') on the individual patient is of critical importance, and the advent of automatic 'shimming' programs have greatly enhanced both the quality of *in vivo* spectra as well as the ease of acquisition. An example of ^{1}H NMR spectra after an autoshimming program was used is shown in Fig. 2.9.

Radiofrequency subsystem (B_1)

The second magnetic field is delivered by the radiofrequency subsystem. This system consists of a stable frequency source to deliver rf energy at the Larmor frequency, a modulation circuit that allows programmed amplitude, phase and/or frequency variations of the rf signal, amplification stages (to several kilowatts of power), and a transmit–receive switching system. The rf signal is delivered to rf antennas in the main magnet. These

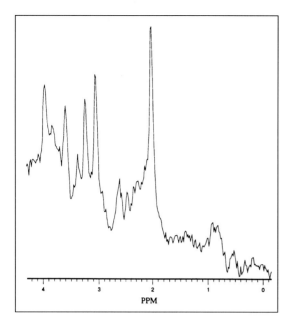

Fig. 2.9 A short echo (TE 20 ms) STEAM spectrum (8 cc) obtained from human brain using an automated shimming program for homogeneity adjustment.

antennas can be constructed in a variety of physical conformations, ranging from simple flat loops to cylindrical volume antennas to flexible arrays of antennas. The requirements of the second rf field are somewhat stringent, with the frequency source being stable to 1 part in 10^{10}. The modulation signals require stability to $\leq 1\%$ in amplitude and $\leq 0.5°$ in phase. The dynamic range requirement of the modulation circuit can be quite large for water-suppressed ^{1}H spectroscopy where the amplitude of the CHESS (Frahm) pulses can be a factor of 1000 less than the slice selection pulses in a Stimulated Echo Acquisition Method (STEAM)[1] (Frahm) or PRESS[2] sequence. If multinuclear capabilities are required, the excitation channel must allow broadband or multifrequency operation and the rf amplifier must operate at more than one frequency (either as a broadband unit or a tunable system). The receiver section of the rf system begins with a preamplifier with a low noise figure (≤ 0.5 dB) and adequate gain (nominally 20–30 dB) protected by the transmit–receive switching system. The output of the preamplifier is demodulated, detected, and digitized in the receiver. As with the transmitter, if multinuclear studies are required, the receiver section must operate at more than one frequency.

The gradient subsystem (G)

In addition to the static magnetic field and rf magnetic field, a magnetic field gradient system is required to provide controlled, usually linear, variations in the magnetic field yield spatial encoding. Relevant characteristics of the gradient subsystem include maximum strength (measured in mT/m), slew rate (T/s) or rise time (5 to 90% of peak), linearity (% deviations of FOV), stability (% variation), duty cycle (%) and degree of eddy current suppression. Examples of gradient specifications required for MRS are shown in Table 2.2. Eddy current suppression can be accomplished with active or passive shielding, pre-emphasis (analog or digital), and post-processing. Current trends include combinations of any or all of these suppression techniques. Suppression of eddy currents are critical to the quality of clinical magnetic resonance spectroscopy, since temporal and spatial variations in the magnetic field will not only distort the localization of the spectra but also the observed linewidths and chemical shift dispersion.

THE IMAGE-GUIDED NMR SPECTROSCOPY EXAMINATION

The first step in an image-guided spectroscopy examination is the choice of an imaging protocol

Table 2.2. Suggested gradient specifications.

Parameter	Suggested specification
Gradient strength	≥ 10 mT/m
Gradient rise time	≤ 500 us to full scale
Gradient slew rate	≥ 20 T/s
Gradient linearity	<2% full field distortion (all axes)
Duty cycle	$\simeq 100\%$
Eddy current suppression	$< 0.5°$ signal phase variation at full gradient strength applied for 50 ms (all axes)
Stability	$< 0.5°$ signal phase variation for repeated pulses (all axes)

Table 2.3. A general scheme for localized spectroscopy, with time frames for the various stages.

Stage of examination	Time
Patient positioning	5–10 minutes
Scout scanning	1–5 minutes
T_2 imaging	8–17 minutes
Coil change (if necessary)	5–10 minutes
Spectroscopy scout image	1–5 minutes
Automatic shimming	3 minutes
Spectroscopy prescan	2–10 minutes
Spectroscopy scan	5–40 minutes
T_1 imaging	4–8 minutes
T_1 imaging plus contrast	8–16 minutes

that will provide image contrast that is adequate for locating the structure or region of interest. The entire examination must be completed in a time frame such that the patient does not become uncomfortable, since patient motion degrades the quality of the study. While there is clearly no absolute rule for the length of such studies, general experience suggests 2 hours as a practical maximum for the total examination. A general scheme for localized spectroscopy is shown in Table 2.3 with approximate times for the various stages.

Coil changes may be made during the procedure much in the same manner that surface coils are added to a scan protocol after a scout image in MRI examinations. In the case of multinuclear studies, the use of dual-tuned or multiple-frequency arrays can minimize the time required for such coil changes. In all cases, patient comfort remains an important consideration to ensure good data quality.

Localization techniques for clinical NMR spectroscopy

Numerous tactics have been developed for localized spectroscopy and have been nicely reviewed by Aue.[3] The majority of clinical MRS applications have been accomplished with three general types of localization: surface (or local) coil localization, single voxel spectroscopy, and chemical shift (or spectroscopic) imaging. Often, combinations of these three techniques are utilized in the same study.

Surface coil localization

Surface coil localization depends on the inherent inhomogeneity of the rf field of the coil, which may be a simple single loop or a more complicated array.[4] Localization can be accomplished by varying the rf power applied to the coil (Fig. 2.10). Each image in Fig. 2.10 represents a change of 2 dB in transmitter power. The major limitation to such studies is contamination of the spectrum by intervening tissue if the lesion is below the surface. Nevertheless, surface coil localization remains useful in studies (such as exercise studies) where some temporal intervention is tracked (Fig. 2.11) or in clinical cases where the metabolic effect is considered diffuse and uniform (e.g. in myopathies).

Voxel localization

The clinical evaluation of focal lesions requires localization beyond that of simple local rf coils. Two classes of techniques address the problem of

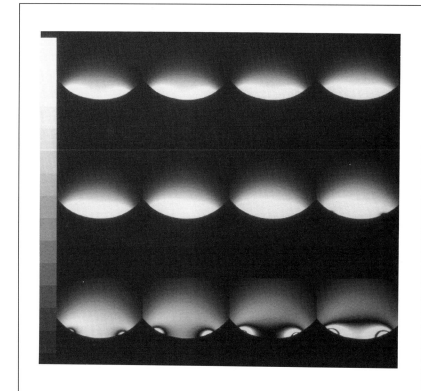

Fig. 2.10 Surface coils, due to their inherent inhomogeneous fields, can be used to effect some degree of localization. Each image represents a change in transmit power of 2 dB moving from the upper left image to the lower right. Notice that some degree of depth localization is obtained, although at increasing power, signal close to the coil begins to increase again. While this form of localization is poor, the use of surface coils in conjunction with other localization (SVS or CSI) is common.

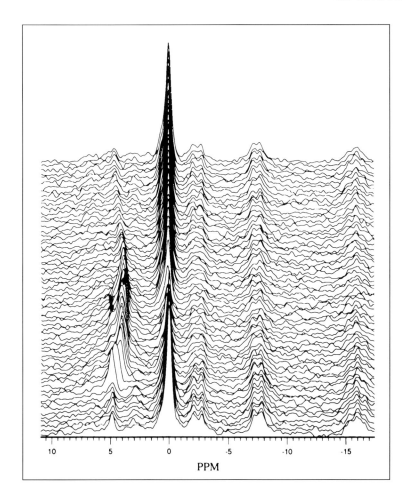

Fig. 2.11 An example of surface coil localized phosphorous spectra from gastrocnemius muscle in a subject with peripheral vascular disease (prior to therapy). Each spectrum represents 32 seconds of signal averaging. After 10 baseline scans, the subject performed a voluntary isometric contraction of the lower leg muscles at a level of minimum discomfort. Notice the rapid shift of the inorganic phosphate signal (5 ppm) indicating decreasing pH. Notice also the presence of different populations of muscles represented by multiple inorganic phosphate signals (arrows). While this spatial contamination is somewhat problematic, the temporal resolution allows determination of recovery times for metabolic indices such as pH.

focal lesions – single voxel spectroscopy and chemical shift (spectroscopic) imaging. In general, single voxel spectroscopy (SVS) operates by selecting three orthogonal slices (or slices that are not orthogonal) in such a manner that the region at the intersection of the slices is the only source of signal (Fig. 2.12). The three SVS techniques in general use are STEAM,[1] PRESS,[2] and ISIS,[5] with STEAM and PRESS providing localization in a single T_R period and ISIS providing localization by encoding the voxel in a minimum of 8 T_R periods.

STEAM selects three orthogonal slices with 90° pulses and acquires the resulting stimulated echo from the confluent voxel. Since several other echoes are generated at the same time, gradient suppression is required to obtain localization in a single acquisition. The PRESS technique acquires the second echo from a 90°–180°–180° pulse train. For the same echo time, the PRESS technique yields twice the single to noise ratio of the STEAM technique. However, shorter echo times are possible with the STEAM technique: it allows detection of more complicated spin systems such as glutamate and glutamine. Examples of SVS spectra at short (20 ms) and long (270 ms) echo times are shown in Figs. 2.9 and 2.4, respectively.

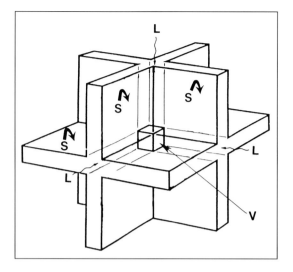

Fig. 2.12 The general principle of single voxel spectroscopy is demonstrated in this schema. Three intersecting slices (S) are arranged such that at the mutual intersection a voxel (V) is created. In the process three lines (L) are also created. The challenge of a SVS technique is eliminating all signal in the object other than that from the desired voxel.

The disadvantage of single voxel spectroscopy is simply that, for a given scan time, only one spectrum is obtained. If the image contrast is adequate to guide the placement of the voxel and if no information regarding metabolic heterogeneity is required, then SVS is an appropriate tool for spectroscopic evaluation. In contrast, chemical shift (spectroscopic) imaging[5] allows coverage of a field of view with a prescribed number of volume elements. Chemical shift imaging is similar to conventional imaging with the exception that, as generally practised, only one point in k-space is encoded per T_R period, whereas an entire line of k-space is encoded per T_R period in conventional Fourier imaging (Fig. 2.13). After the study is completed, the data are reconstructed (usually by FFT although other tactics are possible) to yield spatially resolved spectra as well as spectrally resolved images. The point spread function represents the shape of the individual volume elements and approximates a sinc(x) function in each sampled dimension. Spatial contamination is most apparent from adjacent volume elements (Fig. 2.14), which is analogous to the situation in SVS selection owing to chemical shift arrows. An intrinsic advantage to CSI localization is the presence of adjacent volume elements (Fig. 2.15), which allow easy comparison of anatomic areas.

DATA RECONSTRUCTION AND ANALYSIS

Data reconstruction in MRS varies somewhat from routine image reconstruction, owing to the

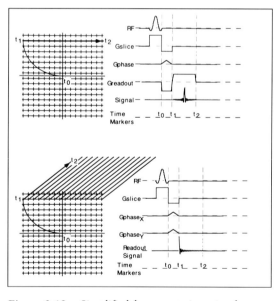

Figure 2.13. Simplified k-space trajectories for conventional imaging where an entire line k-space is read out at a time compared with conventional chemical shift imaging where after a point in k-space has been encoded, the spectrum is read out in the chemical shift dimension.

The whole body NMR spectroscopy examination

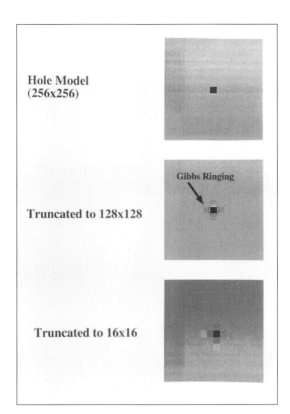

Fig. 2.14 Spatial contamination in sparsely sampled data sets such as CSI is shown in the model sampling a 'hole' in a bright object with a matrix resolution of 256 × 256. When the data are truncated to 128 × 128 and reconstructed, Gibbs ringing close to the hole is seen. This is a familiar artefact in images where bright objects are close to darker objects. While the artefact is similar, truncation to a 16 × 16 matrix shows Gibbs ringing extending further spatially in the object.

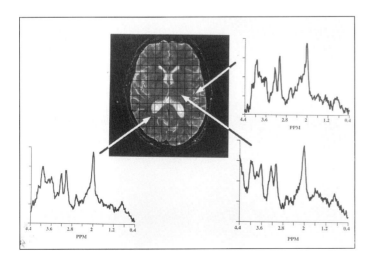

Fig. 2.15 The advantage of CSI is evident is a study such as this where regional comparisons of spectra (eg, subcortical gray matter) as well as contralateral comparisons can be easily made.

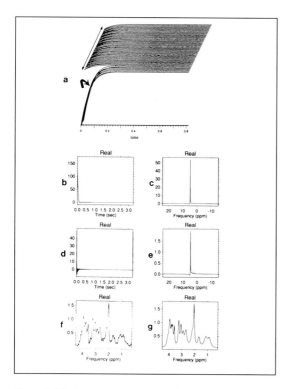

Fig. 2.16 The processing scheme for PROBE (GEMS, Milwaukee, WI) is shown below. The data are acquired in bins (a) with non-water suppressed reference scans at the beginning (curly arrow) and water suppressed data following the reference scans. The data are summed into two bins, the reference scan (b) and water suppress scan (d). The reference scan is Fourier transformed (c) and phase information is determined for use with the water suppressed scan. The water suppressed scan is transformed, either without (e) or with a high pass filter (f). The expanded spectrum (f) can be fitted on a workstation yielding results such as (g) as well as a table of peak parameters for other calculations and reference normalization.

additional dimension. Figure 2.16 shows an example of each stage of the reconstruction process with a single voxel STEAM ^{1}H MRS taken using the PROBE software (GE Medical Systems, Milwaukee, Wisconsin, USA) package as an example. The PROBE software acquires a series of water reference scans followed by several block acquisitions of the water suppressed data. The water reference blocks and the water suppressed blocks are combined, zerofilled, filtered and Fourier transformed. The water reference scans are used to generate the phase correction parameters needed for the spectral data and can be used for water reference intensity normalization (referenced normalized quantitation).

While data reconstruction can be automated, significant effort has been directed at spectral analysis and parameterization. One of the tasks that remains in order to move the method into the clinical domain is the automation and quantitative treatment of both CSI and SVS data. The data need to be presented in a clinically relevant manner and referenced to a specific anatomic location. Some degree of quantification is necessary for comparison with other studies. This can be achieved simply, by taking data at the same time from normal tissue. It can also be referenced to an internal standard, for example the water resonance from the same location or an external reference placed close to the region under study.

REFERENCES

1 **Frahm J, Merboldt K, Hanicke W.** Localized proton spectroscopy using stimulated echoes. *J Mag Res* 1987; 72: 502–8

2 **Bottomley PA,** Spatial localization in NMR spectroscopy in vivo. *Ann NY Acad Sci* 1987; 508: 333–48.

3 **Aue WP.** Localization methods for *in vivo* nuclear magnetic resonance spectroscopy. *Rev Magn Reson Med* 1986; 1(1): 21–72.

4 **Ackerman J, Grove TH, Wong GG, et al.** Mapping of metabolites in whole animals by 31P NMR using surface coils. *Nature* 1980; 283: 167–70.

5 **Ordidge R, Connelly A, Lohman JAB.** Image selected in-vivo spectroscopy (ISIS). A new technique for spatially selective NMR spectroscopy. *J Mag Res* 1986; 66: 283–94.

3 Problems in making useful measurements of *in vivo* spectra

Ian R Young, Martyn Paley

INTRODUCTION

Whereas diagnostic imaging is essentially qualitative in character and depends on the interpretive skill and experience of a radiologist, magnetic resonance spectroscopy (MRS) is more a quantitative science with the target of determining actual concentrations of metabolites. This chapter reviews the problems that can arise in obtaining accurate values from *in vivo* whole body MRS studies. These problems include those that would be experienced even under the ideal conditions of high resolution spectroscopy of biochemical samples and those that are introduced solely by the nature of the human subjects being studied. The obvious predilection that humans have for movement, their size, and the difficulty of placing the receiver coil in the ideal position for making an observation all lead to difficulties. Furthermore, human studies are constrained by examination time, particularly where the subject is a sick patient.

In keeping with the rest of the book, it is not proposed to give an exhaustive mathematical analysis of MRS problems and their potential solutions, but rather to provide an overview of the forms and extent of artifacts encountered in various *in vivo* spectroscopy strategies. High resolution MRS has seen the development of an enormous range of sophisticated methods, many of which could potentially be applied to *in vivo* studies. This chapter confines itself to methods that have been used, or justifiably suggested for use, in whole body experiments. Many of the methods described in this chapter are specific to certain nuclei such as hydrogen or phosphorus, and this is noted where appropriate.

Any spectral experiment involves a series of time-dependent processes:
- preparation;
- excitation;
- spatial localization;
- manipulation;
- acquisition;
- processing; and
- display.

Artifacts may arise during any of these steps, and these will be reviewed with respect to both single voxel and chemical shift imaging methods.

BASIC PRINCIPLE OF SPECTROSCOPIC QUANTIFICATION

The principle of spectroscopic quantitation is very simple:

The free induction decay (FID) signal detected is proportional to the quantity of metabolite that has been excited, and the area of the peak in the resultant frequency spectrum is proportional to the FID signal.

Both these assumptions are questionable in real experiments. The true value of the consant of proportionality between quantity of metabolite and the signal is not usually known. Under ideal circumstances, the system may be calibrated by introducing a sample of known composition, size and concentration. Reproducibility of high-resolution spectrometers using relatively small, homogeneous samples is often good enough to allow the use of a constant calibration factor.

However, this is not possible for *in vivo* studies, as the human body itself may change the sensitivity of the radiofrequency (rf) receiver coil. In detail, this depends on the physical characteristics of each patient. In an attempt to provide a reference value, phantoms with known quantities of a characteristic metabolite are frequently mounted on the rf receiver coil. The ratio of the signal from the phantom compared with the signal from an unknown component is measured from the final spectrum. The quantity of the unknown component can thus be determined by elementary algebra. Reasons why this approach might not be applicable include:

1. The two volumes selected using gradient and/or rf spatial encoding may not be identical in size or shape for the phantom and the tissue sample, resulting in false metabolite concentrations.
2. In large objects, the rf excitation pulse amplitudes may vary with position.
3. Most *in vivo* studies require averaging of several hundred acquisitions in order to achieve the desired spectral quality; this may take many minutes. Movement of the patient relative to the magnet (thereby causing movement of the localized voxel) during the total acquisition period can result in the signal coming from a random mix of tissues. Furthermore, motion of the patient can also modify the sensitivity of the rf coils, introducing additional complication.
4. Because of the low overall metabolite concentrations found *in vivo*, the region from which signal is acquired is usually quite large (>1 ml) and partial volume errors can be substantial (i.e. the implicit assumption of tissue homogeneity is incorrect).
5. One class of spectroscopic acquisition technique, known as chemical shift imaging (CSI), acquires data from multiple locations simultaneously by using a gradient encoding technique. Use of small imaging matrices means that there can be substantial blurring artifacts arising in the data processing.
6. The vastly increased amplifier power requirements needed for large-scale whole-body magnetic resonance systems usally means that rf pulses are longer and gradient pulses have slower rise times than in smaller high-resolution systems. This can result in a variety of contaminating artifacts.
7. Different metabolites vary in the time that they take to return to magnetic equilibrium after each excitation (T_1 recovery time). In order to avoid errors due to incomplete recovery, it is vital to wait until all the metabolites have reached equilibrium. This leads either to long examination times or the need to use estimated correction factors.
8. The size of the patient means that the region being studied may be distant from the reference phantom. The assumption that the reference phantom and the region of interest are in the same MR environment is thus often untrue.

PREPARATION ERRORS

Elimination of unwanted signals

Spatial presaturation

Spatial presaturation is used in spectroscopy in much the same way as it is used in MRI – to eliminate signals from material outside the region of interest which might otherwise fold back on top of the desired data, or to eliminate signals from material (such as blood) which might subsequently move into the region of observation (Fig. 3.1). Errors in presaturation can take two forms – errors of position or errors of completeness. If the presaturated region overlaps the region from which data is to be acquired, then it reduces the effective volume of the voxel of interest and thus causes signal loss.

Spectral presaturation

Failure to achieve complete spectral saturation can be very problematic. Presaturation of the water peak for solvent suppression in proton spectroscopy is the most obvious instance. Typically the chemical shift selective (CHESS) pre-saturation method[1] uses Gaussian pulses with a bandwidth that might be $\delta\Delta$ to saturate a line of width $\delta\omega$, where $\delta\Delta \approx n\delta\omega$ and n is typically in the range $1 < n < 10$. If the main field deviates from the system resonance frequency ω_0 locally by $\gamma\delta B_0$, where $\gamma\delta B_0 > n\delta\omega$, suppression may change by up to a large factor of up to several hundred percent. Effective bandwidth of the spectral saturation pulse is the critical factor. If spatial presaturation is also used, the relevant deviation in B_0 is that across the preselected region of interest. When there is no spatial presaturation, the field error of concern during preparation is across the whole volume within the sensitive region of the receiver coil. All frequency selective presaturation methods are vulnerable to frequency errors. Some methods, e.g. homonuclear presaturation where a narrow spectral line is selectively saturated,[2] are at particular risk.

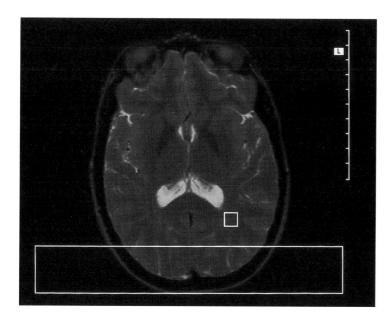

Fig. 3.1. Axial brain image showing localization of the voxel of interest and the placement of a spatial presaturation band to minimize the effects of flow artifacts.

Preinversion of spins

A number of *in vivo* spectroscopy methods use the preinversion of spins. Preinversion can be used to cancel specific unwanted signals and usually exploits differences in T_1 relaxation time of the various tissues. Differences between water and metabolite relaxation T_1 time constants allow water to be effectively suppressed, and cancellation of fat may also be achieved (compare with the short Tau inversion recovery (STIR) sequence[3] used for MR imaging). A further use of preinversion is in methods such as ISIS,[4] where spatially selective inversion pulses are applied as part of more complex spin manipulation processes, which is discussed later.

Preinversion for time constant selectivity

Inversion recovery techniques (selective or otherwise) are prone to errors in rf amplitude. The most reliable method of ensuring a complete inversion is the adiabatic fast passage pulse (AFP probe), which has been exploited in imaging.[5] This involves sweeping the pulse excitation frequency through the resonance, which is helpful in inverting spins in regions of field inhomogeneity, as well as ensuring that the pulse has sufficient bandwidth to invert all the lines in the spectrum equally efficiently.

Spatially selective preinversion

Spatially selective preinversion is vulnerable to several additional artifacts. Firstly, a spatially selective pulse must have a well-defined bandwidth in order to predict the width of the region that it is to affect. When the spectrum to be inverted is relatively broad, the bandwidth of the pulse has to cover the range with an equal amplitude–time product, leading to a requirement for increased field gradient strength. This is desirable, since the applied gradient field should be much greater than any field inhomogeneity. Field inhomogeneity distorts the shape of a selected region, meaning that a nominally straight slice is actually bent. Another incidental virtue of using large gradient fields is that they dominate the effective 'field shifts' between individual spectral lines; this means that all the parts of a spectrum come from the same selected volume. If the frequency shift between two spectral lines is $\delta\omega$, the effective 'field' difference between them is $\frac{\delta\omega}{\gamma}$, and in a gradient field of G for a voxel located at position (x,0,0), the effective shift between them is $\frac{\delta\omega}{\gamma G} \cdot \frac{1}{x}$.

Advanced preparation methods

Although a large number of preparation techniques have been suggested, including the whole repertoire of multiple quantum experiments, relatively few have been used in whole-body studies, owing to limited sensitivity. This section considers in more detail those advanced methods that are applicable to whole-body methods.

Image selective in vivo *spectroscopy (ISIS)*

ISIS works by applying three selective 180° inversion pulses according to all possible binary combinations from 000 to 111. For instance, if the binary pattern is 000, then no rf pulses are transmitted; if the pattern is 010 then only the middle selective pulse is transmitted. In processing the data from an ISIS sequence, it is assumed that the signal amplitude that would be generated using all three pulses (i.e. the 111 pattern) for each of the eight required excitations remains constant. If this amplitude varies because of fluctuations in the rf system, then the additions and subtractions required for processing the data are

Nuclear Overhauser enhancement

Perhaps the most extensively used advanced method is nuclear Overhauser enhancement (nOe). In some situations, one nucleus (e.g. ^{1}H) magnetically couples with another type of nucleus (e.g. ^{31}P or ^{13}C) within the same molecule or on a different molecule. Irradiation at the proton Larmor frequency, for example, can result in an improved spectrum obtained from the second nuclear species. Since the enhancement can be quite pronounced (up to 60% with phosphorus and 200–300% with ^{13}C) the motivation for using this method is apparent.

A number of strategies have been developed for nOe *in vivo*, including quasi-continuous wave irradiation by multiple pulses. Since exposure of the patient to rf dose should be minimized nOe must be used with as few pre-pulses as possible. The bandwidth of the pulses and the energy distribution of the rf power throughout the range of frequencies used must be carefully adjusted to ensure all the coupling nuclei are properly excited. Often additional gradient spoiling is used to dephase coherence in the excited signals.

Errors in the nOe preparation lead to less effective decoupling and thus to loss of signal-to-noise ratio in the resultant spectrum. In many studies, it has simply been assumed that the nOe is equally effective across the spectrum when assessing changes in metabolite concentrations from the relative amplitude of the signals. Some artifacts result in significant variations across the width of the spectrum, thus leading to errors in quantitation.

EXCITATION ERRORS

Amplitude errors

Problems with excitation parallel those of rf errors in the preparation period. The most obvious difference is that any variation in a pulse designed to excite magnetization into the x–y plane with a nominal flip angle, α, is reflected directly in the signal as a linear function in $\sin \alpha$. Errors in α are less likely to be pronounced when $\alpha \approx 90°$.

Radiofrequency excitation errors have an additional significance in chemical shift imaging (CSI) studies, where fluctuations in the flip angle can result in significant artifacts.

Patient motion effects on excitation

After errors in the rf system hardware, the most significant source of rf variation is patient motion. As the patient's body expands and contracts inside the transmitter, the effective load on the transmitter system varies and thus the rf field is modulated. Rf systems must be calibrated to ensure that the waveform generated by the computer is accurately reproduced. Although there is usually feedback within the electronic circuitry to ensure that the output tracks the input, in conventional machines there is generally no feedback around the complete system, including the patient.

Abdominal studies in a closely fitting whole-body coil can result in substantial variations. Motion relative to a surface coil transmitter which is mounted on the patient bed can be even more severe (Fig. 3.2). The impact of the effects may also change with the operating frequency.

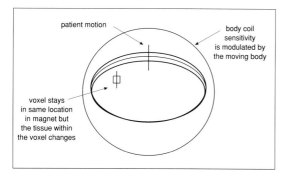

Fig. 3.2. Effect of patient motion on the voxel tissue composition and on the radiofrequency coil sensitivity.

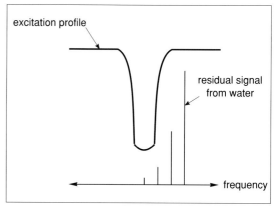

Fig. 3.3. Effects of B_0 errors on the water suppression – as the field deviates from the set value the water suppression becomes less effective.

Chemically selective water suppression pulses

In proton spectroscopy, the exciting rf pulse can be designed in such a way to avoid exciting the water peak. Sequences like multiple-echo selective acquisition (MESA)[6] use 1 $\bar{3}$ 3 $\bar{1}$ pulse groups to excite only metabolites that are not close to the water resonance in frequency (as described by Hore[7]). The 1 $\bar{3}$ 3 $\bar{1}$ pulse is relatively insensitive to rf amplitude variations. However this applies more to water suppression capability than to net rotation of the magnetization of the desired metabolites. The method is very sensitive to deviations in main B_0 field homogeneity. The 1 $\bar{3}$ 3 $\bar{1}$ pulse is designed to minimize excitation at the water peak and maximize excitation at a metabolite like N-acetyl aspartate (NAA) (in human brain spectroscopy), which has a chemical shift of 2.02 ppm. The excitation profile on either side of the suppressed water region changes rapidly and variations in B_0 field mean that different positions in the object may receive a different level of excitation (Fig. 3.3). Patient motion may also cause susceptibility changes which modify the B_0 field and thus the degree of water suppression.

Stimulated echoes

Stimulated echoes are potentially present as soon as the repeat time TR is reduced towards T_2 in sequences using multiple rf pulses. In sequences such as stimulated acquisition mode (STEAM),[8] these echoes are deliberately created and used for data acquisition. With multiple rf pulses, two or more echoes are formed, which may overlap during the data acquisition period, so resulting in artifacts.

In imaging, stimulated echoes mixing with the desired spin echo result in patterns of bright and dark 'zipper-like' signal with quasi-random locations. In spectroscopy, they result in spurious peaks and apparently enhanced noise levels in the data. Stimulated echoes may be removed by using gradient and rf spoiling pulses in methods that are analogous to those used in MR imaging.

Position and size of localized volume

The major problems are:
- overall positioning errors in the region selected;
- width errors in the region;
- shape errors in the selected volume; and
- chemical shift effects where the same volume is not excited for each line of the spectrum.

Overall positioning and width errors are directly analogous to those found in imaging. Region shape artifacts are dependent on sequence type and repetition time. A further complication with spectroscopy is that individual lines have different values of T_1 and T_2, and so the impact of region shape artifacts varies from one line to the next. Errors are worse when data is acquired with relatively short TR compared with T_1. In practice, this is most likely to be a significant problem with ^{31}P studies where a number of metabolites have T_1 values in the range of 4–6 seconds.[9]

^{13}C has very large chemical shifts (200 ppm or so) and low sensitivity. This poor sensitivity means that large voxels must be used, resulting in relatively low gradient amplitudes. Differences in voxel position for different spectral lines can thus be significant, although the actual gradients applied must be scaled from those used for protons by the ratio of gyromagnetic ratios to maintain the correct physical dimensions.

As an example, suppose it was desired to form a 20 mm wide slice for a ^{13}C experiment (as in a depth resolved selective spectroscopy (DRESS)[10] study using this nucleus) with an rf pulse duration of 2 msec at 1.5 T. We would need a gradient of 6 mT/metre. The chemical shift range of 200 ppm represents a frequency range (δf_n) of 3300 Hz, and the relative spatial shift of the two lines is about 5 cm. Thus the two extreme lines of the spectrum can come from distinctly different regions, with the rest of the spectrum being observed from regions between the two. All other methods using selective excitation are similarly affected. With a multistep differencing method such as ISIS, it can be very difficult to evaluate the errors introduced by this effect.

Diffusion effects

Diffusion of molecules through different gradient fields produces signal attenuation. The use of multiple selective rf pulses, particularly with the addition of a number of large gradient spoiling pulses, can thus introduce signal amplitude changes dependent on diffusion. Diffusion anisotropy may also be present, and it may vary for each spectral line.[11] As a result, gradient direction and order of application of multiple spoiler pulses can influence the apparent signal amplitudes.

Magnetization transfer effects

Another possible set of artifacts arises owing to the application of multiple rf pulses. Magnetization transfer occurs when off-resonance irradiation is applied to a complex system, such as tissue-containing free water and water more tightly bound to macromolecules.

Similar effects are found in MR imaging. Multislice imaging results in varying quantities of off-resonance irradiation being applied to a given region during excitation of other slices as well as the required on-resonance excitation for the desired slice.[12] This can result in saturation of magnetization in the 'invisible' pool of tightly bound protons. Resultant change in signal from the free proton pool is observed as the magnetization exchanges and the saturation varies.

This problem is most likely to affect proton spectroscopy and to be complicated by the different degree of off-resonance excitation experienced by each spectral line. However, as the repeat time TR is normally long in spectroscopy, the effects are usually quite small. Increasing TR to reduce magnetization transfer effects is of course beneficial in reducing the risk of selected region distortions and stimulated echoes, and it offers the possibility of quantitation from fully recovered magnetization.

SPATIAL ENCODING ARTIFACTS

Chemical shift imaging

Chemical shift imaging (CSI)[13] is a very efficient localization method. The technique may be used in conjunction with other localization techniques and can be applied in one, two, or three spatial directions. CSI uses the form of spatial encoding used in the phase-encoding direction of the classic spin-warp sequence. All the artifacts found in spin-warp imaging are thus expected to be associated with CSI.

Patient motion artifacts in CSI

The impact of motion effects (or any other data instability) on a set of spatially encoded spectra can be demonstrated quite simply. A small phantom containing water was moved regularly in the magnet along the axis of a surface coil while data was recovered using a one-dimensional phase encoding with its direction at right angles to that of the motion. A simple single peak spectra was acquired from the phantom. An image of the phantom in motion is shown in Fig. 3.4(a). The corresponding sets of 16 phase encoded spectra are shown in Fig. 3.4(b) (no motion) and Fig. 3.4(c) (with motion). Whereas the impact of motion on the images is very apparent, the effect on the CSI spectra is much less predictable.

The total available signal is basically the same in the two experiments shown. Redistribution of data away from the central peak in Fig. 3.4(c) means that the apparent signal-to-noise ratio of this line is reduced. Since in an *in vivo* study motion is always present to some degree, poor signal-to-noise ratio may be the only sign of the presence of artifact. If a measurement that is repeatable in phantom studies

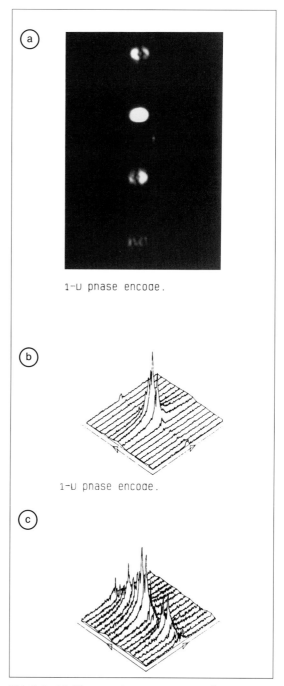

Fig. 3.4. (a). Phantom images in motion. (b). Spectrum at rest. (c). Spectrum in motion illustrating spurious spectral peaks and degraded resolution and signal to noise ratio.

shows signs of significant performance degradation when patients are scanned, then motion artifact is to be suspected and strategies should be adopted to minimize its impact.

Motion relative to the radiofrequency coils in CSI

Motion effects can be much more complex than the obvious movement of signal sources relative to the spatial matrix. This is particularly true when surface coils are in use. If an object moves relative to a surface coil, the signal from a given region will vary in accordance with the sensitivity profile of the coil (see Fig. 3.3). Even if the region of interest moves exactly orthogonal to a phase encoding matrix but with varying sensitivity in the coil, the sort of artifact shown in Fig. 3.4(c) will result. If the coil is of transmit–receive design, the effects of motion are likely to be worse.

Motion in an inhomogeneous field with CSI

Another effect occurs where an extended object of interest is moving in an inhomogeneous field. The spectrum from a particular region of tissue will vary in line shape, width, and amplitude as it moves to different points in the field. This effect contributes artifact in a complicated manner.

Motion artifact reduction techniques for CSI

Strategies available for ameliorating motion artifacts in spectroscopy are as yet largely undemonstrated. Respiratory ordered phase encode (ROPE),[14] centrally ordered phase encode (COPE),[15] motion artifact suppression technique (MAST),[16] and the multiple acquisition method[17] are all possible. Because the T_1 values of many metabolites are long, classic respiratory gating can also be applied.[18] Pseudogating methods are not usually appropriate, because in order to achieve adequate spectral resolution for proton and phosphorus nuclei, data acquisition typically takes up to one second.

Baseline errors with CSI

Another problem with CSI, which is much more apparent with spectroscopy than with imaging, is the time needed between excitation and the beginning of data collection. In carbon and phosphorus spectroscopy where values of T_2 are frequently short, it is not always possible to use a spin echo acquisition. The time needed for the phase-encoding gradient pulses means that the peak of the free induction decay is not sampled. This is tantamount to applying a high-pass filter to the data. Low-frequency signal components are lost, resulting in a baseline which oscillates with a low frequency (Fig. 3.5). Figure. 3.5(a) shows a spectrum collected without an acquisition delay (in this case an unlocalized ^{31}P spectrum of an adult human head), while Fig. 3.6(b) shows a spectrum obtained from a phase encoding experiment with an acquisition delay of 3.1 milliseconds.

Several strategies have been proposed and evaluated for correcting baseline roll.[19,20] Most techniques are useful in limited circumstances but encounter problems when used on data sets with poor signal-to-noise ratio (which is usually the case with ^{31}P spectroscopy!). An example of a frequency domain filter correction technique is shown in Fig. 3.6. Figure 3.6(a) shows the uncorrected spectrum, Fig. 3.6(b) shows the result of the process, and Fig. 3.7(c) is the frequency-

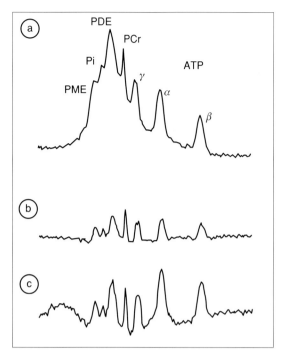

Fig. 3.5. (a). Unlocalized spectrum acquired with zero delay from an adult human head and a surface coil. (b). Rolling baseline due to 3.1 msec time delay for phase encoding between excitation and data acquisition.

dependent roll correction factor. The filter attempts to reconstruct the effects of missing data points.

ACQUISITION AND PROCESSING ERRORS

Small sampling matrices in CSI

It is usually possible to acquire only a few data points in each spatial direction (typically between eight and 32) for the spatial encoding used in CSI, owing to limits of examination time. Performing a Fourier Transform results in ripples in the output data (known as Gibb's ringing). Efforts to avoid this using heavy filtering or zero filling result in data being smoothed and smeared. Signal from

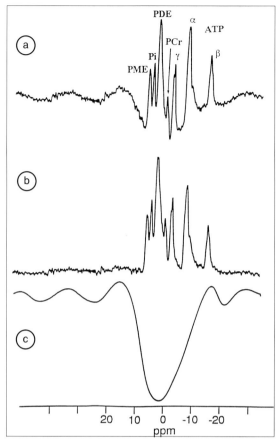

Fig. 3.6. (a). Uncorrected ^{31}P spectrum. (b). Corrected ^{31}P spectrum. (c). Frequency dependent baseline roll correction function.

the voxel of interest is thus contaminated with signal from neighbouring voxels.

A number of different processing and acquisition strategies have been studied (e.g. the Hadamard transform with suitable input data structure,[21] but they have so far not yielded usefully better performance. The consequence of the heavy filtering required to control ripple formation is that data is not quantifiable in absolute terms, although relative quantitation is still possible. Large regions of uniform tissue, particularly from locations that are far from a tissue boundary, are likely to be easier to quantify. This suggests that CSI methods are less likely to be useful for quantification of focal lesions than for non-focal disease.

Convolution differencing and resolution enhancement

Convolution differencing is a method in which a low-pass filtered version of the input spectrum is subtracted from the original spectrum, leaving a quasi-high pass data set. Resolution enhancement is a more direct high-pass filtering process aimed at sharpening the line shape. Many filtering operations are effectively cosmetic in character and are designed to make *in vivo* spectra look more like those obtained from high resolution machines.

Correcting errors due to field inhomogeneity

Haemorrhagic deposits close to tissue structures of interest give rise to complex susceptibility effects. Because of the large voxels required for *in vivo* spectroscopy, the field inhomogeneity leads to broad, distorted lines with reduced peak amplitudes. Single lines may even appear as doublets in extreme cases. Distorted peaks may also be a result of contributions from several different metabolites.

Efforts have been made to map the field error distribution and use this data to correct the spectra from individual voxels for line shape and height.[22] A maximum-entropy processing method has been suggested as a method of improving signal-to-noise ratio without introducing artifact.[23] This has been shown to work satisfactorily with simple spectra, but it is less adequate when applied to low signal-to-noise ratio spectra containing multiple peaks.

Peak area can still be regarded as a measure of the total metabolite concentration present even when it is not possible to resolve the individual components. Failure to resolve the different metabolites present in a peak means that the individual concentrations are unknown.

An example of this problem is provided by the phospho-monoester (PME) and diester (PDE) peaks of the *in vivo* phosphorus spectrum from human liver or brain. Multinuclear decoupling has been demonstrated to improve the situation[24] although the reproducibility of this method in the clinical environment is still not proven.

Relaxation time errors

An additional problem with handling multiplet peaks is that the individual components may have very substantially different T_1 relaxation behaviour. Examination time limits mean that *in vivo* measurements of these time constants have to be made with the minimum number of time points (usually only two points). The results are best described as 'saturation factors'.[25] Precise measurements using more sophisticated sequences such as inversion recovery[26] have been described. However, these are usually only completed in volunteers who can be persuaded to spend very long periods in the magnet.

T_2 measurements are similarly abbreviated. This is particularly problematic as it is most likely that the time constant will be multiexponential for a peak containing more than one component. Further, many spectral lines in spectroscopy are actually multiples produced by interactions between neighbouring nuclei. This is known as hyperfine splitting or J-coupling. If the individual lines are not resolved, then there is an additional amplitude modulation of the peak as they move in and out of phase with each other. The modulation rate is determined by the magnitude of the J-splitting factor. For lactate in proton spectra this J-splitting is 7 Hz, resulting in a modulation period of 1/7 = 135 milliseconds. This is why lactate is inverted in a TE = 135 milliseconds spectrum and positive in a TE = 270 milliseconds spectrum.

CONCLUSION

The variation from examination to examination and the close parallel of the acquisition techniques with MRI suggests that artifacts are a still a major concern for *in vivo* spectroscopy. Currently, the limits of accuracy and reproducibility of spectroscopy in clinical practice are still unknown, although it is encouraging that a number of multi-centre studies are now underway to address this question.[27] *In vivo* spectroscopy has already been demonstrated to produce statistically significant results in large patient groups but whether a diagnostic decision can be reached for an individual patient is still not clear. However, despite all the difficulties described in this chapter, *in vivo* spectroscopy is undoubtedly leading to new insights into the biochemistry of life.

REFERENCES

1 **Haase A, Frahm J, Hanicke W, Matthaei D.** ^{1}H HMR chemical shift selective (CHESS) imaging. *Phys Med Biol* 1985; **30**:341
2 **Anderson WA, Freeman R.** Influence of a second radiofrequency field on high resolution nuclear magnetic resonance spectra, *J Chem Phys* 1962: **37**:85.
3 **Bydder GM, Young IR.** MR imaging: clinical use of the inversion recovery sequence. *J Comput Assist Tomogr* 1985; **9**(4):659–75.
4 **Ordidge RJ, Connelly A, Lohman JAB.** Image-selected in vivo spectroscopy (ISIS). A new technique for spatially selective NMR spectroscopy. *J Magn Reson* 1986; **66**:283–94.
5 **Silver M, Joseph R, Hoult D.** Selective spin inversion in nuclear magnetic resonance and coherent optics through an exact solution of the Bloch–Ricatti equations. *Phys Rev A* 1985; **31**:2753–5.
6 **Lampman D, Murdoch J, Paley M.** In vivo proton metabolite maps using the MESA 3D technique. *Magn Reson Med* 1991; **18**:169–80.
7 **Hore PJ.** Solvent suppression in Fourier transform nuclear magnetic resonance. *J Magn Reson* 1983; **55**:283–300.
8 **Frahm J, Bruhn H, Gyngell ML, Merboldt KD, Hanicke W, Sauter R.** Localized high resolution proton NMR spectroscopy using stimulated echoes: Initial applications of human brain in vivo, *Magn Reson Med* 1989; **9**:79.
9 **Bottomley PA, Foster TB, Darrow RA.** Depth resolved surface coil spectroscopy (DRESS) for in vivo ^{1}H, ^{31}P and ^{13}C NMR. *J Magn Reson* 1984; **59**:338–42.
10 **Bottomley PA, Hardy CJ, Roemer PB.** Depth resolved surface-coil spectroscopy (DRESS) for in vivo ^{1}H, ^{31}P and ^{13}C NMR. *Magn Reson Med* 1990; **14**:425–34.
11 **Doran M, Hajnal JV, Van Bruggen N, King MD, Young IR, Bydder GM.** Normal and abnormal white matter tracts shown by MR imaging using directional diffusion weighted sequences. *J Comput Assist Tomogr* 1990; **14**:865–73.
12 **Hinks RS, Martin D.** Bright fat, fast spin echo and CPMG. *Magn Reson Med, Proceedings of 11th Annual Scientific Meeting, Berlin*, 1992; **S**:4503.
13 **Brown TR, Kincaid BM, Urgubil K.** NMR chemical shift imaging in three dimensions, *Proc Natl Acad Science U S A* 1982; **79**:3523–6.
14 **Bailes DR, Gilderdale DJ, Bydder GM, Collins AG, Firmin DN.** Respiratory ordered phase encoding (ROPE): A method for reducing respiratory motion artifacts in MRI. *J Comput Assist Tomogr* 1985; **9**:835–8.
15 **Haake EM, Patrick JL.** Reducing motion artifacts in two-dimensional Fourier transform imaging. *Magn Reson Imaging* 1985; 359–76.
16 **Pattany PM, Philips JM, Chiu LC et al.** Motion artifact suppression technique (MAST) for MRI. *J Comput Assist Tomogr* 1987; **11**(3): 369–77.
17 **Stark DD, Wittenberg J, Edelman RR et al.** Detection of hepatic metastases; analysis of pulse sequence

performance in MR imaging. *Radiology* 1986; **159**:365–70.

18 **Runge VM**. Respiratory gating in magnetic resonance imaging. In: FA Mettler, LR Muroff, MV Mulkarni eds. *Magnetic Resonance Imaging and Spectroscopy.* New York NY, Churchill Livingstone Inc, 1986:133–46.

19 **Saeed N, Menon DK. A knowledge based approach to minimize baseline roll in chemical shift imaging.** *Magn Reson Med* 1993; **29**:591–8.

20 **McKinnon GC, Burger C, Boesiger P**. Spectral baseline correction using CLEAN. *Magn Reson Med* 1990; **13**:145–9.

21 **Goelman G, Leigh JS**. B1-insensitive Hadamard spectroscopic imaging technique. *J Magn Reson* 1991; **91**:93–101.

22 **Bailes DR, Bryant DJ, Case HA, et al**. In vivo implementation of three-dimensional phase-encoded spectroscopy with a correlation for field inhomogeneity. *J Magn Reson* 1988; **77**:460–70.

23 **Hore PJ**. NMR data processing using the maximum entropy method. *J Magn Reson Imaging* 1985; **62**:561.

24 **Merboldt KD, Chien D, Hanicke W, Gyngell ML, Bruhn H, Frahm J**. Localized ^{31}P NMR spectroscopy of the adult human brain in vivo using stimulated echo (STEAM) sequences. *J Magn Reson* 1990; **89**:343–61.

25 **Bachert P, Belleman ME, Layer G, Koch T, Semmler W, Lorenz WJ**. In vivo ^{1}H, ^{31}P -[^{1}H] and ^{13}C-[^{1}H] magnetic resonance spectroscopy of malignant histiocytoma and skeletal tissue in man. *NMR in Biomed* 1992; **5**:161–70.

26 **Englander SA, Bolinger, L, Leigh JS**. Changes in In-Vivo ^{31}P T1. *Magn Reson Med* 1992, 11th Annual Scientific Meeting, 773.

27 **Alonso J, Cozzone P, Wicklow K, Sauter R, Paley M**. Single volume MR spectroscopy in patients with AIDS: Multicenter Pilot Study. *Radiology* 1993; **189(P)[supp]**:275.

4 MRS of Muscle

Peter A Martin, Henry Gibson, Richard HT Edwards

INTRODUCTION

Technological advances with the development of medium and large bore magnet systems for magnetic resonance spectroscopy (MRS) in the early 1980s gave the first opportunities for the non-invasive study of living human tissue bioenergetics. The easy access and large bulk of skeletal muscle made it the obvious choice for early studies. Muscle is a unique biological machine allowing the conversion of energy into force, movement, and power. Skeletal muscle constitutes 40% of total human body mass, and together with the skeleton form the machinery of locomotion. The requirement to perform delicate precise movements, punches, sprints, or marathons highlights the tremendous functional versatility of skeletal muscle, which may lead to a 100-fold change in metabolic demand. Moreover, the rapid changes of performance seen with training and disease demonstrate an unrivalled potential for adaptation encompassing complex interactions of metabolism and physiological processes.

Phosphorus (^{31}P) MRS offers a unique opportunity to study cell energetics. The importance of MR-visible phosphorus metabolites in cell energetics means that MRS provides much information on the regulation and function of the vital metabolic pathways utilized in contraction. Until the 1980s, the only means of examining tissue chemistry directly was via conventional chemical analysis of tissue samples obtained by needle biopsy. This gave a metabolic 'snapshot' but provided little information on the rapid time courses of energy transduction. Nevertheless, biopsy techniques did contribute much to the understanding of human muscle bioenergetics during exercise and in disease. MRS provides an alternative method of carrying out these studies and has the advantage of being non-invasive and thus more readily acceptable in the clinical setting.[1]

Early MRS studies of muscle, though relatively crude, added to our understanding of muscle metabolism. However, these early studies relied heavily on metabolite ratios, rather than on true quantification, to provide an indication of the time course of alterations in metabolism. Early attempts at quantification used assumptions about the total phosphorus concentration based on biopsy samples. It is only recently that direct measurements from MR of the absolute concentrations of metabolites have been possible.[2] The complex structure and variable composition of muscle requires us to take a critical view when interpreting data obtained from muscle studies. The replacement of muscle fibres with non-contractile fat and connective tissue through disease demands particular care. The potential for artefacts due to partial volume effects is present for both needle biopsy and MRS. One of the benefits of studying human muscle is that not only can a variety of different types of activity be investigated, but it is also possible to study both voluntary and involuntary (stimulated) contractions. In addition by using ergometers and force measuring devices, it is possible to control the work done and to relate muscle performance to metabolism as measured by MRS. As a consequence – under a variety of conditions, e.g. rest, exercise, and recovery – MRS can give information on the physiological and biochemical status of muscle.

The object of this chapter is to provide an overview of MRS as applied to the study of human skeletal muscle metabolism. The final part of this chapter reviews the application of MRS in disease and examines the potential of MRS as a diagnostic tool.

MUSCLE STRUCTURE, FUNCTION AND METABOLISM

The basic structure of a typical skeletal muscle is shown in Fig. 4.1. The functional unit of the muscle is a group of fibres innervated by a nerve, the 'motor unit', although each fibre is a separate working unit in its own right. Under the microscope, the muscle appears striated, hence the use of the terminology 'striated muscle' to describe skeletal muscle. Muscle fibres are single cells filled with fibrils, which contain the contractile proteins, many mitochondria (the site of energy production) and multiple nuclei. A fine layer of connective tissue sheathes the fibres, forming a matrix of connective tissue that blends together with the tendons, finally attaching to anchorage points on the bones. The amount of connective tissue (collagen fibres, elastic fibres, and other cells) varies from muscle to muscle.

The arrangement of fibres within a muscle is functionally important. The force that a muscle generates is dependent on the cross-sectional area of the muscle fibres (i.e. the number of parallel units), and not its length (i.e. the number of serial units) or its mass. By arranging the fibres at an angle to the direction of pull (the pennation angle), it is possible to achieve an effectively larger cross-sectional area and hence generate more force. MR imaging is a very effective technique for the determination of the pennation angle.[3]

An electrical signal that originates in the brain and travels to the muscle along nerve fibres initiates skeletal muscle contraction. This signal or 'action potential' must successfully cross the neuromuscular junction, travel along the cell surface and into a complex network of transverse tubes, releasing stored calcium from the sarcoplasmic reticulum. The calcium deactivates the inhibitory factor (troponin) allowing the formation of cross-bridges between the two contractile proteins, actin and myosin, resulting in force generation. Relaxation involves the active uptake of calcium and the consequent dissociation of cross-bridges. Both contraction and relaxation processes are dependent on energy: depletion of muscle ATP results in a state of 'rigor' in which the muscle fails to relax. The energetics of muscle metabolism are described more fully later in the chapter.

Most human muscles comprise a heterogeneous population of cells, falling mainly into two distinct fibre types with different functional and metabolic characteristics (Table 4.1). The compositional distribution of type I and type II fibres in a healthy muscle often reflects the physiological role of that muscle. Postural muscles such as the soleus have a predominance of type I fibres (slow twitch, oxidative type). Muscles involved in rapid movement,

MRS of Muscle

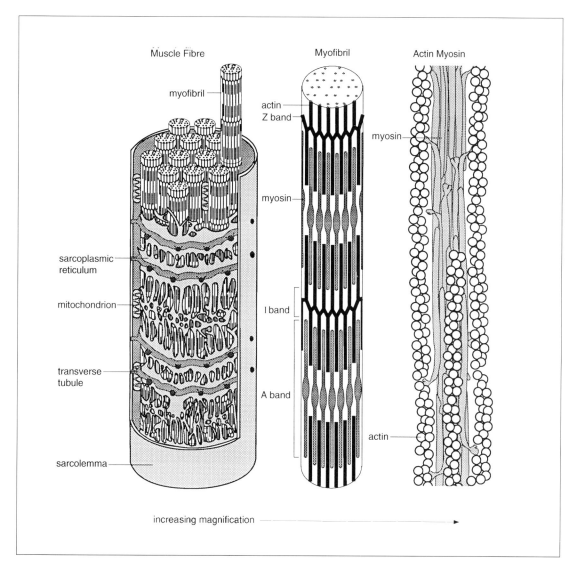

Fig. 4.1 *Sub-divisions of skeletal muscle, indicating the structural network involved in the conversion of relative molecular movement into muscular force.*

such as the quadriceps, are normally mixed, consisting of both type I and type II (fast twitch) fibres arranged in a checkerboard pattern. Type I fibres suit muscles exposed to prolonged periods of aerobic activity, while type II fibres adapt to high intensity, largely anaerobic activity. Intermediate fibres also occur, and different metabolic capacities further sub-divide type II muscle fibres into types IIa and IIb. Except in rare instances of disease, no human muscles are of wholly type II fibres. In other species, we find pure muscles composed only of type II fibres, for example, in the rat extensor digitorum longus. The recognition of two main fibre-type populations in human muscle is impor-

Table 4.1. Summary of the characteristic properties of Type I and Type II fibers in man

Property	Type I (slow twitch)	Type II (fast twitch)
Contraction time	long	short
Relaxation time	long	short
Myosin ATPase activity	low	high
Fatiguability	low	high
Phosphocreatine content	low	high
Oxidative enzyme activity	high	low
Capillary density	high	low
Mitochondria	numerous	few
Glycogen content	no difference	no difference
Fat content	high	low
Myoglobin content	high	low

tant. Depending on the nature of the muscle activity, inhomogeneities in muscle chemistry develop owing to different populations of fibres becoming fatigued to different extents.

Muscle bioenergetics

The focus of most MRS studies of muscle has been on the biochemical pathways for the supply and utilization of energy (Table 4.2). Perhaps the most important reaction to be considered is the hydrolysis of ATP to ADP, catalysed by myosin ATPase; this reaction produces the energy driving muscular contraction. Only a small amount of cytosolic ATP is present in normal muscle, which, as a consequence, can only sustain contractile activity for a very short period of time before requiring fresh supplies of ATP. These supplies come from an energy reservoir in the form of phosphocreatine (PCr) rather than stores of ATP itself. In the cytosol, the creatine kinase reaction uses PCr to recycle ADP back to ATP again. The resulting creatine (Cr) is transported into the mitochondria where 'oxidative phosphorylation' (also known as mitochondrial respiration) occurs. Phosphorylation is simply the reverse of the creatine kinase reaction, where the creatine kinase catalyses the conversion of Cr back into PCr using mitochondrial ATP as the phosphate source.

When the oxygen supply from the blood is good, that is, under aerobic conditions, mitochondrial respiration produces most of the muscle's energy requirement (ATP). When the energy demand exceeds the mitochondrial capacity or when the muscle is ischaemic (anaerobic), only glycogenolysis can supply the ATP for contraction. Thus, during anaerobic metabolism large amounts of lactate are produced, leading to a decrease in intracellular pH (pHi).[4]

MR SPECTROSCOPY TECHNIQUES APPLIED TO SKELETAL MUSCLE

The last 15 years has seen a spectacular improvement in the use of MR imaging which has not been matched by a comparable improvement in *in vivo*

Table 4.2. Energy sources for muscular activity

Short term (anaerobic) sources of energy

ATP hydrolysis (work)

$$\text{Adenosine triphosphate (ATP)} + H_2O \xrightarrow{\text{myosin-ATPase}} \text{adenosine diphosphate (ADP)} + \text{inorganic phosphate (Pi)} + \text{energy}$$

Creatine kinase reaction

$$\text{Phosphocreatine (PCr)} + ADP + 0.9\ H^+ \xrightleftharpoons{\text{creatine kinase}} \text{creatine (Cr)} + ATP$$

Anaerobic glycolysis

$$\text{Glycogen} + 3\ Pi + 3\ ADP \rightarrow 2\ \text{lactate} + 3\ ATP$$

Long term (aerobic) sources of energy

Oxidative phosphorylation

$$\text{Glycogen/glucose/free fatty acids/free amino acids} \rightarrow NADH$$
$$NADH + 1.5\ H^+ + 0.5\ O_2 + 3\ ADP + 3\ Pi \rightarrow H_2O + NAD^+ + 3\ ATP$$

MR spectroscopy. Many modern studies still use the same acquisition techniques as were used on the first small bore *in vivo* spectroscopy systems. Far from becoming a routine diagnostic tool, MR spectroscopy has remained very much in the research arena.

The *in vivo* spectra of Fig. 4.2 show a comparison between a normal forearm and that of a patient with Duchenne muscular dystrophy.[5] The non-water-suppressed hydrogen (^{1}H) spectra are the simplest; they show, for the dystrophic forearm, a decrease in the water peak and an increase in the fat peak compared to those for the normal forearm. These changes reflect the increase in fat content that is the main feature of muscular dystrophy. The more complex phosphorus spectra show a large peak due to PCr and three peaks due to ATP, as well as a small peak due to inorganic phosphate (Pi). The dystrophic ^{31}P spectrum shows a decrease in the total signal of phosphorus metabolites as the fat displaces the muscle. The ^{13}C spectra, on the other hand, show even more peaks, but at the resolution achievable with the 1.5–2.3T of most *in vivo* systems, they are difficult to interpret owing to many overlapping peaks. Peaks due to CH$_3$ and CH$_2$ groups are clearly visible however, and the dystrophic forearm shows a general increase in the ^{13}C signal that is most obvious in fat signals in the CH$_2$ region.

The quality of the spectra that can be obtained sets the limits on the types of muscle studies that can be undertaken. Quantitative analysis requires good signal-to-noise ratios (SNRs) for all the peaks of interest in a spectrum. Qualitative studies are still possible with lower SNRs, but the results need careful interpretation and we can only make inferences as to the presence and direction of any changes in the spectra. Many factors interact to influence spectral quality and thus the design of any muscle study. The size and location of the muscle under study determines the volume from which we obtain a signal, and the type of muscular

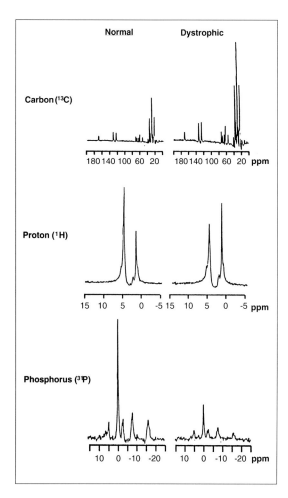

Fig. 4.2 ^{13}C, ^{1}H, and ^{31}P NMR spectra from the human forearm. A comparison between a 12-year-old boy with Duchenne dystrophy and a normal control.

activity decides the importance of motion artefacts. The nucleus to be used (e.g. ^{31}P, ^{13}C, ^{1}H) sets the MR sensitivity, and the magnet field strength and homogeneity governs the resolution. Finally, a combination of the type of radiofrequency coil and the MR localization technique controls the quality of localization.

The first MRS studies of muscle used 'surface coil localization'.[6] This is suitable for all nuclei and is still in wide spread use today. Surface coil localization relies entirely on the localized, but inherently non-uniform, response of a simple loop antenna – the surface coil – to obtain signals from a restricted region. When the surface coil acts as both a transmitter and receiver, it has a sensitive volume that forms a hemisphere penetrating to a depth of roughly one coil radius.[7] Supplementation of surface coil localization by magnetic field localization techniques improves overall quality of localization.

The earliest of these to be widely used for muscle studies was topical magnetic resonance (TMR).[8] TMR uses static high order (Z4) gradients to spoil deliberately the magnet homogeneity over all but a small central region of the magnet; the diameter of the sensitive volume is variable and adjusted by varying the Z4 gradient strength. Any signals coming from within this uniform central volume produce narrow peaks, whereas any signals coming from outside this volume appear as a broad hump in the baseline. Processing via a convolution-difference technique removes the baseline hump from the final spectra.[9] The big advantage of TMR is that a single acquisition gives good localization – superior to that of a surface coil alone. Thus it is suitable for studies requiring time resolution of the order of 1 or 2 seconds.

The big disadvantage of TMR is that it relies on static, non-switching, magnetic field gradients that only produce a uniform field over a small region at the centre of the magnet. The small size of this homogeneous region makes TMR systems unsuitable for general MR imaging purposes. Anatomical constraints often mean that the muscle of interest cannot be positioned at the centre, thereby preventing the use of TMR. For these reasons, the TMR method rapidly fell out of favour once the combined, switched gradient imaging and spectroscopy systems became available. These combined systems held the promise of spectroscopy via pulsed gradient techniques, allowing localization to any chosen volume within the (larger) homogenous region in the centre of the magnet.

Of the many gradient localization techniques introduced over the last 10 years, few of them have found practical application in muscle studies. This is because most muscle studies have used ^{31}P spectroscopy. Natural abundance carbon (^{13}C) spectroscopy is possible, but the most useful results are obtained with the use of (expensive) ^{13}C labeled compounds.[10,11] Simple proton (^{1}H) spectroscopy can only provide crude spectra that are of little value, since they contain only a water and a fat peak. In order to see potentially more interesting metabolites, it is necessary to suppress both the water and the fat signals. The application of water-suppressed ^{1}H spectroscopy is commonplace in head studies but not in muscle studies. Techniques that adequately remove both the water and the fat peaks are not available at present, though we might expect them to become available in the near future.

The gradient-mediated localization techniques generally fall into two categories, single voxel techniques and chemical shift imaging (CSI). The use of CSI techniques is not widespread and they have inherently long acquisition times ($\geq$10 min for phosphorus) making them unsuitable for many dynamic studies. An alternative localization strategy that has been used by some groups is to use one-dimensional, 'rotating frame' chemical shift imaging.[12] This has not been widely adopted, however, largely because of the T_2 distortion to which it is susceptible.

Of the single voxel techniques, two that have found application in proton head studies are: Pixel RESolved Spectroscopy (PRESS) and the STimulated Echo Acquisition Method (STEAM). Compared to hydrogen (protons), phosphorus nuclei tend to have short T_2 relaxation times.[13] This makes them unsuitable for most such gradient mediated localization techniques, as these tend to rely on spin or stimulated echoes with long (>10 msec) echo times. One usable gradient-mediated technique is Depth REsolved Surface coil Spectroscopy (DRESS). This relies on a simple slice selective excitation pulse followed by acquisition of the free induction decay (FID).[14] The necessity of a refocusing gradient after the excitation pulse means that there is some delay between excitation and acquisition, which results in loss of some signal and baseline distortions and makes accurate quantification difficult. The signal loss in DRESS is particularly severe for the β-ATP peak, which is unfortunately the peak often chosen as an internal standard.

The best true volume localization technique for phosphorus is the Image Selected *In vivo* Spectroscopy technique (ISIS). This has the potential advantage that it does not use spin echoes and thus is capable of giving good undistorted spectra suitable for quantification. In practice, ISIS suffers from the serious disadvantage that it is a differencing technique requiring a minimum of eight signals for full volume localization. This makes it highly susceptible to motion artefacts and thus completely unsuitable for exercise studies. In addition, the quality of localization via ISIS is subject to T_1 effects. To avoid these in muscle studies, repetition times (TRs) should be 15 seconds or more, making ISIS unsuitable for short duration time course studies.

Because of the problems associated with volume localization techniques, most muscle studies still use surface coil localization alone. This presents some difficulties for quantification, as the non-uniform response of the surface coil results in a variation of the flip angle throughout the sensitive volume. The resultant signal varies in strength with spatial position, if values of muscle T_1 and T_2 also change across the volume, then quantification of the signal becomes impossible. These effects are ameliorated a little by the use of 'designer' pulses such as BIR-4,[15] which give uniform flip angles over a larger proportion of the coil's sensitive volume.

In vivo spectroscopy is currently capable of seeing only the narrow lines produced from long T_2 metabolites, i.e. those in solution in the cytoplasm. When in a viscous medium or bound to the mitochondrial matrix, these metabolites become MR invisible or only give broad lines that appear as a hump in the baseline. This is actually an advantage, as only the MR-visible cytoplasmic metabolites are important in assessing the thermodynamics of muscle biochemistry.

In vivo quantification has always been difficult. As a result, many muscle studies report their results in terms of ratios of peak areas. To achieve a semblance of quantification, we must make appropriate assumptions about the MRS-visible total phosphorus or ATP concentrations, which enables estimates to be made of the ^{31}P-containing metabolite concentrations. A common reference method is by comparison with β-ATP. To obtain improved quantification requires some form of external or internal standard. Internal standards have to be inherent in the muscle under study; thus one method is to collect a ^{1}H spectrum and to use the water peak as a reference. One problem with this is that the water content of muscle varies, especially during exercise, and we cannot use assumptions about the water concentration.[16] One approach to overcoming this problem is to collect several water spectra at different TRs and to calculate the true H_2O concentration. An acceptable alternative method is to use an external phantom as a reference. Attaching the standard to the surface coil provides a means of determining the effects of coil loading on MR signal. The coil loading information provides correction factors between the *in vivo* spectra and those from an *in vitro* phosphate solution.

The man–machine interface

The ease of access to muscle and the simplicity of MRS surface coil techniques makes muscle studies seem deceptively simple. It is of little value performing such investigations without considering an objective measure of muscle performance: resting spectra rarely show abnormalities except under certain circumstances, e.g. in arterial disease,[17] injury,[18] or myopathy.[5]

Interventional studies of the kinetics of energy supply, fatigue mechanisms, and consequences of disease and myopathy require exercise with its consequent pronounced changes in metabolism. The small size of the bore of the magnet and the deleterious effects of the magnetic environment upon electronic instruments hampers the design of exercise regimes. The adoption of a critical approach to the design of MRS studies is essential. The observed changes in muscle metabolism depend on many factors, including the nature of the contraction (dynamic or isometric), the type of activity (repetitive or continuous), and the strength of contraction.

Early studies in medium bore magnets (15–30 cm diameter) involved simple exercise protocols, for example repetitively squeezing a rubber ball at a given rate and monitoring the metabolic changes within forearm flexor muscles. This approach was particularly useful for examining metabolic processes in disease, but it gave no information on how the functional characteristics of muscle altered with metabolism; neither did it give an objective work standard. One approach adopted to overcome this problem, particularly for electrophysiological investigations,[19] was duplication of the study outside the magnet. This had the disadvantage of necessitating repeat experiments, which was often unacceptable to subjects.

The advent of large bore magnets (1.5 and 2.0T, 50–60 cm free bore diameter) enabled more complex studies involving dynamic activity to be realized. Simultaneous measurement of force and chemistry for larger limb movements became possible. Although many ergometers are commercially available, they are of limited use in conjunction with *in vivo* MR systems. The horizontal design of the magnet means that the subject must lie either supine or (less often) prone. Moreover, the servo-braked devices used in the laboratory are often bulky and expensive, and they disrupt the homogeneity of the magnetic system. This has spurred the development of simple MR-compatible ergometers constructed from non-magnetic materials. These ergometers rely on performing a known amount of work by lifting weights via pulley systems or pushing against resistive pneumatic devices using pedal or lever systems.[20] Securing the subject to immobilize limbs and/or the abdomen helps prevent muscle groups other than those under study from contributing to the exercise. The constraints of the magnet still mean that it is difficult to perform very large limb movements and this restricts MRS studies to the smaller muscle groups such as tibialis anterior or triceps surae (although by lying the subject prone, hamstring flexion[21] is possible).

The easiest exercise to study, particularly for large muscle groups, is isometric exercise (when the muscle contracts without altering its length). During isometric exercise, external power output is zero; however the force–time integral provides an indicator of muscular activity.[22] The force produced is recorded using a non-magnetic force

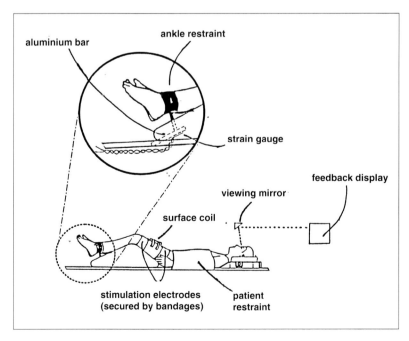

Fig. 4.3 Diagram of the Liverpool Apparatus for combined MR and exercise studies of the quadriceps.

transducer, constructed with the minimum of metal components in order not to affect the MR measurements. Our own studies of the quadriceps function use this approach, and by having the knee extended by up to 120°,[23] sufficient flexion of the knee joint is generally possible to enable maximum force generation. Skeletal and biomechanical considerations together with the force length properties of muscle[24] determine the optimum angle of 120°.

Figure. 4.3 illustrates the apparatus in which the subject is lying supine, with the lower limb placed over a polystyrene foam block for support and the ankle attached to a force transducer. A wide belt across the waist prevents the subject from using the back and abdominal muscles to aid contraction. An LED bar graph indicator placed behind the subject and visualized with the aid of a mirror fixed above the subject's head displays the force measurements. This visual feedback informs the subject of the target force required for submaximal contractions. LED 'traffic light' indicators and a tone generator controlled by the spectrometer computer tell the subject when to contract a muscle and when to rest. Verbal encouragement contributes markedly to the success of obtaining maximum efforts. When appropriate, ischaemia is induced in the limb by simply inflating a thigh sphygmomanometer cuff to 100 mmHg above systolic blood pressure, in order to stop the blood supply. This ischaemia produces a closed system, trapping metabolites and preventing oxidative recovery processes.

More recent studies have utilized electrical stimulation to induce contraction, based on well established laboratory techniques.[24] This permits a controlled means of contraction development, giving objective measures of function independent of motivation. This provides more reliable interpretations of effort than the subjective measures of voluntary effort. To obtain an objective measure of maximum voluntary effort, we superimpose single twitch stimuli on the voluntary effort:[25] the presence of the stimulated twitch force indicates a submaximal effort. The application of stimulation techniques within the magnet has not proved difficult, particularly for percutaneous stimulation of large muscle groups. We have used large (12 cm^2) gel-based conducting electrodes (defib-rillator electrodes) connected to a commercial stimulator via small (2 cm^2) copper conducting pads attached to screened leads and rf filters.[22] Crepe bandages strap the electrodes in place over the proximal and distal portions of the thigh and a surface coil is placed between the gel electrodes, thereby minimizing interference in field homogeneities. Small (5 mm) diameter tubes filled with copper sulfate solution provide MRI-visible markers that ensure the correct placing of the surface coil over the region of interest. This arrangement provides functional measures such as force, relaxation and contraction rates, and frequency–force characteristics.[26] Recent studies by Vestergaard-Poulsen and colleagues[27] have also utilized electromyographic recording techniques during voluntary activity, providing a further handle for the interpretation of chemical and electrophysiological events during muscle contraction.

Of note is a recent study by Adams et al,[28] who have shown that percutaneous stimulation does not necessarily stimulate a homogeneous region of muscle under the electrodes, even for development of 75% of the maximum force. This raises the possibility of partial volume effects, both for the MRS measurements of stimulated muscle and for submaximal voluntary activity where not all the fibres within a muscle are active.

Despite these reservations, MR studies have made it possible to obtain a detailed analysis of the chemistry, physiology, and function of muscle in a single non-invasive investigation. It is likely that future developments may allow more complex exercise protocols within the magnet system, with the subject standing or possibly even running or cycling in movements akin to everyday activity. Even with improvements in magnetic shielding, it is unlikely that the need for specially designed non-magnetic ergometers will disappear; indeed the opportunities opened up by this new field will require new and more sophisticated designs.

Applications of MRS in skeletal muscle

Resting muscle

The wide range of physiological states found in normal human populations hampers the study of resting muscle. Normal human populations have large intrinsic variations in the levels of individual training, ranging from those who are completely sedentary to the highly trained athlete. Furthermore, the variety of types of training produces individuals who have opposing characteristics, such as marathon runners and sprinters.

One of the first and perhaps the most important finding from studies of resting muscle is the observation that the concentration of PCr is consistently higher when measured by ^{31}P MRS than when measured by freeze-clamped needle biopsy. This is almost certainly due to rapid hydrolysis of PCr to Pi during the freeze clamping process. Support for this interpretation comes from the observation that the total phosphorus concentration [PCr + Pi] obtained by both methods is approximately the same.

Measurements of ATP show no such discrepancies between the techniques, possibly owing to the rapid equilibration of the creatine kinase reaction during the extraction procedures. Most quantitative ^{31}P MRS studies on resting muscle have relied not on true quantification, but on the assumption that the ATP levels are reasonably constant. This assumption requires care, since there is evidence that changes in the resting levels of metabolites depend on the prior history of the muscle. Muscle exercise protocols involving lengthening contractions[18] significantly reduce the PCr:Pi ratio for up to 1 hour after exercise. This reduction continues with the ratio reaching a minimum at 1 day after exercise and remaining low for between 3–10 days after exercise. Similarly, studies show abnormal spectra persisting for up to 2–3 days after short-term exercise, even when there is no evidence of muscle fibre damage. Much evidence is available to show that type I and type II fibres have different PCr:Pi ratios, with type II fibres having elevated PCr:ATP ratios compared to type I fibres. It is probable that the variety of different fibre type ratios found in normal subjects explains the resting ratios and changes discussed above.[29]

Measurements of pH

Inorganic phosphate (Pi) *in vivo* has a pK$_a$ of about 6.75 at physiological pH and exists in an equilibrium between two forms $H_2PO_4^-$ and HPO_4^{2-}. These give rise to two separate phosphate resonances, 2.3 ppm apart, which undergo rapid chemical exchange (10^9 to 10^{10} s^{-1}). This results in the spectrum containing a single resonance whose frequency depends on the relative amounts of the two moieties. Since the equilibrium between the two forms of Pi is pH dependent, the chemical shift of the Pi peak acts as an indicator of pH. Measurement of the chemical shift of Pi is relative to an external standard such as methylene diphosphonate (MDP) or to an internal standard such as the pH-insensitive PCr peak. In resting human gastrocnemius muscle, the chemical shift of Pi from PCr is 4.88 ppm, corresponding to a pH of 7.07.

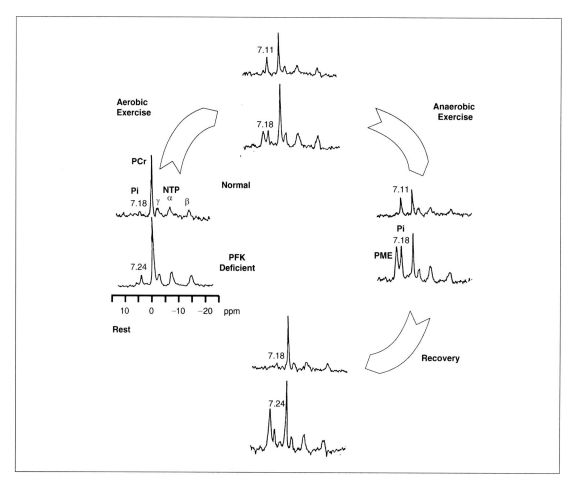

Fig. 4.4 *A series of spectra showing the effects PFK deficiency during an exercise regime.*

During exercise (Fig. 4.4), the chemical shift of Pi is 4.27 ppm corresponding to a pH of 6.62, owing to the presence of lactate. Equation 1 gives the relationship between chemical shift and pH:

$$\mathrm{pH} = 6.75 + \log\left[\frac{(\delta - 3.27)}{(5.69 - \delta)}\right] \quad \text{(Equation 1)}$$

The buffering capacity of muscle

Since the equilibrium involves H^+ ions, a knowledge of the buffering capacity of muscle is necessary for determination of ATP fluxes (see Table 4.2). During the initial part of aerobic exercise, we can assume that the proton efflux is negligible, enabling the glycogenolytic rate to be estimated in the same way as for ischaemic exercise. This is no longer true after exhaustion of the muscle's buffering capacity when the pH starts to fall. The cytosolic buffering capacity of skeletal muscle depends on Pi (pK = 6.57), bicarbonate (pK = 6.1), and other buffers, largely imidazole groups in histidine residues. From the work of Kemp and Radda,[30] it can be shown that in a closed system (i.e. muscle during ischaemia), where the total CO_2 is constant, the buffer capacity β is given by:

For Pi:

$$\beta = \frac{2.3 \cdot [Pi]}{\{[1 + 10^{(pH-6.75)}] \cdot [1 + 10^{(6.75-pH)}]\}}$$

(Equation 2)

where β is measured in slykes (i.e. mmol/litre/pH unit).

For bicarbonate:

$$\beta = \frac{2.3 \cdot S \cdot pCO_2 10^{(pH-6.1)}}{\{[1 + 10^{(pH-6.1)}] \cdot [1 + 10^{(6.1-pH)}]\}}$$

(Equation 3)

where S is the solubility of CO_2. Taking pCO_2 as 5 kPa and S as 0.3 mmol/litre/kPa then at its resting pH in closed muscle, $\beta < 5$ slykes.

For other buffers:

$\beta = 20$ to 30 slykes – inferred by analysis of ^{31}P MRS data and from measurements in muscle homogenates.

Exercise and fatigue

While MRS studies of resting muscle give much information about its energy status, it is during exercise that the most dramatic changes occur, owing to the high metabolic demand of muscle contraction. Particular interest lies in the study of individuals with metabolic defects, which can give information about metabolic processes normally not accessible. Exercise studies using a variety of workloads enable the observation of changes in metabolic demand due to physiological changes in function. The improvement and maintenance of muscle performance are important both in sport and in the treatment of disease. Muscle fatigue – the decline in force or power output with prolonged activity – has received much attention in this respect. The type of muscular activity undertaken and the many cellular physiological factors related to the chemical changes taking place complicates the interpretation of the results.

Comparison of the force output between alternate periods of voluntary contractions and peripheral electrical nerve stimulation provides a measure of the central contribution to fatigue. If the voluntary force output falls more than the stimulated force output, the difference is due to fatigue of central motor control mechanisms rather than that of the muscle itself.

Except in the rare neuromuscular disease myasthenia gravis, fatigue due to failure of the neuromuscular transmission is rare. Over the last 50 years the direction taken by much of the research in this field has been towards gaining an understanding of the balance between impaired energy supply and electromechanical coupling failure in a particular form of muscular activity.[31,32] The response of the muscle to different frequencies of electrical stimulation provides a means of classifying muscle fatigue. An example of fatigue produced by high frequency electrical stimulation (high frequency fatigue – HFF) occurs in myasthenia gravis. Low frequency fatigue (LFF), in which there is selective impairment of force generation at low frequency (i.e. a reduced 20:50 Hz force ratio) can occur for a long time after ischaemia[24] or an eccentric muscular contraction,[33] or when force generation at high frequency has recovered. Needle biopsy and MR studies early in recovery from severe ischaemic exercise have shown that, from the energetics point of view, HFF and LFF can be distinguished. Rapid recovery of the membrane excitation – Compound Muscle Action Potential (CMAP) – with HFF means recovery of force may occur before full recovery of PCr:β-ATP or Pi:PCr ratios. With LFF, force is still reduced 1 hour after exercise, when both CMAP and the PCr:β-ATP ratio have recovered.[34]

Ischaemic exercise

During exercise, the contraction of the muscle itself can result in the occlusion of the blood supply producing ischaemia. For the purposes of exercise studies, the use of an inflatable cuff can enforce ischaemia, resulting in anaerobic exercise. During anaerobia, ATP production comes from two sources, the hydrolysis of PCr catalysed by creatine kinase (i.e. PCr depletion) and glycogenolysis (the breakdown of glycogen to lactic acid). Knowledge of [PCr] and pH enables us to calculate the rate of ATP.[35] Since MRS provides direct measurement of the decay of the PCr peak, we obtain a direct measure of the hydrolysis of PCr (D) and the rate of ATP synthesis calculated:

$$D = \frac{-\delta[\text{PCr}]}{\delta t} \quad \text{(Equation 4)}$$

Hydrolytic ATP synthesis also results in a net proton consumption at the rate:

$$\text{Net proton consumption} = \frac{D}{[1 + 10^{(\text{pH}-6.75)}]} \quad \text{(Equation 5)}$$

The effect of lactic acid production on pH allows estimation of ATP production rates via glycogenolysis (L). Since during ischaemic exercise protons cannot escape from the system, and by anaerobic glycolysis, 1 glycosyl unit (1 mole) produces 2 moles of lactic acid and 3 moles of ATP:

$$L = \left(\frac{3}{2}\right)\left(\frac{D}{[1 + 10^{(\text{pH}-6.75)}]} - \frac{\beta\,\delta\text{pH}}{\delta t}\right) \quad \text{(Equation 6)}$$

where β is the cytosolic buffering capacity of the muscle (derived as above).

Since the total rate of ATP synthesis (F) is:

$$F = D + L \quad \text{(Equation 7)}$$

The total turnover of ATP is:

$$F = D + \left(\frac{3}{2}\right)\left(\frac{D}{[1 + 10^{(\text{pH}-6.75)}]} - \frac{\beta\,\delta\text{pH}}{\delta t}\right) \quad \text{(Equation 8)}$$

Aerobic exercise

In aerobic (oxidative) exercise, oxidative synthesis of ATP from PCr occurs in the mitochondria and, as before, the rate of depletion of PCr and thus the rate of hydrolytic ATP production is readily accessible by MRS. The presence of oxidative ATP synthesis and the fact that the system is no longer closed, i.e. there is a net efflux of protons from the cells, complicates the assessment of total ATP production. One strategy has been to compare aerobic with anaerobic exercise, making use of the power output relationships derived from graded ischaemic exercise to provide the 'missing data' from aerobic exercise studies at the same power output.

At the start of aerobic exercise, we can neglect proton efflux and the glycogenolytic rate estimated as per ischaemic exercise. At the end of 'aerobic' exercise, the muscle is actually anaerobic, the pH is falling, and the effect of the large proton efflux from the system requires that a correction be applied to the estimates of glycogenolytic ATP synthesis rates. The correction can be obtained by assuming that the proton efflux rate has the same linear relationship to pH during recovery.

Recovery

The study of recovery after exercise gives information on the pH dependence of proton efflux, which, as described above, enables calculation of a correction factor for the analysis of aerobic exercise.[36]

After the completion of exercise, the accumulated ADP continues to stimulate mitochondrial respiration to resynthesize ATP; as a consequence, the creatine kinase equilibrium replenishes the cytoplasmic PCr pool. Making the muscle ischaemic after exercise by rapid inflation of a sphygmomanometer cuff does not produce any evidence of metabolic recovery. This shows that the PCr replenishment during recovery from exercise must be entirely dependent on mitochondrial respiration.[37] Thus, we can consider the rate of PCr resynthesis to be a good index of mitochondrial function.[38]

The recovery of PCr after exercise appears to follow an approximately exponential time course and experimentally is usually treated as such. The true course of PCr recovery is biphasic,[39] with a rapid initial rise followed by a much slower rate of recovery back to resting level. The rate of PCr resynthesis is dependent on the extent of intracellular acidosis, which in turn depends on the work rate during exercise. The final rate of PCr recovery is dependent on the rate of pH recovery and a direct linear relationship that exists between the value of intracellular pH (at the end of exercise) and the rate of PCr recovery.[40]

At the end of exercise, intracellular pH (pHi) continues to fall until a minimum which, depending on the work rate during the exercise, occurs approximately 1 minute after the end of the exercise period. The higher the work rate the later pH recovery begins (see Fig. 4.4). Although they are likely to involve active transport mechanisms,[41] we do not understand the mechanisms that control pHi. During recovery, [Pi] generally mirrors [PCr], and transport of the Pi accumulated during exercise into the mitochondria produces phosphorylation of ADP to ATP. Like PCr, Pi recovery is biphasic[42] after the end of exercise and during the period in which pH is still falling Pi recovery is fast. Subsequently, as the pH recovers, Pi recovery slows down and [Pi] decreases to undetectable levels for several minutes before it again reappears and recovers to resting levels. During this time, a temporary decrease in the total [PCr + Pi] occurs; [PCr + Pi] otherwise remains constant throughout exercise. When performing exhausting exercise that results in a reduction of [ATP], e.g. strenuous aerobic exercise, then impairment of all recovery processes occurs.[43]

Studies of muscle disease

A considerable body of knowledge of muscle energy metabolism in disease has arisen from needle biopsy studies. However, muscle behaves in a heterogeneous manner during exercise, and therefore needle biopsies, while being cheap and quick to perform, have the disadvantage of sampling only very small and not very representative portions of muscle. Additionally, needle biopsies are invasive and thus impractical for multiple sampling, e.g. during exercise studies. MRS affords an alternative non-invasive, non-intrusive approach to the study of metabolism in disease.

The advantage of continuously monitoring metabolite levels for muscle research studies is clear, but the diagnostic value of MRS in muscle remains in doubt. Resting spectra can provide valuable metabolic information in certain myopathies such as in dystrophy[44] and in muscle injury.[18] The question of whether metabolic abnormalities are secondary to other disease processes is becoming of increasing interest clinically, particularly because of the consequences of disease on muscle pain, performance, and fatigue.

A further application of MRS has been to study the effects of therapy, e.g. in altering muscle energetics where enzyme defects have limited energy supply for contraction, such as glucose infusion in McArdle's disease[45] (glycogenolysis impairment caused by myophosphorylase deficiency).

MRS monitoring of drug trials in Duchenne dystrophy[44] and observation of the consequences of insulin infusion on phosphate metabolism[46] have also been performed.

Metabolic abnormalities

The early applications of MRS in disease, in particular the patients with clear metabolic defects, provided a novel means to study the normal physiological mechanisms that limit exercise. Such patients have been considered to represent nature's experiments, allowing examination of metabolic processes under conditions that would normally not be possible in human muscle. McArdle's disease is an example, in which a defect in metabolism may manifest itself as a failure of membrane excitation. There is no lactate production in these individuals, and hence no acidic shift in the Pi peak occurs, making it possible to measure initial rates of PCr resynthesis and the Km for ADP control of oxidative phosphorylation.[45] Figure 4.4 illustrates the metabolic changes seen during exercise in a patient with phosphofructokinase (PFK) deficiency[5] which is perhaps metabolically more significant than McArdle's disease,[47] in that oxidation of blood-borne glucose is not possible. The PME peak increases abnormally during exercise.[48,49] Accumulation of fructose-6-phosphate traps Pi, resulting in a relatively small rise in Pi, and again no change in pH occurs. In these PFK-deficient patients, Pi or H^+ increases cannot explain the rapid onset of fatigue. Similarly, abnormally raised PME peaks during exercise with limited acidosis are reported for phosphoglycerate mutase defects.[50]

The mitochondrial abnormalities represent a further group of the metabolic myopathies. Those patients affected show abnormalities in oxidative phosphorylation and consequently a high Pi, a low PCr,[51] and slow PCr resynthesis rates.[52,53] Various studies of patients with defects in mitochondrial function, show reduced PCr and ADP recovery rates, although the most reliable indications come from the resting spectra. Interestingly, encephalomyopathy induces changes as a result of mitochondrial enzyme defects in which an abnormal transfer function and slow recovery of PCr and pH occur;[54,55] encephalomyopathy also causes abnormalities in brain energy metabolism.

Dystrophies

Abnormalities in MRS parameters in Duchenne muscular dystrophy have been discussed earlier. The question of whether there is a reduction in energy state of the dystrophic tissue has been difficult to answer, owing to the increasing proportion of muscle tissue replaced with fat and connective tissues as the disease progresses.[44] This highlights the inherent partial volume effects that are always a problem when acquiring a signal from a heterogeneous tissue. Several studies have attempted to correct the acquired phosphorous signal for dilution affects from non-contractile tissue using data from biopsy samples, and indeed several studies have shown a reduced PCr:ATP ratio in dystrophic muscle. Moreover, a reduction in PCr:Pi ratio suggests impairment of mitochondrial function,[56] which is likely to be secondary to the dystrophic process. As greater changes in the PCr:Pi ratio occurs in patients with enzyme defects in whom histological evidence of damage is not apparent; this impairment is probably due to disuse of the muscle, rather than being a contributing factor to the damage process. Indeed, ^{31}P MRS is sufficiently sensitive to show abnormalities in metabolism in patients with minimal or no muscle weakness,[57] and Duchenne carriers demonstrate altered energy metabolism.[58]

Other disorders

The most marked changes in muscle metabolism demonstrated in disease are probably secondary in origin, particularly when impairment of blood flow is evident. Examples include peripheral vascular disease,[17,59] where the relative ischaemia results in an acid shift and fall in the Pi:PCr ratio with exercise; and sickle cell anemia,[60] with a consequential ischaemia in muscle leading to a reduced total ^{31}P; congenital heart disease,[61] in which the impaired oxygenation and cyanosis leads to elevated resting pH and Pi with abnormal PCr depletion and acidification during exercise with prolonged recovery times; and anaesthetic-induced malignant hyperthermia, which results in subjects having a high resting Pi:PCr ratio with slow recovery following exercise.[62] It is likely that many more disease states where some degree of fatigue or muscle impairment occurs secondary to the disease may also demonstrate abnormalities in metabolism of muscle.

Future perspectives

Although MRS has found application in the study of muscle disease for purely scientific purposes, it has yet to find any great application as a diagnostic test for muscle disease. This is primarily due to the specialized nature of the MRS examination and the great expense involved when many less expensive procedures are available (e.g. needle biopsy followed by chemical and genetic analysis). Furthermore, these other procedures are capable of providing information on a greater range of metabolites or gene products than is currently possible by MRS. There is little doubt however that MRS will continue to provide a useful tool for the scientific elucidation of muscle biochemistry, particularly when combined with the use of labeled compounds such as ^{13}C labeled glucose tracers.

REFERENCES

1 Radda GK. Control of bioenergetics: from cells to man by phosphorus NMR, 19th Ciba Medal Lecture. *Biochem Soc Trans* 1986; **14**:517–25.

2 Boska M. Estimating the ATP cost of force production in the human gastrocnemius/soleus muscle group using 31P MRS and 1H MRI. *NMR Biomed* 1991; **4**:173–81.

3 Narici MV, Landoni L, Minetti AE. Assessment of human knee extensor muscles stress from in vivo physiological cross-sectional area and strength measurements. *Eur J Appl Physiol* 1992; **65**:438–44.

4 Sahlin K. Intracellular pH and energy metabolism in skeletal muscle of man, with special reference to exercise. *Acta Physiol Scand (Suppl)* 1978; **445**:1–56.

5 Edwards RHT, Dawson MJ, Wilkie DR, Gordon RE, Shaw D. Clinical use of nuclear magnetic resonance in the investigation of myopathy. *Lancet* 1982; i:725–31.

6 Hoult DI, Busby SJW, Gadian DG, Radda GK, Richards RE, Seeley RJ. Observation of tissue metabolites using 31P nuclear magnetic resonance. *Nature* 1974; **252**:285–6.

7 Ackerman JJH, Grove TH, Wong GG, Gadian DG, Radda GK. Mapping of metabolites in whole animals by 31P NMR using a surface coil. *Nature* 1980; **283**:167–70.

8 Gordon RE, Hanley PE, Shaw D. Tropical magnetic resonance. *Prog Nucl Magn Reson Spectrosc* 1981; **15**:1–47.

9 Campbell ID, Dobson CM, Williams RJP, Xavier AV. The convolution difference method for baseline correction. *J Magn Reson* 1978; **11**:172–6.

10 Price TB, Rothman DL, Avison MJ, Buonamico P, Shulman RG. 13C-NMR measurements of muscle glycogen concentration during low intensity exercise. *J Appl Physiol* 1991; **70**:1836–40.

11 Taylor R, Price TB, Katz LD, Shulman RG, Shulman GI. Direct measurement of change in muscle glycogen concentration after a mixed meal in normal subjects. *Am J Physiol* 1993; **265**:224–9.

12 Dunn JF, Kemp GK, Radda GK. Depth selective quantification of phosphorus metabolites in human calf muscle. *NMR Biomed* 1992; **5**:154–60.

13 Thomsen C, Jensen KE, Henriksen O. 31P NMR measurements of T2 relaxation times of 31P metabolites in human skeletal muscle in-vivo, *Magn Reson Imaging* 1989; **7**:557–9.

14 Bottomley PA, Foster TB, Darrow RD. Depth REsolved Surface coil Spectroscopy (DRESS) for in-vivo H-1, P-31 and C-13 NMR. *J Magn Reson* 1984; **59**:338–46.

15. Sakuma H, Nelson SJ, Vigneron DB, Hanliala J, Higgins CB. Measurement of T1 relaxation times of cardiac phosphate metabolites using BIR-4 adiabatic pulses and variable mutation method. *Magn Reson Med* 1993; **29**:688–91.

16. Rodenburg JB, de Boer RW, Schiereck P, van Echteld CJA, Bar PR. Changes in phosphorus compounds and water content in skeletal muscle due to eccentric exercise. *Eur J Appl Physiol* 1994; **68**:205–13.

17. Hands LJ, Bore PJ, Galloway G, Morris PJ, Radda GK. Muscle metabolism in patients with peripheral vascular disease investigated by 31P nuclear magnetic resonance spectroscopy. *Clin Sci* 1986; **71**:283–90.

18. McCully KK, Argov Z, Boden BP, Brown RL, Bank WJ, Chance B. Detection of muscle injury in humans with 31-P magnetic resonance spectroscopy. *Muscle Nerve* 1988; **11**: 212–16.

19. Cady EB, Jones DA, Lynn J, Newham DJ. Changes in force and intracellular metabolites during fatigue of human skeletal muscle. *J Physiol* 1989; **48**:311–25.

20. Gonzalez-de-Susa JM, Bernus G, Alonso J, et al. Development and characterization of an ergometer to study the bioenergetics of the human quadriceps muscle by 31P NMR spectroscopy inside a standard MR scanner. *Magn Reson Med* 1993; **29**:575–81.

21. Yoshida T, Watari H. 31P-nuclear magnetic resonance spectroscopy study of the time course of energy metabolism during exercise and recovery. *Eur J Physiol* 1993; **66**:494–9.

22. Edwards RHT, Hill DK, Jones DA, Merton PA. Fatigue of long duration in human skeletal muscle after exercise. *J Physiol* 1977; **272**:769–78.

23. Martin PA, Gibson H, Hughes S, Edwards RHT. Electrical stimulation and force recording of human quadriceps muscle during 31P MRS. *Proc Soc Magn Reson Med* 1991; **10**:547.

24. Gordon AM, Huxley AF, Julian FJ. The variation in isometric tension with sarcomere length in vertebrate muscle fibres. *J Physiol* 1966; **184**:170–92.

25. Belanger AY, McComas J. Extent of motor unit activation during effort. *J Appl Physiol* 1981; **51**: 1131–5.

26. Martin PA, Gibson H, Saugen E, Vollestad NK, Edwards RHT. Combined MRS voluntary and stimulated muscle exercise studies demonstrate individual pathways for electromechanical coupling and energy utilisation in submaximal exercise. *Proc Soc Magn Res Med* 1993; **3**:1133.

27. Vestergaard-Poulsen P, Thomsen C, Sinkjaer T, Stubgaard M, Rosenfalck A, Henriksen O. Simultaneous electromyography and 31P nuclear magnetic resonance spectroscopy with application to muscle fatigue. *Electroencephalogr Clin Neurophysiol* 1992; **85**:402–11.

28. Adams GR, Harris RT, Woodard D, Dudley GA. Mapping of electrical muscle stimulation using MRI. *J Appl Physiol* 1993; **74**:532–7.

29. Kushmerick MJ, Moreland TM, Wiseman RW. Mammalian skeletal muscle fibers distinguished by contents of phosphocreatine, ATP, and Pi. *Proc. Natl Acad Sci U S A*, 1992; **89**:7521–5.

30. Kemp GK, Taylor DJ, Styles P, Radda GK. The production, buffering and efflux of protons in human skeletal muscle during exercise and recovery. *NMR Biomed* 1993; **6**:73–83.

31. Porter R, Whelan J (eds). Human muscle fatigue: physiological mechanisms. *Ciba Foundation Symposium* 82, Pitman Medical: London, 1981.

32. Sargeant AJ, Kernell D (eds). *Neuromuscular Fatigue.* Royal Netherlands Academy of Arts and Science: Amsterdam, 1993.

33. Newham DJ, Mills KR, Quigley BM, Edwards RHT. Pain and fatigue after concentric and eccentric muscle contractions. *Clin Sci* 1983; **64**:55–62.

34. Miller RG, Giannini D, Milner-Brown HS, et al. Effects of fatiguing exercise on high energy phosphates, force, and EMG: evidence for three phases of recovery. *Muscle Nerve* 1987; **10**:810–21.

35. Kemp GK, Thompson CH, Barnes PR, Radda GK. Comparisons of ATP turnover in human muscle during ischemic and aerobic exercise using 31P magnetic resonance spectroscopy. *Magn Reson Med* 1994; **31**:248-58.

36. Kemp GK, Taylor DJ, Radda GK. Control of phosphocreatine resynthesis during recovery from exercise in human skeletal muscle. *NMR Biomed* 1993; **6**:66–72.

37. Taylor DJ, Bore PJ, Styles P, Gadian DG, Radda GK. Bioenergetics of intact human muscle: a 31P nuclear magnetic resonance study. *Mol Biol Med* 1983; **1**:77–94.

38. Kemp GK, Taylor DJ, Thompson CH, et al. Quantitative analysis by 31P magnetic resonance spectroscopy of abnormal mitochondrial oxidation in skeletal muscle during recovery from exercise. *NMR Biomed* 1993; **6**:302–10.

39. Arnold DL, Matthews PM, Radda GK. Metabolic recovery after exercise and the assessment of mitochondrial function in vivo in human skeletal muscle

by means of 31P NMR. *Magn Reson Med* 1984; **1**:307–15.
40 Bendahan D, Confort-Gouny S, Kozak-Reiss G, Cozzone PJ. Heterogeneity of metabolic response to muscular exercise in humans. New criteria of invariance defined by in vivo phosphorus NMR spectroscopy. *FEBS Letters* 1990; **272**:155–8.
41 Nadshus IH. Regulation of intracellular pH in eukaryotic cells. *Biochem J* 1988; **250**:1–8.
42 Iotti S, Funicello R, Zaniol P, Barbiroli B. Kinetics of post-exercise phosphate transport in human skeletal muscle: an in vivo 31P-MR spectroscopy study. *Biochem Biophys Res Commun* 1991; **176**:1204–9.
43 Taylor DJ, Styles P, Matthews PM, et al. Energetics of human muscle: exercise induced ATP depletion. *Magn Res Med* 1986; **3**:44–54.
44 Griffiths RD, Cady EB, Edwards RHT, Wilkie DR. Muscle energy metabolism in Duchenne dystrophy studied by 31P-NMR: controlled trials show no effect of allopurinol or ribose. *Muscle Nerve* 1985; **8**:760–7.
45 Lewis SF, Haller RG, Cook JD, Nunnally RL. Muscle fatigue in McArdle's disease studied by 31P-NMR: effect of glucose infusion. *J Appl Physiol* 1985; **59**: 1991–4.
46 Taylor DJ, Coppack SW, Cadoux-Hudson TAD, et al. Effect of insulin on intracellular pH and phosphate metabolism in human skeletal muscle in vivo. *Clin Sci* 1991; **81**:123–8.
47 McArdle B. Myopathy due to a defect in glycogen breakdown. *Clin Sci* 1951; **10**:13–35.
48 Duboc D, Jehenson P, Tran Dinh S, Marsac C, Syrota A, Fardeau M. Phosphorus NMR spectroscopy study of muscular enzyme deficiencies involving glycogenolysis and glycolysis. *Neurology* 1987; **37**:663–71.
49 Argov Z, Bank WJ, Maris J, Leigh JS Jr, Chance B. Muscle energy metabolism in human phosphofructokinase deficiency as recorded by 31P nuclear magnetic resonance spectroscopy. *Ann Neurol* 1987; **22**: 46–51.
50 Argov Z, Bank WJ, Boden B, Ro YI, Chance B. Phosphorus magnetic resonance spectroscopy of partially blocked muscle glycolysis. *Arch Neurol* 1987; **44**:614–17.
51 Gadian DG, Radda GK, Ross BD, et al. Examination of a myopathy by phosphorus nuclear magnetic resonance. *Lancet* 1981; **ii**:774–5.
52 Radda GK, Bore PJ, Gadian DG, et al. 31P-NMR examination of two patients with NADH-CoQ reductase deficiency. *Nature* 1982; **295**:608–9.
53 Edwards RHT, Griffiths RD, Cady EB. Topical magnetic resonance for the study of muscle metabolism in human myopathy. *Clin Physiol* 1985; **5**:93–109.
54 Hayes DJ, Hilton-Jones D, Arnold DL, et al. A mitochondrial encephalomyopathy: a combined 31P-magnetic resonance and biochemical investigation. *J Neurol Sci* 1985; **71**:105–18.
55 Barbiroli B, Montagna P, Cortelli P, Martinelli P, Sacquegna, T, Zaniol P. Complicated migraine studied by phosphorus magnetic resonance spectroscopy. *Cephalalgia* 1990; **10**:263–72.
56 Chance B, Eleff S, Leigh JS Jr, Solilow D, Sapega A. Mitochondrial regulation of phosphocreatine/phosphate ratios in exercising human limbs: gated 31P NMR study. *Proc Natl Acad Sci U S A* 1981; **78**:6714–18.
57 Barbiroli B. 31P MRS of human skeletal muscle. In: Certaines JD, Bovee WMM, Podo F (eds). *Magnetic Resonance Spectroscopy in Biology and Medicine*. Pergamon Press: Oxford, 1992:369–86.
58 Barbiroli B, Funicello R, Ferlini A, Montagna P, Zaniol P. Muscle energy metabolism in female DMD/BMD carriers: a 31P-MR spectroscopy study. *Muscle Nerve* 1992; **15**:344–8.
59 Zatina MA, Berkowitz HD, Gross GM, Maris JM, Chance B. 31P nuclear magnetic resonance spectroscopy: non-invasive biochemical analysis of the ischemic extremity. *J Vasc Surg* 1986; **3**:411–20.
60 Norris SL, Gober JR, Haywood LJ, Halls J, Boswell W, Colletti P, Terk M. Altered muscle metabolism shown by magnetic resonance spectroscopy in sickle cell disease with leg ulcers. *Magn Reson Imaging* 1993; **11**:119–23.
61 Adatia I, Kemp GJ, Taylor DJ, Radda GK, Rajagopalan B, Haworth SG. Abnormalities in skeletal muscle metabolism in cyanotic patients with congenital heart disease: a 31P nuclear magnetic resonance spectroscopy study. *Clin Sci* 1993; **85**:105–9.
62 Olgin J, Rosenberg H, Allen G, Seestedt R, Chance B. A blinded comparison of non-invasive in vivo phosphorus nuclear magnetic resonance spectroscopy and the in vitro halothane/caffeine contracture test in evaluation of hyperthermia susceptibility. *Anesth Analg* 1991; **72**: 36–47.

5 Human cardiac NMR spectroscopy

Paul A Bottomley

INTRODUCTION

The key to the utility of spectroscopy for studying patient populations is the identification of NMR-visible moieties that play an important role in, or are able to characterize, disease processes. In the heart, which is the largest consumer of energy per unit of tissue mass, defects affecting the supply and demand of energy are commonly implicated in dysfunction. Adenosine triphosphate (ATP) is the fundamental energy currency of the body. It provides energy for muscular contraction via cleavage of the high-energy phosphate bond forming inorganic phosphate (Pi):

$$\text{ATP} \rightarrow \text{adenosine diphosphate (ADP)} + \text{Pi} + \text{energy}.$$

Phosphorus (^{31}P) spectroscopy can directly measure naturally abundant ATP. It can also monitor phosphocreatine (PCr), a reservoir of cellular energy that can supply high-energy phosphate to maintain ATP levels during short anaerobic episodes via the creatine kinase reaction:

$$\text{PCr} + \text{ADP} \leftrightarrow \text{ATP} + \text{creatine}.$$

These compounds exist in the heart at concentrations of only about 6 μmol/g and 11 μmol/g (wet weight), and therefore their NMR signals cannot match the proton (^{1}H) signal from 90 mmol/g tissue water used for NMR imaging (MRI) in sensitivity, spatial resolution and scan-time.[1] Nevertheless, with coarser spatial resolution of 8 ml or more,[2] and longer scan times of (typically) 5–30 min, PCr and ATP can be observed in the anterior myocardium.[3] The critical questions are:

- what do such measurements from normal subjects and patients tell us about human myocardial energy metabolism in normal and disease states; and
- could the technique find a role in the clinic?

In the case of ischemic heart disease, the effect of a transient imbalance in oxygen supply and demand in regions of the heart that are perfused by stenosed arteries is anticipated by much prior work on animals with coronary occlusion or low-flow ischemia:[4] excess PCr consumption occurs,[5] accompanied by elevations in Pi. This manifests itself as transient reductions in PCr:ATP ratio which can be seen, for example, by localized ^{31}P NMR performed during stress-testing in the NMR system.[5] In dilated cardiomyopathy (DCM), hypertrophy, and coronary artery disease (CAD),[6,7] chronic reductions in the levels of PCr and Cr, and of creatine kinase activity, have also been measured in biochemical assays of biopsies taken at surgery. These observations link changes in energy metabolism to contractile dysfunction in the absence of evidence of active ischemia.[8] PCr, ATP, and their ratio are thus useful indices of supply-side myocardial energy metabolism.

In this chapter, ^{31}P NMR studies of patients with cardiomyopathies, including DCM, hypertrophic cardiomyopathy (HCM) and pressure overload left ventricular hypertrophy (LVH), transplanted hearts, and ischemic heart disease (including myocardial infarction) are reviewed (see elsewhere for a more detailed review[9]) and the outlook for potential clinical applications are considered. The precision with which spectroscopic localization is achieved and how the ^{31}P results are quantified are critical factors that affect the outcome of the patient studies, and provide an appropriate place to start. Emerging opportunities that may provide access to metabolites in the glycolytic and citric acid cycle, or to deoxymyoglobin in the human heart, via ^{1}H, carbon (^{13}C) or ^{13}C/^{1}H double resonance NMR spectroscopy, are also covered.

METHODS

NMR coils and optimizing sensitivity

To date, ^{31}P NMR spectroscopy of the intact human heart has invariably deployed surface detection coils placed externally on the chest adjacent to the anterior myocardium. Surface coils generally provide better sensitivity than larger volume coils, provided that they are optimized so that sample (magnetic) losses represent the dominant noise source and the coil diameter is approximately equal to the depth of the tissue of interest.[10] Excitation is preferably accomplished with a separate large NMR coil to ensure uniform excitation and to avoid spatially dependent spectral distortion.[11]

The limited range of sensitivity of surface coils necessitates their careful positioning relative to the heart, which is most conveniently visualized by MRI,[3] although ultrasound imaging has also been used to guide positioning.[12] The coil positioning problem can be alleviated by use of phased-arrays of ^{31}P surface detection coils for the heart,[13] as was pioneered for conventional MRI.[10] Phased arrays can also provide an additional (quadrature) sensitivity improvement of up to about $\sqrt{2}$ relative to the best-positioned single-surface coil of the same diameter as one of the phased array coils. A four-coil ^{31}P cardiac phased array is pictured in Fig. 5.1.[13]

Substantial sensitivity gains can also be realized from the use of a prone versus a supine patient orientation, which brings the heart closer to the detection coil,[11] and from gating acquisitions synchronous to the heart cycle. Finally, at 1.5 T, a significant proton nuclear Overhauser effect (nOe) can be realized for PCr ($\eta \approx 0.6$) and ATP ($\eta \approx 0.4$–0.6) in the human heart (Table 5.1).[14,15] The requisite narrow or broadband ^{1}H irradiation for nOe can be provided via an additional ^{1}H surface coil or a body MRI coil, care being taken not to exceed radio frequency (rf) power deposition limits. Because η can differ for PCr and ATP, nOe may distort the observed PCr:ATP ratio, necessitating a correction for quantitative intersite comparisons.[14]

Localization

Localization methods based on the application of MRI gradients, and gradients in the rf excitation field, have been successful in distinguishing myocardial spectral information from chest wall contributions. The first of these, depth resolved surface coil spectroscopy (DRESS), used a single

Human cardiac NMR spectroscopy

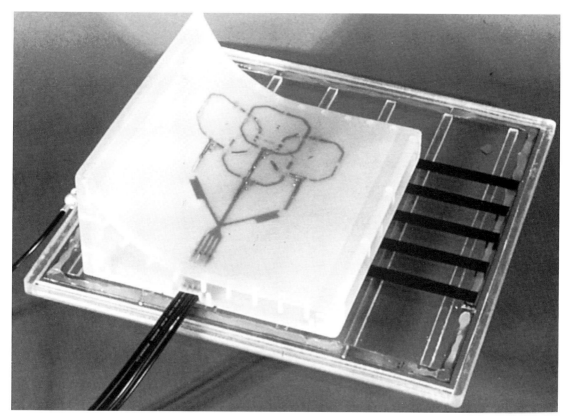

Fig. 5.1. A ^{31}P cardiac phased array of four 6.5-cm diameter surface detector coils. The subject lies prone on the coil set, with the heart near the center. ^{31}P NMR is excited with a large 40-cm square transmitter coil seen below it: the black stripes are velcro attachment strips. The coils are overlapped and fed to low input impedance preamplifiers to minimize interactions, and a four-channel quadrature receiver. For a CSI study, the four data sets must be first reconstructed and phased prior to combination into a single data set with improved signal-to-noise.[13] (Reproduced with permission.)

slice-selective excitation pulse to limit excitation to a plane parallel to the surface coil intersecting the anterior myocardium.[3] It is still in use today.[25] This and other one-dimensional (1-D) localization methods rely on the coil's sensitivity profile to provide localization in the other two dimensions coplanar with the coil. Full 3-D localization to single volume elements (voxels) in the heart has been achieved with the ISIS (image selected *in vivo* spectroscopy) technique.[26] ISIS uses a sequence of up to three slice-selective inversion pulses applied prior to excitation and detection of a free induction decay following an additional non-selective excitation pulse. The sequence must be applied at least eight times, with each combination of selective pulses. The large resultant signals are added and subtracted to generate the heart spectrum. The method is susceptible to signal artefacts from outside the selected volume because of motion[27] or partial saturation

Table 5.1 ^{31}P NMR measurements of phosphate metabolites in normal volunteers.[2,14–24] CK = creatine kinase; γ-ATP, β-ATP = gamma- and beta-phosphates of ATP. *95% confidence intervals.

Parameter	Value	Reference	Parameter	Value	Reference
PCr/ATP	1.80 ± 0.21	16	pH	7.15 ± 0.2	22
	1.93 ± 0.21	17		7.15 ± 0.03	19
	1.8 ± 0.1	2	CK forward rate	0.5 ± 0.2 s^{-1}	23
	1.95 ± 0.45	18	CK forward flux	6 ± 3 μmol(g wet wt s)$^{-1}$	23
	1.65 ± 0.26	19	T_1(PCr)	4.37 ± 0.48*	24
[PCr]	11 ± 3 μmol/g wet wt	20	T_1(γ-ATP)	2.52 ± 0.45*	24
	11.3 ± 3.7 μmol/g wet wt	21	T_1(β-ATP)	2.28 ± 0.54*	24
[ATP]	6.9 ± 1.6 μmol/g wet wt	20	nOe of PCr at 1.5 T	0.61 ± 0.25	14
	7.4 ± 1.6 μmol/g wt wt	21		0.6 ± 0.1	15
Pi/PCr	< 0.25	22	nOe of ATP at 1.5 T	0.6 ± 0.3 (γ-ATP)	14
	0.14 ± 0.06	19		0.3 ± 0.2 (β-ATP)	14
				0.4 ± 0.2 (γ & β-ATP)	15

effects.[28] A significant disadvantage of single voxel methods for dynamic studies or for studies of focal abnormalities that are not easily identified by other modalities is that the location of the suspected abnormality must be correctly selected *a priori*.

Techniques that permit the acquisition of signals from multiple voxels in the heart and chest in one dimension are the 1-D chemical shift imaging (CSI) method, which utilizes a single hard rf pulse followed by a phase-encoding gradient pulse to encode a contiguous series of slices parallel to the surface coil,[5,11] and the rotating frame zeugmatography (RFZ) method which encodes with an rf pulse applied by a transmitter field with a near-linear gradient.[12,29] The CSI method is easily extended to 3-D, either by using concurrent phase-encoding gradients in all three dimensions,[30] or by replacing the hard excitation pulse by a slice-selective pulse in one dimension, followed by phase-encoding in the other two.[11,20] While all of these multiple voxel techniques yield essentially the same signal-to-noise ratio from all excited voxels as that achievable from a single voxel method with the same spatial resolution and total scan time, the total scan time in CSI is constrained by the requirement that at least $n_x.n_y.n_z$ different phase encoding steps must be applied, where n_x, n_y, and n_z are equal to the dimensions of the field-of-view in the X, Y, and Z directions, divided by the desired spatial resolution in those dimensions. In practice, scan times will be limited by patient tolerance as, for example, in stress-testing for myocardial ischemia.[5]

Correcting distortion

While myocardial PCr, ATP, and conceivably the PCr:ATP ratio from CSI data sets could be displayed as image intensity, findings and the testing of hypotheses invariably require the derivation of numerical quantities from the spectra, representing metabolite concentrations or ratios. Here, the convention that all metabolite ratios and concentrations are based on the integrated signal of the corresponding resonances in the spectrum is adopted throughout.[31] Quantification should involve correction for experimental distortions, such as the distortion introduced by partial saturation and contamination from blood in the ventricles. How these are or are not dealt with probably represents the major source of inter- and intra-laboratory variability, with published values from normal volunteers ranging from 0.9 ± 0.3[30] to 2.1 ± 0.4.[32]

Partial saturation distortion of the PCr or ATP signals occurs when the NMR pulse sequence repetition periods, T_R, are comparable or less than the spin-lattice relaxation times, T_1, of PCr or ATP. For patient heart studies using ^{31}P, signal-to-noise ratio considerations and the lengthy PCr T_1 of 4–5 seconds dictate partial saturation conditions in virtually all cases.[33] The myocardial PCr:ATP ratio is also distorted because the T_1 values of the two moieties are different.[24,33] The 95% confidence intervals for published values of T_1 are shown in Table 5.1. The ratio of T_1(PCr):T_1(ATP) is about 1.9,[24] which leads to a distortion factor by which the short-T_R PCr:ATP ratio must be multiplied to yield the true myocardial PCr:ATP ratio, of $\sim$1–2 depending on T_R and the NMR pulse flip-angles.[9,33] Uncorrected measurements of PCr:ATP ratios can thus differ from corrected values by up to about twofold.

The saturation distorted PCr and ATP signals can be corrected if myocardial T_1 values and the experimental NMR pulse flip-angle, α, are known:

$$S_0 = \frac{S[1 - \cos \alpha \exp(-T_R/T_1)]}{[1 - \exp(-T_R/T_1)] \sin \alpha},$$

where S and S_0 are the distorted and undistorted signals, respectively.[33] As T_1 measurements are time consuming, T_1 values must be measured on a separate group of individuals (usually healthy volunteers) if published values are not adopted. The assumption that the values of T_1 are the same in the study group of interest (for example, patients) as in the T_1 controls is implicit. An alternative approach for correcting the PCr:ATP ratio for saturation distortion is to measure the saturation factor directly from the ratio of signals acquired under partially saturated and fully relaxed conditions on each subject. To provide adequate sensitivity for a measurement that can be accommodated in several minutes of a study protocol, the unlocalized, uniformly-excited surface coil signal will suffice.[5,33] This correction rests on the assumption that the T_1(PCr):T_1(ATP) ratio is substantially the same in the chest and heart tissue that contribute to the unlocalized spectrum.[33] This assumption is consistent with the available data.[24] For both methods, errors resulting from the use of incorrect values for T_1 or failures in the underlying assumptions, will be minimized by using small-angle or Ernst-angle ($\cos \alpha = \exp[-T_R/T_1]$) excitation.[9] A complete set of data from a ^{31}P cardiac exam showing unlocalized surface coil chest spectra used for correcting saturation by the second method, a 1-D CSI ^{31}P set of spectra, and corresponding annotated surface coil MRI, are exemplified in Fig. 5.2.

Blood contamination is a second common source of scatter and potentially erroneous myocardial PCr:ATP values. Blood contains ATP but no PCr. It therefore can reduce the observed PCr:ATP ratio in voxels intersecting the ventricles.[16] Blood also contains 2,3-diphosphoglycerate (DPG) which exhibits characteristic resonances at about 5.4 and 6.3 ppm relative to PCr. The [ATP]:[DPG] ratio is about 0.3.[19,25,34,35] The amount of ATP contributed from the blood pool can be estimated by measuring the blood DPG signal in the spectrum:

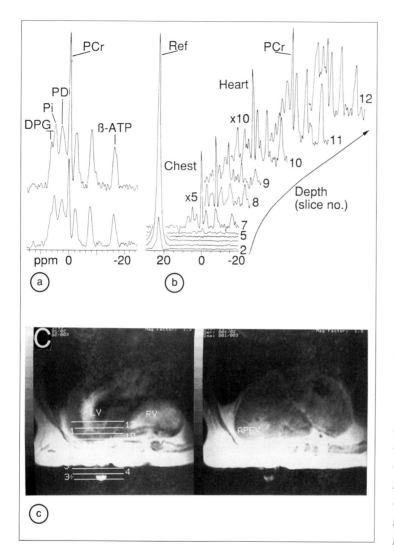

Fig. 5.2. Data from a complete cardiac ^{31}P exam of an elderly female CAD patient.[33] (a) Unlocalized cardiac-gated ^{31}P spectra obtained from the chest and heart with a 6.5-cm diameter surface detection coil, and a separate uniform transmitter. The lower spectrum was acquired at the heart-rate (60 excitations, $T_R = 0.8$ seconds); the upper spectrum at $T_R = 15$ s (20 excitations). The peak ratios from the two spectra can be used to correct partial saturation effects.[33] (b) A 1-D CSI data set of 6-mm thick coronal slices as a function of depth (vertical) extending into the myocardium, acquired in 12 minutes. The reference peak (Ref) is phosphonitrilic chloride trimer embedded in the surface soil.
(c) Corresponding axial surface coil images from a multislice data set on the surface coil axis (left, annotated to show 1-D CSI slices) and 1-cm inferiorly (right). The images were acquired prior to spectroscopy to confirm coil location, without disturbing the subject. (Reproduced, with permission, from Bottomley et al.[33])

$$(\text{corrected ATP}) = (\text{contaminated ATP}) - 0.5([\text{ATP}]/[\text{DPG}]) \times (\text{DPG}),$$

recalling that DPG has two phosphates.

Typically, the effect of blood ATP corrections on raw PCr:ATP values are fairly small at $13 \pm 6\%$.[16–19,25] This is fortuitous as there are potential problems for the correction with short T_R protocols if there are differences between the values of T_1 for blood ATP and DPG that are not discounted by the effect of inflowing unsaturated blood or the use of small flip angles. The apparent blood ATP:DPG ratio could also vary a little with the timing of the localization sequence and the study group,[36] and there is the possibility that DPG estimates are contaminated during spectral analysis by neighbouring Pi or phosphomonoester (PME) resonances, which would result in overestimation of the amount of blood contamination.

Normal myocardial PCr:ATP, [PCr], and [ATP]

While published PCr:ATP ratios for the human heart range from about 1 to 2, when those studies in which both saturation effects and blood ATP contamination were not accounted for are excluded, there is a consensus of around 1.8 for the normal human myocardial PCr/ATP ratio (see Table 5.1).[2,9,16–19] As ATP and PCr are likely to be 100% visible by ^{31}P NMR,[37,38] and because PCr loss may be unavoidable when human surgical biopsy specimens are assayed by traditional biochemical means,[6] this should represent a best current estimate of the true ratio.

Fully corrected measurements of metabolite ratios can be extended to absolute concentration measurements in a voxel by including a determination of the NMR signal intensity of a known concentration reference in the study protocol, and a measurement of the tissue volume or mass which contributes to the heart spectrum from the voxel.[20] The concentration reference can be included in the CSI data set, for example, at a fixed location within the field-of-view during the patient study. In calculating concentrations, any differences in detection coil sensitivity at the location of the reference and at the voxel being assayed, and any non-uniformity in the excitation field must be compensated for.[20] Alternatively, NMR measurements of a reference concentration at the same location as the heart may be performed in a separate experiment in which the subject is absent, provided that any differences in coil loading are taken into account.[21] Tissue volumes can be estimated by multislice MRI acquired without moving the subject. Human myocardial PCr and ATP concentrations reported to date are preliminary, but agree with each other (see Table 5.1).

Pi, pH, and metabolic flux

The quantification of Pi in the normal heart is problematic, owing to its low concentration and the presence of neighboring, often overlapping, DPG and PME resonances.[22] It is probably best quantified in ^{31}P NMR as an upper bound of a concentration or a ratio to PCr or ATP (see Table 5.1). Although ^{1}H decoupling can narrow DPG linewidths and thereby enhance spectral resolution in the region of the spectrum where the Pi resonance falls, myocardial Pi may still be undetectable in a large fraction of normal subjects:[19] it may in fact be partially NMR-invisible.[37,38] The chemical shift of Pi also varies depending on intracellular (pHi), from about 3.7 to 5.2 ppm for a pH range of 6.0 to 7.3.[39] The pH of the normal heart in which Pi is detectable appears to be about 7.15 (see Table 5.1). The possibility that blood Pi contaminates these measurements of myocardial Pi and pH[40] is unlikely, as [Pi] may be as low as 0.08 mM[41] in whole human blood, and is imperceptible in blood ^{31}P spectra.[25,36]

The forward rate of turnover of PCr through the creatine kinase reaction in the heart can be determined with saturation transfer NMR experiments that have been modified to incorporate spatial localization.[23] The localized saturation transfer experiment involves acquisition of a localized ^{31}P data set in the presence and absence of saturating NMR irradiation applied to the γ-phosphate resonance of ATP. Saturation may be achieved simply by applying a steady low-level RF signal at the resonant frequency of the γ-phosphate. With γ-ATP saturated, PCr decreases as it reacts to form ATP in the absence of unsaturated phosphates being supplied by the reverse reaction. The pseudo-first-order forward rate constant, k, in units of the T_1 of PCr in the presence of the saturating radiation, T_1', is given by the ratio of PCr signals measured with (S') and without (S) the γ-ATP saturation:[23]

$$kT_1' = 1 - S'/S.$$

Measurements at 4 T suggest that about half of the PCr is turned over per second (see Table 5.1), consistent with data from animals.[23]

Patient studies

Some myocardial PCr/ATP data obtained by ^{31}P NMR from the human heart are summarized in Table 5.2.

Cardiomyopathy

^{31}P NMR studies of animal models of cardiomyopathy,[46–51] and biochemical analyses of biopsies

Table 5.2 Summary of myocardial PCr/ATP data from patients.[5,16–19,22,25,29,30,32,42–45] Values are mean ±SD. MD = muscular dystrophy; BB = cardiac beri-beri; CA = cardiac amyloidosis; MI = myocardial infarction. Controls were normal, or #LVH patients not in heart failure,[29] or DCM patients with NYHA class < III failure,[18] or patients with no transplant rejection,[17] or the same patients tested after intervention.[5] *$P < 0.001$, §$P < 0.01$, and †$P < 0.05$ vs controls, or, for ischemia patients, values at rest.[5,45]

PCr/ATP				PCr/ATP			
Patients	Controls	Comments	Ref	Patients	Controls	Comments	Ref
Cardiomyopathy studies showing significant changes				Heart transplant patients			
1.3 ± 0.3*	2.1 ± 0.4	MD, BB, CA	32	1.6 ± 0.5§	1.9 ± 0.2	all patients	17
1.4 ± 0.4*	2.1 ± 0.4	HCM	32	1.6 ± 0.5†	1.9 ± 0.4#	rejection with myocyte necrosis	17
1.3 ± 0.3†	1.7 ± 0.3	HCM, NYHA class 0–II	19	Myocardial infarction			
1.1 ± 0.4§	1.7 ± 0.3	HCM	25	1.7 ± 0.4	1.6 ± 0.4	elevated Pi at ≤ 9 days post-MI	22
1.1 ± 0.3	1.6 ± 0.2#	LVH, NYHA class II–III	29				
1.5 ± 0.31*	1.8 ± 0.2	DCM, NYHA class II–IV	16	(normal)		[PCr]*, [ATP]† lower, old MI	44
1.4 ± 0.5†	1.9 ± 0.4#	DCM, NYHA class ≥III	18	1.8 ± 0.5	2.0 ± 0.5	0.5–24 mo post-MI	18
Cardiomyopathy studies showing no significant changes				Myocardial ischemia			
1.7 ± 0.2	1.8 ± 0.2	HCM	5	0.9±0.2*	1.7 ± 0.2	during exercise stress	5
0.8 ± 0.3	0.9 ± 0.3	LVH, severe	30	1.0 ± 0.3†	1.6 ± 0.2#	during exercise stress	5
1.9 ± 0.2	2.1 ± 0.4	LVH	32	0.9 ± 0.3*	1.6 ± 0.2	during exercise stress	45
0.7 ± 0.3	0.9 ± 0.3	DCM	30				
1.5 ± 0.3	1.5 ± 0.1	DCM, NYHA class I–III	42				
1.9 ± 0.4	2.1 ± 0.4	DCM	32				
1.6 ± 0.2	1.9 ± 0.2	DCM, NYHA class I–III	43				
1.5 ± 0.6	1.7 ± 0.3	DCM, NYHA class II–III	19				

taken during surgery from patients with DCM and LVH[6,7] provide a basis for expecting that myocardial high-energy phosphates are reduced in human cardiomyopathy *in vivo*. Despite this, human *in vivo* ^{31}P NMR studies have yielded mixed results with respect to the question of whether, and how, a reduced PCr:ATP ratio is involved in the disease process. Thus, some studies report significant reductions in anterior myocardial PCr:ATP ratios in patients with cardiomyopathy due to specific diseases such as muscular dystrophy, beri-beri, and amyloidosis,[32] patients with HCM,[19,25,32] and patients with LVH[29] and DCM[16,18] who were in heart failure, while others have found no statistically significant changes in LVH[30,32] or DCM.[19,30,32,42,43]

In general, PCr:ATP ratios have correlated poorly with functional indices of disease severity such as left ventricular ejection fraction or fractional shortening[16,18,42] and the etiology of disease.[16] However, in two reports on cardiomyopathy, changes in myocardial PCr:ATP ratios have been directly linked to the presence of advanced heart failure as distinct from cardiomyopathy *per se*. One showed that the PCr:ATP ratio was significantly lower in LVH patients with valve disease being treated for heart failure, but not in patients not being treated for failure.[29] The other found a significant negative correlation between the myocardial PCr:ATP ratio and the New York Heart Association (NYHA) clinical classification for heart failure. In addition, in cases where intervention improved the NYHA classification, the myocardial PCr:ATP ratio also recovered.[18]

In assessing these results, it is likely that two factors account for the bulk of the difference between studies of cardiomyopathy that have found reduced PCr:ATP ratios and those that have not. First, it appears that the presence and severity of heart failure in the study population may be an important factor. In the study that found a correlation between the PCr:ATP ratio and NYHA failure classification, when PCr:ATP data from patients in mild and severe failure were averaged, only those patients in mild failure were considered, significant PCr:ATP ratio changes were not manifest.[18] The second factor is statistical sensitivity. Thus, although the statistical findings vary, in all studies of HCM, DCM, and LVH published to date, the mean myocardial PCr:ATP ratios in patients were lower than in the corresponding normal control groups.[9]

Those studies linking PCr:ATP ratio changes to heart failure suggest that reduced myocardial PCr:ATP ratios do not become detectable by ^{31}P NMR until the more advanced phases in which physical activity becomes limited by fatigue with symptoms such as palpitation and dyspnea (NYHA classes II and III).[18,29] This would be consistent with prior work that concluded that the ability of myocytes to sustain normal levels of ATP is compromised only in advanced stages of failure.[52] It has been hypothesized that the reduced myocardial energy reserve in heart failure evidenced by the lower myocardial PCr:ATP ratio, seen both by *in vivo* ^{31}P NMR and by decreases in myocardial creatine and creatine kinase activity assayed at surgery, may limit the heart's ability to do work and lead to contractile dysfunction.[8] However, it remains unclear why the metabolic changes are not manifest, at least in ^{31}P NMR spectra, earlier in the failure process. Furthermore, in some studies of HCM,[19,53] there is no clear link between PCr:ATP ratio reductions and the severity of heart failure, so the possibility that other factors may contribute to metabolic change cannot be ruled out.

It has been suggested that ^{31}P NMR might find clinical utility in identifying patients with heart failure in cases where diagnosis is complicated by other conditions such as age[29] and lung disease. The identification of metabolic abnormalities by ^{31}P NMR might also find use for managing surgical intervention in valve disease so as to achieve maximum benefit with minimum permanent injury.[54] Further work is needed to ascertain precisely when and where benefits may be realized.

Heart transplant patients

A number of ^{31}P NMR studies of animals with non-working transplanted hearts have shown significant, even dramatic, deterioration in myocardial PCr:Pi and/or PCr:ATP ratios in the first week or two post-transplantation.[55–59] Because metabolic changes preceded histological evidence for acute rejection, it was proposed that the metabolic changes may prove useful as predictors of histological rejection in human heart transplant recipients. The accepted standard, and the most reliable means of detecting significant rejection to date, is by histological examination of endomyocardial biopsies sampled at regular cardiac catheterization procedures. The presence of myocyte necrosis is generally regarded as a warning of a bout of rejection requiring augmented immunosuppressive therapy.

Studies to test whether the ^{31}P NMR findings from animals were applicable to human transplant patients, and whether metabolic changes might predict rejection and thereby help to limit the number of catheterization procedures on these patients, appeared in conference reports from 1988 onwards. While virtually all of these studies confirmed the presence of PCr:ATP and/or PCr:Pi ratio reductions in transplanted hearts, the reliability with which ^{31}P NMR could predict the histological findings was mixed. In 1991, a published report of 19 cases studied up to 5.5 years after transplantation found significantly lower resting anterior myocardial PCr:ATP ratios relative to normal controls, but agreement between ^{31}P NMR abnormalities and histological evidence for necrosis could be achieved in only about 60–70% of exams.[17] This suggested that ^{31}P NMR was not a precise predictor of significant histological rejection in many transplant patients. This finding is supported by the most recent conference reports.[60,61]

The overall changes in PCr:ATP ratios in heart transplants cannot presently be linked to hypertrophy or to CAD.[17] The poor correlation of metabolic indices with the histological outcome probably reflects fundamental differences in the techniques: myocyte necrosis should not cause an altered PCr:ATP ratio, since dead cells have neither PCr nor ATP ratio. The ratio should therefore not index necrosis *per se*, although a lower PCr:ATP ratio might warn of impending damage if a large enough proportion of cells were involved for them to be detected by the ^{31}P NMR exam. An elevated Pi might be expected to index necrosis better, but the PCr:Pi ratio data from animals do not strongly support the view that Pi might be a useful index of the severity of histological rejection or for differentiating rejection, that involves necrosis from milder rejection not involving necrosis (Fig. 5.3). To test the hypothesis that PCr:ATP ratio changes are indicative of an earlier phase of the rejection process, a much more frequent schedule of NMR exams and biopsies than the conventional biopsy schedule would be needed, which may not be practical. Nevertheless, the cause and effect of altered myocardial energy metabolism in transplant patients merits further investigation.

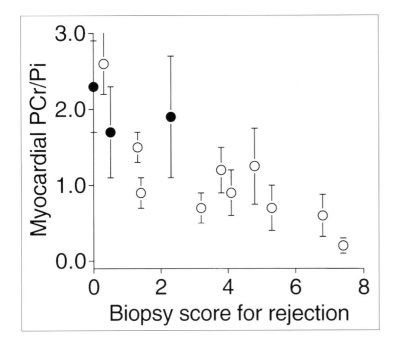

Fig. 5.3. Canine myocardial PCr:Pi ratio (means ±SD) as a function of histological score for rejection in groups of seven and 10 non-immunosuppressed (empty symbols) and immunosuppressed (filled symbols) heterotopic cardiac allografts implanted in the neck, adapted from data in reference 58. Biopsy scoring is via Billingham's criteria[58] (1–3, mild rejection, perivascular and interstitial mononuclear cell infiltration; 4–6, moderate, mononuclear cell infiltration plus myocyte necrosis; 7–8, severe, polymorphonuclear leukocyte infiltration plus myocyte necrosis).

Myocardial infarction

The PCr:ATP ratios in myocardial infarction at rest generally exhibit no significant alterations.[18,22,44] The possible role played by other factors such as heart failure and/or cardiomyopathy in reducing resting PCr:ATP ratios in some patients with myocardial infarction remains to be seen.[9,45] In recent myocardial infarction, there is evidence from canine[62] and human [22] studies to suggest that elevated Pi produced by PCr and ATP consumption may persist for several days after onset. Animal studies using biochemical methods also show that ATP and PCr are completely depleted within the first few hours of an ischemic injury that results in infarction.[4] Because high-energy phosphates are absent from necrotic myocytes, the normal resting PCr:ATP ratios observed by ^{31}P NMR in the absence of symptoms of heart failure in recent myocardial infarction must derive from a mixture of surrounding or interspersed, metabolically normal, scarred and possibly jeopardized, myocardium.

The fact that metabolite concentrations may be reduced in infarction is suggested by observations of a significant negative correlation between ATP levels and the size of perfusion deficits in thallium (^{201}Tl) radionuclide images of the heart.[44] The findings of altered metabolite levels measured by localized ^{31}P NMR, however, might not translate into altered tissue concentrations, since concentration measurements must account for the tissue volume present in the voxel. Thus, reductions in tissue volume caused by any thinning of the ventricular wall associated with the infarction will tend to offset infarction-induced reductions in metabolite levels when concentrations are calculated. Nevertheless, the observations[44] are consistent with a model for infarction in which the PCr and ATP levels are reduced in the heart spectrum proportionate to the volume of infarction in the voxel.[9]

Myocardial ischemia

As in patients with myocardial infarction, the resting myocardial PCr:ATP ratio in patients with ischemic heart disease involving severe stenosis of the anterior vessels, may be normal,[18] or nearly so.[5] The observation of reversible ischemia requires the development of stress protocols that can be conveniently performed by patients lying in the NMR magnet for the duration of a localized ^{31}P NMR acquisition. An aerobic exercise involving leg weight-lifting can produce a sustained increase in cardiac work-load, as indexed by the heart-rate–blood-pressure product (HRBP), of up to about 70% for periods of 30 minutes in normal volunteers.[63] With somewhat less motion during spectral acquisition, a continuous isometric hand-grip exercise at 30% of the subject's maximum force produces about a 30% increase in HRBP and coronary vasoconstriction in the presence of critical levels of coronary stenosis as well.[5] Exercise stress can be immediately terminated should complications arise, and the hand-grip exercise is well-tolerated by CAD patients for NMR acquisitions of 5–8 minutes or so. Stress induced by pharmaceutical agents such as dobutamine has also been used in ^{31}P NMR studies of volunteers with DCM.[43] This can increase HRBP by up to 2.3 times.

In healthy volunteers who are free of known heart disease, stress-testing by ^{31}P NMR does not alter myocardial PCr:ATP ratios significantly.[5,43,45,63] In groups of 16 and 15 patients with ischemia and significant CAD involving ≥ 70% stenosis of the left main or left anterior descending coronary arteries, the handgrip exercise protocol produced a reversible 40% decrease in the myocardial PCr:ATP ratio, while in patients with non-ischemic heart disease, including cardiomyopathy and valve disease and CAD patients

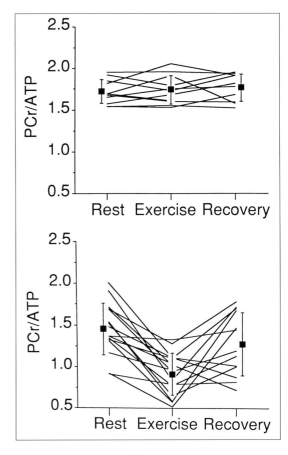

Fig. 5.4. Results from ^{31}P NMR hand-grip exercise stress-testing of disease-free normal controls (top), and patients with ischemia and significant CAD involving the anterior vessels (bottom).[5] The data represent the anterior myocardial PCr/ATP measured by localized spectroscopy before, during and after continuous isometric exercise at 30% of the subjects maximum force. Lines connect values for each individual: squares with error bars are means ±SD. Scan times were 9–14 minutes before exercise, and 5–8 min thereafter. (Reprinted, by permission of the New England Journal of Medicine *from Weiss et al.*[5])

with fixed anterior ^{201}Tl radionuclide imaging defects, myocardial PCr:ATP was unaltered by the same exercise (Fig. 5.4). Five[5,45] CAD patients were studied before and after successful revascularization therapy: a significant 33% reduction in the PCr:ATP ratio seen prior to therapy resolved in the post-therapy study.[5]

All of these findings, and the additional one that dobutamine stress-testing does not significantly alter myocardial PCr:ATP ratios in DCM patients,[43] suggest that stress-induced reductions in PCr:ATP ratios may be specific to ischemic disease. The potential clinical utility of ^{31}P NMR stress-testing may depend to some extent on whether studies can be extended to the inferior and posterior walls, for example with the assistance of nOe and phased-array ^{31}P cardiac detection coils, as well as on the outcome of studies comparing the sensitivity and cost efficacy relative to current diagnostic procedures.

Spectroscopy of other nuclei

The strongest candidate nuclei, other than ^{31}P, for human cardiac spectroscopy are ^{13}C and ^{1}H. Studies to date have been limited to a few normal volunteers. Only 1.1% of carbon is ^{13}C, and its sensitivity is lower than ^{31}P so millimolar-level energy metabolites such as glucose, glutamate, or even lactate are probably not accessible to natural abundance ^{13}C NMR in realistically tolerable human exams. The natural abundance normal human heart ^{13}C spectrum is dominated by fatty acid resonances, which are probably contaminated heavily by pericardial fat, given the inevitably coarse spatial resolution.[64] Glycogen may be visible with ^{1}H decoupling and nOe.[64] Animal ^{13}C spectroscopy studies with infused ^{13}C-enriched glucose have suggested that lactate and/or

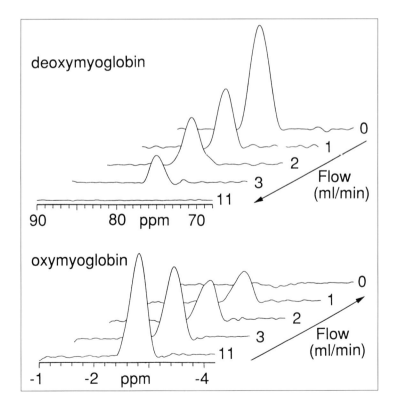

Fig. 5.5. Unlocalized ^{1}H spectra of deoxymyoglobin (top) and oxymyoglobin (bottom) resonances in an isolated perfused rat heart as a function of perfusion flow rate. The spectra were acquired in about 10 minutes each at 7 T. (Adapted with permission, from Kreutzer et al.[69])

glutamate levels may be more sensitive to low blood flow than ^{31}P measures.[65] With new ^{1}H-observe, ^{13}C-excite double resonance detection methods[66] such studies may be possible in humans.

Water-suppressed ^{1}H cardiac spectra are also dominated by fat resonances.[67] This is more problematic than in ^{13}C NMR, because the ^{1}H chemical shift dispersion is more than ten-fold smaller, and many metabolites of interest have resonances falling close to, if not overlapped by, the fat resonances. The use of multiple quantum NMR techniques[68] might be one method of overcoming some of these problems. Meanwhile, ^{1}H could provide access to measurements of total creatine, whose resonance is a little further away from those of fat: biochemical assays show altered creatine levels in cardiomyopathy and heart failure.[7,8] Finally, in rat hearts, the resonances of oxy- and deoxy-myoglobin have been identified in ^{1}H spectra at chemical shifts far from water and fat, providing another potential route for the study of ischemia (Fig. 5.5).[69]

ACKNOWLEDGEMENTS

I thank RG Weiss, M Conway, B Rajagopalan and GK Radda for many helpful discussions.

REFERENCES

1 **Bottomley PA**. Human in vivo NMR spectroscopy in diagnostic medicine: clinical tool or research probe? *Radiology* 1989; **170**:1–15.

2 **Menon RS, Hendrich K, Hu X, Ugurbil K**. ^{31}P NMR spectroscopy of the human heart at 4T: detection of substantially uncontaminated cardiac spectra and differentiation of subepicardium and subendocardium. *Magn Reson Med* 1992; **26**:368–76.

3 **Bottomley PA**. Noninvasive study of high-energy phosphate metabolism in human heart by depth-resolved ^{31}P NMR spectroscopy. *Science* 1985; **229**: 769–72.

4 **Jennings RB, Reimer KA**. Lethal Myocardial ischemic injury. *Am J Pathol* 1981; **102**:241–55.

5 **Weiss RG, Bottomley PA, Hardy CJ, Gerstenblith G**. Regional myocardial metabolism of high-energy phosphates during isometric exercise in patients with coronary artery disease. *N Engl J Med* 1990; **323**: 1593–1600.

6 **Swain JL, Sabina RL, Peyton RB, Jones RN, Wechsler AS, Holmes EW**. Derangements in myocardial purine and pyrimidine nucleotide metabolism in patients with coronary artery disease and left ventricular hypertrophy. *Proc Natl Acad Sci U S A* 1982; **79**:655–9.

7 **Ingwall JS, Kramer MF, Fifer MA, Lorell BH, Shemin R, Grossman W, Allen PD, et al**. The creatine kinase system in normal and diseased human myocardium. *N Engl J Med* 1985; **313**:1050–4.

8 **Ingwall JS**. Is cardiac failure a consequence of decreased energy reserve? *Circulation* 1993; **87**(suppl VII): VII-58–VII-62.

9 **Bottomley PA**. MR spectroscopy of the human heart: the status and the challenges. *Radiology* 1994; **191**:593–612.

10 **Roemer PB, Edelstein WA, Hayes CE, Souza SP, Mueller OM**. The NMR phased array. *Magn Reson Med* 1990; **16**:192–225.

11 **Bottomley PA, Hardy CJ**. Strategies and protocols for clinical ^{31}P research in the heart and brain. *Philos Trans R Soc Lond A* 1990; **333**:531–44.

12 **Blackledge MJ, Rajagopalan B, Oberhaensli RD, Bolas NM, Styles P, Radda GK**. Quantitative studies of human cardiac metabolism by P-31 rotating frame NMR. *Proc Natl Acad Sci U S A* 1987; **84**:4283–7.

13 **Hardy CJ, Bottomley PA, Rohling KW, Roemer PB**. An NMR phased array for human cardiac ^{31}P spectroscopy. *Magn Reson Med* 1992; **28**:54–64.

14 **Bottomley PA, Hardy CJ**. Proton Overhauser enhancements in human cardiac phosphorus NMR spectroscopy at 1.5T. *Magn Reson Med* 1992; **24**: 384–90.

15 **Kolem H, Sauter R, Friedrich M, Schneider M, Wicklow K**. Nuclear Overhauser enhancement and proton decoupling in phosphorus chemical shift imaging of the human heart. In: Pohost GM, ed, *Cardiovascular Applications of Magnetic Resonance*. Futura: Mt Kisco NY, 1993:417–26.

16 **Hardy CJ, Weiss RG, Bottomley PA, Gerstenblith G**. Altered myocardial high-energy phosphate metabolites in patients with dilated cardiomyopathy. *Am Heart J* 1991; **122**:795–801.

17. Bottomley PA, Weiss RG, Hardy CJ, Baumgartner WA. Myocardial high-energy phosphate metabolism and allograft rejection in patients with heart transplants. *Radiology* 1991; **181**:67–75.

18. Neubauer S, Krahe T, Schindler R, Horn M, Hillenbrand H, Entzeroth C, et al. ^{31}P magnetic resonance spectroscopy in dilated cardiomyopathy and coronary artery disease. *Circulation* 1992; **86**:1810–18.

19. de Roos A, Doornbos J, Luyten PR, Oosterwaal LJMP, van der Wall EE, den Hollander JA. Cardiac metabolism in patients with dilated and hypertrophic cardiomyopathy: assessment with proton-decoupled P-31 MR spectroscopy. *JMRI* 1992; **2**:711–19.

20. Bottomley PA, Hardy CJ, Roemer PB. Phosphate metabolite imaging and concentration measurements in human heart by nuclear magnetic resonance. *Magn Reson Med* 1990; **14**:425–34.

21. Okada M, Mitsunami K, Yabe T, Kinoshita M, Morikawa S, Inubushi T. Quantitative measurements of phosphorus metabolites in normal and diseased human hearts by ^{31}P NMR spectroscopy. *Proc Soc Magn Reson Med* 1992; **2**:2305 [abstract].

22. Bottomley, PA, Herfkens RJ, Smith LS, Bashore TM. Altered phosphate metabolism in myocardial infarction: P-31 MR spectroscopy. *Radiology* 1987; **165**:703–7.

23. Bottomley PA, Hardy CJ. Mapping creatine kinase reaction rates in human brain and heart with 4 tesla saturation transfer ^{31}P NMR. *J Magn Reson* 1992; **99**:443–8.

24. Bottomley PA, Ouwerkerk R. Optimum flip-angles for exciting NMR with uncertain T_1 values. *Magn Reson Med* 1994; **32**:137–41.

25. Sakuma H, Takeda K, Tagami T, Nakagawa T, Okamoto S, Konishi T, et al. ^{31}P MR spectroscopy in hypertrophic cardiomyopathy: comparison with T1-201 myocardial perfusion imaging. *Am Heart J* 1993; **125**:1323–8.

26. Schaefer S, Gober J, Valenza M, Karczmar GS, Matson GB, Camacho SA, et al. Nuclear magnetic resonance imaging-guided phosphorus-31 spectroscopy of the human heart. *J Am Coll Cardiol* 1988; **12**:1449–55.

27. Bottomley PA, Hardy CJ. PROGRESS in efficient three-dimensional spatially localized *in vivo* ^{31}P NMR spectroscopy using multidimensional spatially selective pulses. *J Magn Reson* 1987; **74**:550–6.

28. Lawry TJ, Karczmar GS, Weiner MW, Matson GB. Computer simulation of MRS techniques: an analysis of ISIS. *Magn Reson Med* 1989; **9**:299–314.

29. Conway MA, Allis J, Ouwerkerk R, Niioka T, Rajagopalan B, Radda GK. Detection of low phosphocreatine to ATP ratio in failing hypertrophied human myocardium by ^{31}P magnetic resonance spectroscopy. *Lancet* 1991; **338**:973–6.

30. Schaefer S, Gober JR, Schwartz GG, Twieg DB, Weiner MW, Massie B. In vivo phosphorus-31 spectroscopic imaging in patients with global myocardial disease. *Am J Cardiol* 1990; **65**:1154–61.

31. Bottomley PA. The trouble with spectroscopy papers. *Radiology* 1991; **181**:344–50.

32. Masuda Y, Tateno Y, Ikehira H, Hashimoto T, Shishido F, Sekiya M, et al. High-energy phosphate metabolism of the myocardium in normal subjects and patients with various cardiomyopathies – the study using ECG gated MR spectroscopy with a localization technique. *Jpn Circ J* 1992; **56**:620–6.

33. Bottomley PA, Hardy CJ, Weiss RG. Correcting human heart ^{31}P NMR spectra for partial saturation. Evidence that saturation factors for PCr/ATP are homogeneous in normal and disease states. *J Magn Reson* 1991; **95**:341–55.

34. Minakami S, Suzuki C, Saito T, Yoshikawa H. Studies on erythrocyte glycolysis. I. Determination of the glycolytic intermediates in human erythrocytes. *J Biochem* 1965; **58**:543–50.

35. Beutler E. The erythrocyte. In: Williams WJ, Beutler E, Erslev AJ, Lichtman MA, eds. *Hematology* 3rd edn. McGraw–Hill: New York, 1983:283–4.

36. Horn M, Neubauer S, Bomhard M, Kadgien M, Schnackerz K, Ertl G. ^{31}P-NMR spectroscopy of human blood and serum: first results from volunteers and patients with congestive heart failure, diabetes mellitus and hyperlipidaemia. *Magn Reson Materials Phys Biol Med* 1993; **1**:55–60.

37. Humphrey SM, Garlick PB. NMR-visible ATP and Pi in normoxic and reperfused rat hearts: a quantitative study. *Am J Physiol* 1991; **260**:H6–H12.

38. Garlick PB, Townsend RM. NMR visibility of Pi in perfused rat hearts is affected by changes in substrate and contractility. *Am J Physiol* 1992; **263**:H497–H502.

39. Gadian DG. *Nuclear Magnetic Resonance and its Applications to Living Systems*. Oxford University Press: Oxford, 1982:30–4.

40. Brindle KM, Rajagopalan B, Bolas NM, Radda GK. Editing of ^{31}P NMR spectra of heart in vivo. *J Magn Reson* 1987; **74**:356–65.

41. Altman DL, Dittmer DA. Blood and other body fluids. *FASEB J* 1961; 21–2,29–31.

42 Auffermann W, Chew WM, Wolfe CL, Tavares NJ, Parmley WW, Semelka RC, et al. Normal and diffusely abnormal myocardium in humans: functional and metabolic characterization with P-31 MR spectroscopy and cine MR imaging. *Radiology* 1991; **179**:253–9.

43 Schaefer S, Schwartz GG, Steinman SK, Meyerhoff DJ, Massie BM, Weiner MW. Metabolic response of the human heart to inotropic stimulation: in vivo phosphorus-31 studies of normal and cardiomyopathic myocardium. *Magn Reson Med* 1992; **25**:260–72.

44 Mitsunami K, Okada M, Inoue T, Hachisuka M, Kinoshita M, Inubishi T. In vivo ^{31}P nuclear magnetic resonance spectroscopy in patients with old myocardial infarction. *Jpn Circ J* 1992; **56**:614–19.

45 Yabe T, Mitsunami K, Okada M, Morikawa S, Inubushi T, Kinoshita M. Detection of myocardial ischemia by ^{31}P-magnetic resonance spectroscopy during hand-grip exercise. *Circulation* 1994; **89**: 1709–16.

46 Markiewicz W, Wu S, Parmley WW, Higgins CB, Sievers R, James TL, et al. Evaluation of the hereditary Syrian hamster cardiomyopathy by ^{31}P nuclear magnetic resonance spectroscopy: improvement after acute verapamil therapy. *Circ Res* 1986; **59**:597–604.

47 Camacho SA, Wikman-Coffelt J, Wu ST, Watters TA, Botvinick EH, Sievers R, et al. Improvement in myocardial performance without a decrease in high-energy phosphate metabolites after isoproterenol in Syrian cardiomyopathic hamsters. *Circulation* 1988; **77**:712–19.

48 Wu S, White R, Wikman-Coffelt J, Sievers R, Wendland M, Garrett J, et al. The preventative effect of verapamil on ethanol-induced cardiac depression:phosphorus-31 nuclear magnetic resonance and high-pressure liquid chromatographic studies of hamsters. *Circulation* 1987; **75**:1058–64.

49 Nicolay K, Aue WP, Seelig J, van Echteld CJA, Ruigrok TJC, de Kruijff B. Effects of the anti-cancer drug adriamycin on the energy metabolism of the rat heart as measured by in vivo ^{31}P-NMR and implications for adriamycin-induced cardiotoxicity. *Biochim Biophys Acta* 1987; **929**:5–13.

50 Kopp SJ, Klevay LM, Feliksik JM. Physiological and metabolic characterization of a cardiomyopathy induced by chronic copper deficiency. *Am J Physiol* 1983; **245**:H855–H866.

51 Afzal N, Ganguly PK, Dhalla KS, Pierce GN, Singal PK, Dhalla NS. Beneficial effects of verapamil in diabetic cardiomyopathy. *Diabetes* 1988; **37**:936–42.

52 Krause SM. Metabolism in the failing heart. *Heart Failure* 1988; 267–73.

53 Whitman JR, Chance B, Bode H, Maris J, Haselgrove J, Kelley R, et al. Diagnosis and therapeutic evaluation of a pediatric case of cardiomyopathy using phosphorus-31 nuclear magnetic resonance spectroscopy. *J Am Coll Cardiol* 1985; **5**:745–9.

54 Anonymous. When to operate in aortic valve disease. *Lancet* 1991; **338**:981 [editorial].

55 Canby RC, Evanochko WT, Barrett LV, Kirklin JK, McGiffen DC, Sakai TT, et al. Monitoring the bioenergetics of cardiac allograft rejection using in vivo P-31 nuclear magnetic resonance spectroscopy. *J Am Coll Cardiol* 1987; **9**:1067–74.

56 Haug CE, Shapiro JL, Chan L, Weil R. P-31 nuclear magnetic resonance spectroscopic evaluation of heterotopic cardiac allograft rejection in the rat. *Transplantation* 1987; **44**:175–8.

57 Fraser CD, Chacko VP, Jacobus WE, Soulen RL, Hutchins GM, Reitz BA, et al. Metabolic changes preceding functional and morphological indices of rejection in heterotopic cardiac allografts. *Transplantation* 1988; **46**:346–51.

58 Fraser CD, Chacko VP, Jacobus WE, Mueller P, Soulen RL, Hutchins GM, et al. Early phosphorus 31 nuclear magnetic bioenergetic changes potentially predict rejection in heterotopic cardiac allografts. *J Heart Transplant* 1990; **9**:197–204.

59 Fraser CD, Chacko VP, Jacobus WE, Hutchins GM, Glickson J, Reitz BA, et al. Evidence from 31P nuclear magnetic resonance studies of cardiac allografts that early rejection is characterized by reversible biochemical changes. *Transplantation* 1989; **48**:1068–70.

60 Evanochko WT, den Hollander JA, Luney DJE, Blackwell G, Pohost GM. ^{31}P MRS in human heart transplants: a clinical update. *Proc Soc Magn Reson Med* 1993; **3**:1092 [abstract].

61 van Dobbenburgh JO, de Jonge N, Klopping C, Lahpor JR, Woolley SR, van Echteld CJA. Altered myocardial energy metabolism in heart transplant patients: consequence of rejection or a postischemic phenomenon *Proc Soc Magn Reson Med* 1993; **3** [abstract]:1093.

62 Bottomley PA, Smith LS, Brazzamano S, Hedlund LW, Redington RW, Herfkens RJ. The fate of inorganic phosphate and pH in regional myocardial ischemia and infarction: a noninvasive ^{31}P NMR study. *Magn Reson Med* 1987; **5**:129–42.

63 Conway MA, Bristow JD, Blackledge MJ, Rajagopalan B, Radda GK. Cardiac metabolism during exercise in healthy volunteers measured by ^{31}P magnetic resonance

spectroscopy. *Br Heart J* 1991; **65**:25–30.
64 Bottomley PA, Hardy CJ, Roemer PB, Mueller OM. Proton-decoupled, Overhauser-enhanced, spatially localized carbon-13 spectroscopy in humans. *Magn Reson Med* 1989; **12**:348–63.
65 Weiss RG, Chacko VP, Glickson JD, Gerstenblith G. Comparative ^{13}C and ^{31}P NMR assessment of altered metabolism during graded reductions in coronary flow in intact rat hearts. *Proc Natl Acad Sci* 1989; **86**: 6426–30.
66 van Zijl PCM, Barker PB, Soher BJ, Gillen J, Bottomley PA, Duyn J, et al. Proton spectroscopic imaging of ^{13}C-labelled compounds on a 1.5 Tesla standard clinical imager. *Proc Soc Magn Reson Med* 1993; **1**:373 [abstract].
67 den Hollander JA, Evanochko WT, Pohost GM. Observation of cardiac lipids by localized ^{1}H magnetic resonance spectroscopic imaging. In: *Book of Abstracts: Society of Magnetic Resonance in Medicine 1992, Works in Progress*. Society of Magnetic Resonance in Medicine: Berkeley, 1992:849.
68 Hurd RE, Freeman DM. Metabolite specific proton magnetic resonance imaging. *Proc Natl Acad Sci U S A* 1989; **86**:4402–6.
69 Kreutzer U, Wang DS, Jue T. Observing the ^{1}H NMR signal of the myoglobin Val-E11 in myocardium: an index of cellular oxygenation. *Proc Natl Acad Sci U S A* 1992; **89**:4731–3.

Paediatric applications of MRS

David G Gadian, J Helen Cross, Alan Connelly

INTRODUCTION

Magnetic resonance spectroscopy (MRS) provides an attractive non-invasive approach to investigating disease mechanisms in childhood, many of which remain poorly understood. The large majority of MRS studies of children have been carried out on the brain, and this chapter is therefore restricted to a discussion on investigations of brain disease.

The two nuclei that have been used for metabolic studies of brain disease in childhood are ^{1}H and ^{31}P. ^{31}P MRS provides a means of probing energy metabolism via the detection of signals from ATP, phosphocreatine (PCr), and inorganic phosphate (Pi). In addition to measuring the relative concentrations of these metabolites, it is also possible to determine the intracellular pH (pHi) from the chemical shift of the inorganic phosphate signal. The most extensive ^{31}P MRS studies have been carried out in neonates with hypoxic ischaemic encephalopathy, and in the first section of this chapter, this work is summarized. Discussion of the role of ^{1}H MRS in investigations of brain disease follows. ^{1}H MRS is becoming increasingly used for such studies, and rather than review the field, we describe some illustrative examples from our own work which highlight the ways in which ^{1}H MRS can lead to a better understanding of disease processes.

^{31}P MRS OF ENERGY METABOLITES IN THE NEONATAL BRAIN

Birth asphyxia is the commonest cause of impaired neurological function in full-term infants,[1] and hypoxic-ischaemic injury is likely to be the main factor involved. ^{31}P MRS of neonates was started as soon as magnets of large enough bore size became available,[2–4] and the sensitivity of ^{31}P MRS to abnormalities of oxidative phosphorylation provides scope for investigating the mechanisms underlying hypoxic-ischaemic injury in newborns. Figure. 6.1 shows a ^{31}P spectrum obtained from the cerebral cortex of a normal full-term infant. The phosphomonoester (PME) signal, which is believed to contain a major contribution from phosphorylethanolamine, is particularly large in the neonatal brain; in fact, several features of the ^{31}P spectra show an age dependence, and this has to be taken into account when comparing the spectra of normal and diseased subjects.[5,6] An extensive range of studies carried out by the group at University College London has shown that the PCr:Pi ratio provides a good prognostic indicator of the likely clinical outcome following birth asphyxia. PCr:Pi ratios within or close to the 95% confidence limits for normal controls were shown to be associated with normal outcome or with only minor impairment at

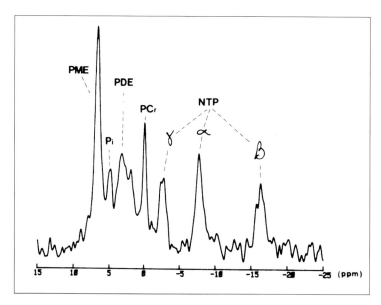

Fig. 6.1. ^{31}P spectrum from the cerebral cortex of a normal full-term infant. The spectrum shows characteristic signals from nucleoside triphosphates (NTP, the major component being ATP), phosphocreatine (PCr), phosphodiesters (PDE), inorganic phosphate (Pi), and phosphomonoesters (PME). (Reproduced, with permission from Cady EB.[34])

12 months, whereas low PCr:Pi ratios were associated with major neuromotor impairment.[7,8] The ratio of nucleoside triphosphates (NTP) to the total NMR-visible phosphate pool was also related to outcome; values below the confidence limits for normal controls were almost always associated with fatal outcome.[7] Hypoxic-ischaemic injury is often associated with increased echo densities on ultrasound scans, but this itself does not necessarily imply a poor neurological outcome. ^{31}P MRS, however, can discriminate on the basis of the PCr:Pi ratio between those infants with increased echo densities who survive, and those who either die or develop cerebral atrophy.[9] In infants with localized cerebral lesions, however, it was not possible to predict on the basis of the PCr:Pi ratios those who would develop neurological impairment.[7]

An important feature of the time course of the ^{31}P spectra is that an apparently normal spectrum (in terms of metabolite ratios) is recorded soon (8–17 hours) after the birth asphyxia episode, but over the next few days the energy status declines (see Fig. 6.2).[1] This raises the possibility that intervention during the intermediate 'normal' stage may be able to ameliorate the clinical outcome.

The development of ^1H MRS for studies of brain metabolism

The ^{31}P studies of neonatal brain metabolism discussed above began some years before the emergence of ^1H MRS as an alternative NMR approach to the investigation of cerebral metabolism in children. The development of ^1H NMR for metabolic studies lagged behind ^{31}P NMR for both technical and biochemical reasons.

Technically, ^1H NMR is more complex than ^{31}P NMR because of the need to suppress the large signals from water and, in some cases, from fats, and because of the large number of metabolites that produce signals in a relatively narrow chemical shift range. However, techniques for solvent suppression and spectral 'editing' are now sufficiently well developed to permit the non-invasive

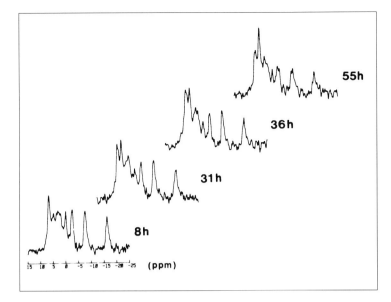

Fig. 6.2. A series of ^{31}P spectra obtained from the cerebral cortex of a birth-asphyxiated infant born at 37 weeks gestation. The post-natal ages at the time of study are indicated (in hours, h). For peak assignments, see Fig. 6.1. The spectrum obtained at 8 hours is similar to those obtained from normal infants, but the subsequent spectra show abnormalities, including a reduced PCr/Pi ratio. (Reproduced, with permission, from Azzopardi et al.[7])

monitoring of many metabolites of interest, provided that the field homogeneity is sufficiently good. For the brain, excellent field homogeneity can often be obtained, so that even at the relatively low (by spectroscopists' standards) field strengths of 1.5–2.0 T that are most commonly used for clinical spectroscopy, adequate spectral resolution can be achieved for many of the signals. In addition, the normal human brain (as opposed to the scalp and bone marrow) generates little, if any, signal from fats, and this facilitates the detection of the co-resonant signal from lactate. For other tissues, these problems are less easily overcome, which explains, at least in part, why most ^1H studies of tissue metabolism have focused on the brain.

From a biochemical viewpoint, ^1H NMR lagged behind ^{31}P studies because of the perceived strength of ^{31}P NMR in monitoring energy metabolism, as described above. However, it has become apparent that there are a number of disease states where ^1H spectroscopy might reveal abnormalities in circumstances where the ^{31}P spectra may appear normal. In other situations, the information available from ^1H MRS can complement that provided by ^{31}P studies. For example, while the ^{31}P metabolite ratios of birth-asphyxiated infants may return to the normal range about two weeks after birth, the ^1H spectra remain abnormal for much longer. These abnormalities include a reduction in the signal intensity ratio of N-acetylaspartate (NAA) to the other main ^1H signals, which presumably reflects irreversible neuronal damage (see below). In a preliminary study, it was shown that this correlated with neurodevelopmental outcome at one year.[10] Thus it seems that for studies of birth asphyxia, ^{31}P MRS can provide important information in the acute stages, while ^1H MRS is likely to be more informative in later examinations.

A key feature of ^1H spectroscopy is the high sensitivity of the ^1H nucleus in comparison with other nuclei. In principle, this means that metabolites could be detected at relatively low concentrations. However, it is not necessarily straightforward to observe metabolites at low concentrations, as their signals may be masked by larger signals from other compounds that are present at higher concentrations. In practice, therefore, the higher sensitivity is generally exploited by trading signal-to-noise ratio with spatial resolution. The higher sensitivity of ^{1}H spectroscopy means that adequate signal-to-noise ratios can be obtained from smaller volume elements. For example, the linear spatial resolution for clinical ^{1}H spectroscopy of the brain is typically 1–2 cm, which is about twofold superior to the resolution achieved with ^{31}P spectroscopy.

In the light of the above comments, it is perhaps not surprising to find that, while the majority of clinical spectroscopy studies of the body and limbs still rely on the ^{31}P nucleus, the proton is now the dominant nucleus for brain investigations. We now discuss the type of information that ^{1}H spectroscopy can provide by means of illustrative examples from the work that we and our colleagues have carried out in recent years.

^{1}H MRS STUDIES OF BRAIN METABOLISM IN CHILDREN

All our clinical studies have been carried out using a 1.5 T Siemens whole body system, with a standard quadrature head coil. Most of the children were examined under sedation according to the protocol of the Hospital for Sick Children, Great Ormond Street, London, UK.[11] Full diagnostic MRI was carried out together with the spectroscopy in each examination. Using the images as a guide, spectra were obtained from 2 × 2 × 2 cm cubes centred on specific regions of interest, using a 90–180–180 spin echo technique[12] with the three selective radiofrequency pulses applied in the presence of orthogonal gradients of 2 mT/m. Water suppression was achieved by pre-irradiation of the water resonance using a 90° Gaussian pulse with a 60 Hz bandwidth, followed by a spoiler gradient. TR was 1600 ms and TE was 135 ms. After global and local shimming, and optimization of the water suppression pulse, data were collected in 2–4 blocks of 128 scans. The time domain data were corrected for eddy–current induced phase modulation using non-water-suppressed data as a reference.[13] Exponential multiplication corresponding to line broadening of 1 Hz was carried out prior to Fourier transformation, and a cubic spline baseline correction was performed. In most examinations, spectra were obtained from two regions of interest.

Figure 6.3 shows a normal ^{1}H spectrum from a 2 × 2 × 2 cm cubic region centred on the basal ganglia of a 12 month old girl, together with an image showing the region selected for spectroscopy. The signal at 2.0 ppm is from N-acetyl-containing compounds, the dominant contribution being from NAA. This signal is therefore labelled NAA, and any changes in its intensity are attributed to changes in the contribution from NAA. The signal at 3.0 ppm is from creatine (Cr) + phosphocreatine (PG), while the signal at 3.2 ppm is from choline-containing compounds (Cho), including phosphorylcholine and glycerophosphorylcholine. The relative intensities of these signals show both age dependence and regional dependence,[10,14–17] which must be considered when making comparisons.

At first sight, it may appear that these signals are not very informative; the role of NAA remains unclear, the Cr signal may remain unchanged, even when phosphocreatine is converted to creatine under conditions of energy failure, while interpretation of the Cho signal is not straightforward because of the range of choline-containing compounds that can contribute to this signal. However, as discussed below, the NAA signal does give useful information; furthermore, additional metabolites of interest can be detected in certain disorders, and also if shorter echo times are used.

Paediatric applications of MRS

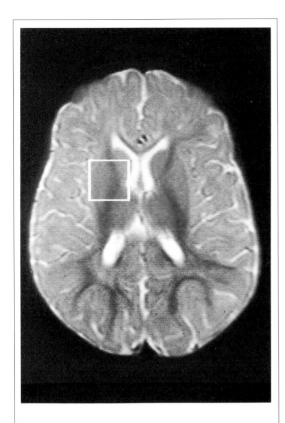

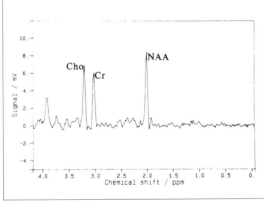

Fig. 6.3. *Magnetic resonance image and a normal spectrum obtained from a 12-month-old girl. The box indicates the position of the 8 ml cubic volume of interest from which the spectrum was obtained. (Reproduced, by kind permission of Kluwer Academic Publishers, Lancaster, England and SSIEM, from Cross et al.[22])*

In practice, our observations have fallen into two broad categories:
- the accumulation of brain metabolites in children with inherited metabolic disease; and
- spectral changes (more specifically a loss of signal from NAA) which are attributed to selective neuronal loss or damage.

Inherited metabolic diseases

Lactic acidosis and mitochondrial disorders

The congenital lactic acidoses form a large group of disorders that are commonly associated with profound neurological dysfunction. Difficulties are frequently encountered in establishing a precise diagnosis, and the mechanisms underlying brain damage are poorly understood. It has been shown that in several disorders of the brain, lactate can be seen by MRS at elevated concentrations.[18,–21] We have performed ^{1}H MRS on 24 patients under investigation for suspected metabolic disorder, and have compared the MRS observations of brain lactate with measurements of cerebrospinal fluid (CSF) lactate.[22]

We obtained spectra from the basal ganglia in all 24 children, and from occipital white matter in 16 of the children. There was good concordance between the MRS and CSF investigations, in that all of the nine children with CSF lactate concentrations of 2.5 mmol/l or less gave no detectable lactate signal on spectroscopy, while the 13 children with CSF lactates above 4 mmol/l all showed an inverted doublet signal characteristic of lactate (Fig. 6.4). This concordance serves to validate both types of measurement, and also suggests that ^{1}H MRS may have a role in the investigation of those children who have neurological dysfunction in whom screening of blood or urine may not be adequate to establish a diagnosis. However, the

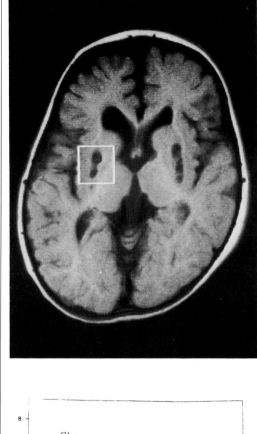

Fig. 6.4. *Magnetic resonance image and ^1H spectrum from a child with a CSF lactate of 7.2 mmol/l. The box indicates the position of the 8 ml cubic volume of interest from which the spectrum was obtained. (Reproduced by kind permission of Kluwer Academic Publishers, Lancaster, England and SSIEM, from Cross et al.[22])*

diagnostic capability of MRS in such disorders still remains to be evaluated.

Of further interest is the distribution of lactate in different areas of the brain. It is well recognized that the basal ganglia are particularly susceptible to damage in disorders of lactate metabolism. Eleven of the 13 patients with CSF lactate concentrations above 4 mmol/l showed the peak in the basal ganglia region, but lactate was also detected in the occipital white matter in all of the eight examinations in which the spectra could be interpreted unambiguously. Regional variations in the lactate signal intensity were observed in some cases. Further investigations of such variations, and of their relationship to focal brain damage, preferably using metabolic imaging methods,[21] should help to explain the neurological patterns that are observed in these disorders.

Ornithine carbamoyl transferase deficiency

Ornithine carbamoyl transferase (OCT) deficiency is one of the five inherited disorders of the urea cycle. Affected patients are prone to recurrent encephalopathy associated with hyperammonaemia, but the underlying mechanisms responsible for this encephalopathy remain uncertain. Suggested aetiologies have included alterations in energy metabolism, neurotransmitter imbalance, astrocyte swelling, seizures, and changes in cerebral blood flow and intracranial pressure. However, many of these phenomena could be secondary to the effects of glutamine, to which ammonia is converted, rather than directly attributable to ammonia itself.

^1H MRS provides a non-invasive approach to the measurement of brain glutamine *in vivo*. However, because of the relatively complex magnetic resonance characteristics of the glutamine signals, and because of spectral overlap with signals from other metabolites, the unambiguous

Paediatric applications of MRS

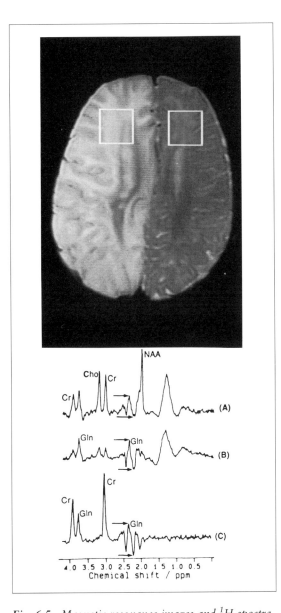

Fig. 6.5. Magnetic resonance images and ^1H spectra from a child with OCT deficiency. The spectra from the damaged (B) and contralateral (A) regions show signals which can be assigned to glutamine. Confirmation of this assignment was obtained by comparison with the spectra of solutions containing creatine and glutamine (C). All of these spectra were obtained with an echo time TE of 135 ms.
(Reproduced with permission from Connelly et al.[24])

detections of brain glutamine using clinical NMR systems poses difficulties. These difficulties were overcome by Kreis et al[23] who, with the use of short echo times, reported the detection of brain glutamine in a series of patients with hepatic encephalopathy. In two children with OCT deficiency,[24] we were also able to detect glutamine signals, even with relatively long (TE = 135 ms) echo times (Fig. 6.5). Confirmation of this assignment was obtained by comparison with the spectra of solutions of creatine and glutamine acquired with the same pulsing conditions. In both children, glutamine was detected in two regions of the brain during episodes of acute hyperammonaemic encephalopathy with focal neurological abnormalities. In one of the children, spectra obtained after treatment showed a marked decrease in these signals, indicating a return of brain glutamine towards normality.

In these studies, we were able to obtain lower limits for the concentration of brain glutamine. The average of the four measurements in the two children was 13 mmol/kg wet weight, which is similar to the values of brain glutamine obtained on post-mortem studies of patients who had died from hepatic failure, and four times the values given for normal human brain tissue.[24] Since this value of 13 mmol/kg wet weight is a lower limit, the actual concentrations within the brain are likely to be considerably higher, possibly by 50–100%, and they could be even greater in the astrocytes, where the glutamine is synthesized. This could result in a rise in osmotic pressure, with consequent astrocytic swelling and hence cerebral oedema and brain damage. These findings are therefore consistent with the hypothesis that intracerebral accumulation of glutamine contributes to the encephalopathy associated with hyperammonaemia. Further work is required to establish why some areas of the brain are more

Canavan's disease

Canavan's disease is an autosomal recessive disorder presenting in the first year of life with developmental regression and an enlarging head circumference. Pathologically, findings include a spongy degeneration of white matter, with extensive histological evidence of vacuolation and demyelination. Recently, a deficiency of the enzyme aspartoacylase has been described in this disorder, and high levels of NAA have been found in the urine and CSF of sufferers, but not in patients with other leukodystrophies. We have described studies of two children with Canavan's disease.[25] MRI showed extensive high signal throughout the white matter, with little evidence of normal myelination. Spectra were acquired from regions centred on the basal ganglia and occipital white matter (Fig. 6.6). Both regions showed very high signal intensity ratios of NAA:Cr and NAA:Cho, consistent with the known enzyme defect in Canavan's disease. When taken together with the demyelination that is demonstrated by MRI in the same examination, they support the suggestion that the metabolic pathways involving NAA may play a role in the process of myelination. Such a role may be related to the finding that NAA has been found, not only in neuronal cell cultures, but also in a glial precursor cell, the O-2A progenitor.[26]

N-acetylaspartate and neuronal loss or damage

We now turn to the second category of observations made with ^{1}H MRS, namely conditions in which there is a loss of signal from NAA. While

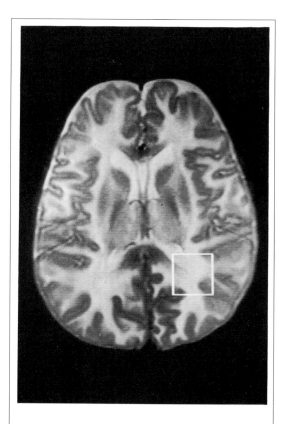

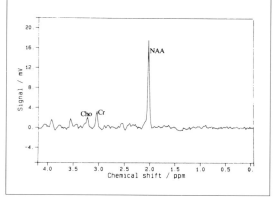

Fig. 6.6. Magnetic resonance image and spectrum obtained from a 9-month-old girl with Canavan's disease. The box indicates the position of the 8 ml cubic volume of interest from which the spectrum was obtained. (Reproduced, with permission of Williams and Wilkins, from Austin et al.[25])

the function of NAA remains uncertain, it is believed to be located primarily within neurons, and therefore an unusually low NAA:Cr ratio is commonly interpreted in terms of selective neuronal loss or damage. However, some caution is required in interpreting NAA loss in young children because:

- the ratio of NAA to the Cr and Cho signals increases with development;[10,14,15,17] and
- NAA is present not only in neuronal cells but also in the glial cell precursor known as the O-2A progenitor.[26] Bearing in mind this cautionary note, a reduced NAA:Cr ratio has been observed in numerous disorders of the brain,[27] including stroke and tumours, multiple sclerosis, and epilepsy, and this is entirely consistent with the neuronal or axonal loss or damage that might be anticipated on clinical grounds. Below we describe some of our studies of patients with epilepsy, in which the NAA, Cr and Cho signals have been used to investigate the pathology associated with this disorder.

Temporal lobe epilepsy

The option of surgical treatment for patients with intractable temporal lobe epilepsy has become increasingly accepted in this disabling disorder. To offer a definitive neurosurgical approach, the epileptogenic region must be determined with a high degree of accuracy. This frequently requires the implantation of intracerebral electrodes, which carries a significant risk of morbidity and can be particularly difficult to carry out and indeed to justify in children. We are using a combination of ^{1}H MRS, MRI, relaxation time mapping, neuropsychology, EEG, and clinical assessment to identify non-invasively the underlying pathological basis of the seizure foci in temporal lobe epilepsy and to increase our understanding of the nature of such focal damage and its relationship to brain function. Here we consider the specific role of ^{1}H MRS in these investigations.

Spectra are obtained, as one part of integrated MRI–MRS examinations, from $2 \times 2 \times 2$ cm cubes within the medial region of each temporal lobe. Many of the spectra from adults and children with epilepsy show reduced NAA:(Cho+Cr) ratios in comparison with control subjects.[28–30] (Fig. 6.7), and the metabolic abnormalities detected by MRS can aid lateralization of the seizure origin.[29] Further information about the underlying cellular pathology is available from measurements of the absolute signal intensities. By multiplying the observed signal intensities by the 90° pulse voltage, it is possible to compensate for differences in radiofrequency coil loading,[31] and thereby to compare signal intensities between different subjects. We have found that, in comparison with control subjects, the patients show a mean reduction in the NAA signal, but also a mean increase in the Cr and Cho signals. The reduction in NAA is interpreted in terms of neuronal loss or damage. The explanation for the increase in the Cr and Cho signals remains unclear, but on the basis of our cell studies showing that the Cr and Cho concentrations (expressed relative to cellular protein content) were much higher in astrocyte and oligodendrocyte preparations than in cerebellar granule neurons,[32] this increase may reflect reactive astrocytosis. Further studies are needed to establish whether this is indeed the basis of the increase in the Cr and Cho signals, but meanwhile it is apparent that the combination of *in vivo* and cellular studies provides a powerful approach to defining the cellular pathology associated with disorders of the brain.

Correlation of these MRS data with MRI, relaxation time mapping, neuropsychology, EEG, and clinical assessment is currently in progress,

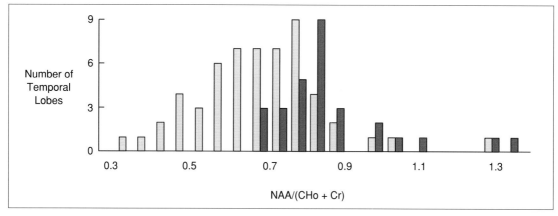

Fig. 6.7. Bar chart showing NAA:Cho+Cr ratios vs. number of temporal lobes for controls (dark bars) and children with complex partial seizures (light bars). (Reproduced, with permission, from Gadian et al.[30])

and the initial analysis appears promising, both in terms of helping in the identification of the seizure focus,[29] and also in investigating the pathological basis of functional deficits.[33]

CONCLUSIONS

Magnetic resonance spectroscopy provides a non-invasive method of investigating brain metabolism in infants and children. Despite its limited spatial resolution, it is providing clinically useful information in a wide variety of diseases of the brain. ^{1}H MRS in particular can be readily incorporated into integrated MRI–MRS examinations, and it is apparent that spectroscopy can provide new insights into the biochemical and pathophysiological basis of neurological disorders.

REFERENCES

1. Hope PL, Reynolds EOR. Investigation of cerebral energy metabolism in newborn infants by ^{31}P NMR spectroscopy. *Clin Perinatol* 1985; **12**:261–75.
2. Cady EB, Costello AMdeL, Dawson MJ, et al. Non-invasive investigation of cerebral metabolism in newborn infants by phosphorus nuclear magnetic resonance spectroscopy. *Lancet* 1983; **i**:1059–62.
3. Hope PL, Costello AMdeL, Cady EB, et al. Cerebral energy metabolism studied with phosphorus NMR spectroscopy in normal and birth-asphyxiated infants. *Lancet* 1984; **ii**:366–70.
4. Younkin DP, Delivora-Papadopoulos M, Leonard JC, et al. Unique aspects of human newborn cerebral metabolism evaluated with phosphorus nuclear magnetic resonance spectroscopy. *Ann Neurol* 1984; **6**:581–6.
5. Azzopardi D, Wyatt JS, Hamilton PA, Cady EB, Delpy DT, Reynolds EOR. Phosphorus metabolites and intracellular pH in the brains of normal and small for gestational age infants investigated by magnetic resonance spectroscopy. *Pediatr Res* 1989; **25**:440–4.
6. Boesch C, Gruetter R, Martin E, Duc G, Wuthrich K. Variations in the in-vivo P-31 MR spectra of the developing human brain during postnatal life. *Radiology* 1989; **172**:197–9.
7. Azzopardi D, Wyatt JS, Cady EB, et al. Prognosis of newborn infants with hypoxic-ischaemic brain injury assessed by phosphorus magnetic resonance spectroscopy. *Pediatr Res* 1989; **25**:445–51.
8. Roth SC, Azzopardi D, Edwards AD, et al. Relation between cerebral oxidative metabolism following birth asphyxia and neurodevelopmental outcome and brain growth at one year. *Dev Med Child Neurol* 1992; **34**:285–95.

9 Hamilton PA, Hope PL, Cady EB, Delpy DT, Wyatt JS, Reynolds EOR. Impaired energy metabolism in brains of newborn infants with increased cerebral echodensities. *Lancet* 1986; **ii**:1242–6.

10 Peden CJ, Rutherford MA, Sargentoni J, Cox IJ, Bryant DJ, Dubowitz LMS. Proton spectroscopy of the neonatal brain following hypoxic-ischaemic injury. *Dev Med Child Neurol* 1993; **35**:502–10.

11 Shepherd JK, Hall-Craggs MA, Finn JP, Bingham RM. Sedation in children scanned with high field magnetic resonance: the experience at the Hospital for Sick Children, Great Ormond Street. *Br J Radiol* 1990; **63**:794–7.

12 Ordidge RJ, Bendall MR, Gordon RE, Connelly A. Volume selection for *in vivo* spectroscopy. In: Gorvind G, Khatrapal C, Saran A, eds. *Magnetic Resonance in Biology and Medicine*. Tata–McGraw-Hill: New Delhi, 1985:387–97.

13 Klose U. *In vivo* proton spectroscopy in presence of eddy currents. *Magn Reson Med* 1990; **14**:26–30.

14 van der Knapp MS, van der Grond J, van Rijen PC, Faber JAJ, Valk J, Willemse K. Age-dependent changes in localized proton and phosphorus spectroscopy of the brain. *Radiology* 1990; **176**:509–15.

15 Connelly A, Austin SJ, Gadian DG. Localised ^{1}H MRS in the paediatric brain: age and regional dependence. *Proceedings of the 10th Annual Meeting of the Society of Magnetic Resonance in Medicine* 1991:379.

16 Frahm J, Bruhn H, Gyngell ML, Merboldt KD, Hanicke W, Sauter R. Localized proton NMR spectroscopy in different regions of the human brain *in vivo*. Relaxation times and concentrations of cerebral metabolites. *Magn Reson Med* 1989; **11**:47–63.

17 Toft PB, Christiansen P, Pryds O, Lou HC, Henriksen O. T1, T2, and concentrations of brain metabolites in neonates and adolescents estimated with H-1MR spectroscopy. *J Magn Reson Imaging* 1994; **4**:1–5.

18 Bruhn H, Frahm J, Gyngell ML, Merboldt KD, Hanicke W, Sauter R. Cerebral metabolism in man after acute stroke: new observations using localised proton NMR spectroscopy. *Magn Reson Med* 1989; **9**:126–31.

19 Detre JA, Wang Z, Bogdan AR, et al. Regional variation in brain lactate in Leigh syndrome by localized ^{1}H magnetic resonance spectroscopy. *Ann Neurol* 1991; **29**:218–21.

20 Grodd W, Krageloh-Mann I, Klose U, Sauter R. Metabolic and destructive brain disorders in children: findings with localized proton MR spectroscopy. *Radiology* 1991; **181**:173–81.

21 Luyten PR, Marien AJH, Heindel W, et al. Metabolic imaging of patients with intracranial tumors: H-1 MR spectroscopic imaging and PET. *Radiology* 1990; **176**:791–9.

22 Cross JH, Gadian DG, Connelly A, Leonard JV. Proton magnetic resonance spectroscopy in lactic acidosis and mitochondrial disorders. *J Inherited Metab Dis* 1993; **16**:800–11.

23 Kreis R, Ross BD, Farrow NA, Ackerman, Z. Metabolic disorders of the brain in chronic hepatic encephalopathy detected with H-1 MR spectroscopy. *Radiology* 1992; **182**:19–27.

24 Connelly A, Cross JH, Gadian DG, Hunter JV, Kirkham FJ, Leonard JV. Magnetic resonance spectroscopy shows brain glutamine in ornithine carbamoyl transferase deficiency. *Pediatr Res* 1993; **33**:77–81.

25 Austin SJ, Connelly A, Gadian DG, Benton JS, Brett EM. Localised ^{1}H NMR spectroscopy in Canavan's Disease: a report of two cases. *Magn Reson Med* 1991; **19**:439–45.

26 Urenjak J, Williams SR, Gadian DG, Noble M. Specific expression of N-acetyl-aspartate in neurons, oligodendrocyte-type 2 astrocyte (O-2A) progenitors and immature oligodendrocytes *in vitro*. *J Neurochem* 1992; **59**:55–61.

27 Gadian DG, Shaw D, Moonen CTW, van Zijl P. Advances in proton magnetic resonance spectroscopy of the brain: A report on a workshop held at the University of Oxford, December 1992. *Magn Reson Med* 1993; **30**:1–3.

28 Gadian DG, Connelly A, Duncan JS, et al. ^{1}H magnetic resonance spectroscopy in the investigation of intractable epilepsy. *Acta Neurol Scand Suppl* 1994; **152**:116–21.

29 Connelly A, Jackson GD, Duncan JS, King MD, Gadian DG. Magnetic resonance spectroscopy in temporal lobe epilepsy. *Neurology* 1994; **44**:1411–17.

30 Gadian DG, Connelly A, Cross JH, et al. Magnetic resonance spectroscopy in children with brain disease. In: Fejerman N, Chamoles NA, eds. *New Trends in Paediatric Neurology*. Elsevier: Amsterdam, 1993: 23–32.

31 Hoult DI, Richards RE. The signal-to-noise ratio of the nuclear magnetic resonance experiment. *J Magn Reson* 1976; **24**:71–85.

32 Urenjak J, Williams SR, Gadian DG, Noble M. Proton nuclear magnetic resonance spectroscopy unambiguously identifies different neural cell types. *J Neurosci* 1993; **13**:981–9.

33 Gadian DG, Connelly A, Cross JH, Jackson GD, Isaacs

EB, Vargha-Khadem F. *Proceedings of the 12th Annual Meeting of the Society of Magnetic Resonance Medicine* 1993:430.

34 Cady EB. *Clinical Magnetic Resonance Spectroscopy*. Plenum Press: New York, 1990:60.

7 Clinical MRS studies of the liver and spleen

Chitta R Paul, Barton M Milestone, Robert E Lenkinski

INTRODUCTION

The purpose of this chapter is to review the published clinical MRS studies that describe results obtained from the liver and spleen. There has been considerable interest in establishing MRS methods for studying liver metabolism and function, but to date there has been only one published report dealing with MRS of the spleen. There are a number of technical challenges that must be overcome in order to make these studies successful. One clear challenge involves obtaining MR spectra that are 'pure' spectra from the organ of interest. In order to achieve this goal all of the studies in the literature employ spatial localization methods to obtain MR spectra that are uncontaminated from resonances arising from surrounding tissues. In ^{31}P studies, the signals from surrounding muscle tissue are the major source of contamination. In ^{13}C studies, contamination can occur from either subcutaneous fat or surrounding muscle. While it is beyond the scope of this chapter to discuss all of the spatial localization methods in detail, it is important to recognize that each of these methods offers advantages and disadvantages in terms of their spatial discrimination and spectral distortion. All of these factors influence the quantitative reliability of the spectra. As will become evident throughout this chapter, quantitation, particularly with regard to the ability to compare spectral data obtained at different sites, remains a challenge.

This chapter focuses on the unique metabolic information available from MRS studies, and attempts to highlight the common findings that have been reported in the literature. It is clear from all of the studies reported that MRS studies of the liver can provide valuable clinical information in a variety of pathologies. However, the majority of the reports reviewed in this chapter have indicated the *potential* clinical applications possible through results obtained from studies carried out on relatively small cohorts of subjects or patients. As will become evident, there are still a number of challenges that must be met before large-scale MRS studies of the liver can be considered to be of routine clinical use.

LIVER

The first published reports of MRS of the liver occurred almost ten years ago.[1] An example of an early ^{31}P spectrum is shown in Fig. 7.1. Note the absence of any phosphocreatine (PCr) in the spectrum. Since the liver has no creatine kinase activity, the presence of PCr in a spectrum obtained from the liver indicates spatial contamination of the spectrum with signals from muscle.

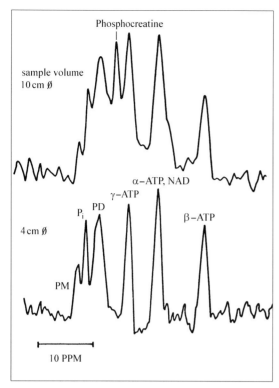

Fig. 7.1 Effect of magnetic field profiling. The radio frequency coil was positioned on the thoracic wall in the region of the liver. The top panel shows the ^{31}P spectrum before magnetic field profiling. The spectrum contains signals from intercostal muscle and liver. The bottom panel shows a normal ^{31}P spectrum of the liver after magnetic field profiling. The repetition rate was 1 second and 1024 scans were acquired. PM phosphomonoester (e.g. sugar phosphates); Pi inorganic phosphate; PD – phosphodiester; ATP – adenosine triphosphate. Reproduced with permission from Oberhaensli et al.[1]

In fairness, it is important to note that the ^{31}P spectrum of normal muscle contains resonances from inorganic phosphate, PCr and adenosine triphosphate. There are only minimal signals from phosphomonoesters (PME) and phosphodiesters (PDE). Thus the contamination of the ^{31}P spectrum of liver with a small contribution from muscle should have little effect on the levels of the resonances arising from PME and PDE.

NMR studies of extracts of normal liver

Bell et al[2] employed high field NMR (11.7 T) to assign and quantify the resonances present in perchloric acid extracts of normal (n = 6) and diseased liver (n = 10). Quantification was accomplished by adding a known amount of PCr to the extract. This compound was employed both as a chemical shift reference and as an internal standard for the determination of relative peak areas. An example of a ^{31}P spectrum of an extract of normal liver tissue is shown in Fig. 7.2. The levels of nucleotide triphosphates (NTP) found in the extracts of normals was 1.732 µg/g wet weight, which is lower than that reported previously.[3,4] However, the authors noted that the NTP:nucleotide diphosphate (NDP) ratio was variable, indicating that periods of ischemia might have taken place during the collection of these samples. The concentrations were determined for PME and PDE in normal liver samples (PME = 0.84 µmol/g wet weight; PDE = 4.62 µmol/g wet weight). There was a marked increase in both phosphoethanolamine (PE) and phosphocholine (PC) in the abnormal tissues. There was also a marked reduction in the concentrations of glycerophosphocholine (GPC) and glycerophosphoethanolamine (GPE) in the abnormal tissue.

A proton spectrum taken from an extract of normal liver is shown in Fig. 7.3. Elevations were observed in the levels of lactate, alanine, and succinate in the extracts of tumors. Bell et al[2] suggested that these changes reflected the fact that tumors were more ischemic than normal liver. There were also increases observed in the levels

Clinical MRS studies of the liver and spleen

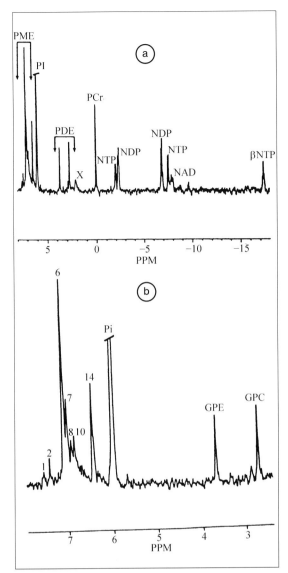

Fig. 7.2 Typical proton-decoupled ^{31}P NMR spectrum of perchloric acid extract prepared from a histologically normal liver biopsy. (a) Full spectrum; (b) PME and PDE regions. PME – phosphomonoesters; PDE – phosphodiesters; NAD – NADH+NAD; NTP – nucleotide triphosphates; NDP – nucleotide diphosphate; X – a PCA extract artefact. PCr (phosphocreatine) was added as an internal standard. Reproduced with permission from Bell et al.[2]

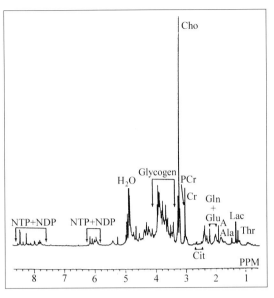

Fig. 7.3 500 MHz proton spectrum of perchloric acid prepared from a control liver biopsy. Abbreviations: Cr, creatine; PCr, phosphocreatine; A, acetate; Lac, lactate; Ala, alanine; Thr, threonine; Gln+Glu, glutamine+glutamate; Cit, citrate; Chol, choline containing metabolites; NTP+NDP: ATP, GTP, ADP, AMP, NADH+NAD. Chol: predominantly GPC, phosphocholine and carnitine. Reproduced with permssion from Bell et al.[2]

of taurine and citrate and decreases observed in the levels of threonine and creatine in tumor tissue.

^{31}P studies of normal liver

Relaxation studies

Blackledge et al[5] employed a radio frequency localization method to obtain estimates of the spin-lattice (T_1) relaxation times of normal liver. These values are given in Table 7.1. Buchtal et al[6] employed a surface coil in combination with a progressive saturation method to obtain estimates of the values of T_1 for normal liver at 1.5 T. One-

Table 7.1 Spin-lattice relaxation times (in seconds) of normal liver.

Peak	T_1^a	T_1^b	T_1^c	T_1^d	T_1^e
PME			0.84 (0.26)	4.0	0.76
Pi	0.44 (0.06)	0.41 (0.1)	0.97 (0.25)	0.5	0.38
PDE		1.4 (0.13)	1.36 (0.37)	7.0	1.16
γ-ATP	0.33 (0.35)	0.43 (0.18)	0.35 (0.06)		
α-ATP		0.68 (0.09)	0.46 (0.09)		
β-ATP		0.40 (0.13)	0.35 (0.05)	1.0	0.47

Standard deviations are given in parentheses.
a. Taken from Blackledge et al[5]
b. Taken from Buchtal et al[6]
c. Taken from Meyerhoff et al[7]
d. Taken from Cox et al[8] calculated from seven TR values
e. Taken from Cox et al[8] calculated from seven SHR values

dimensional chemical shift imaging (CSI) data were acquired. The spectra were analyzed using the peak identification, quantitation, and automatic baseline estimation (PIQABLE) algorithm. After shimming on the free induction decay (FID) of water, the water-proton line width was typically around 25 Hz. A graphical presentation (Fig. 7.4) illustrates the variation in the T_1 of the adenosine triphosphate (ATP) resonance as a function of slice position in the abdomen. As the slice position changed from wall muscle to the liver, the T_1 decreased by a factor of five. This change is probably a reflection of the relatively high concentration of iron in liver tissue. The T_1 values of the phosphorous metabolites in the human liver at 1.5 T obtained by this method are also given in Table 7.1.

Meyerhoff et al[7] employed a three-dimensional single voxel method based on the image-selected *in vivo* spectroscopy (ISIS), modified for use with surface coils, to determine the T_1 values of normal liver at 1.5 T. These values are also given in Table 7.1. Note the relatively good agreement between all these results.

Cox et al[8] reported estimates of T_1 values obtained on both normal controls and patients with liver disease. These estimates were obtained from an analysis of a slice selective two-dimensional CSI sequence carried out with repetition times (TR) of 0.5 sec and 5 sec. In one, normal control spectra were obtained with seven different TRs. The T_1 values based on the analysis of these seven different data sets are given in Table 7.1. The data sets obtained at the two different TRs were analyzed in terms of a signal–height ratio (SHR). The mean values of these ratios in normal controls (n = 6) were: PME 2.0; Pi 1.3; PDE 2.8; and β-ATP 1.4. These ratios can be related to the T_1 values using the following equation:

$$\text{SHR} = \frac{(1 - E^{-5.0/T_1})}{(1 - E^{-0.5/T_1})} \quad [1]$$

Using the values given above for the measured SHRs for PME, Pi, PDE, and β-ATP in equation 1, we estimate the T_1 values of these compounds to be 0.76 sec, 0.38 sec, 1.16 sec, and 0.47 sec respectively. For comparison purposes these values are also given in Table 7.1. Note that the values

Clinical MRS studies of the liver and spleen

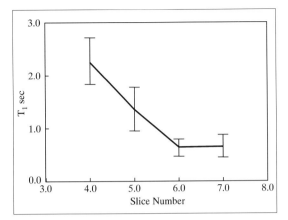

Fig. 7.4 ^{31}P T_1 *values for α-ATP as a function of depth in the abdomen. T_1 values were calculated for α-ATP as described for rows 4–7 of the CSI spectral data. Row 4 corresponds to a slice approximately 1–2 cm deep, and row 7 to a slice approximately 4–5 cm deep in the abdomen. Reproduced with permission from Buchthal et al.*[6]

obtained from an analysis of the SHR are in relatively good agreement with the values given in Table 7.1.

Spectral resolution

Luyten et al[9] showed that improvements in the spectral resolution of ^{31}P liver spectra at 1.5 T were possible through broad-band proton decoupling. An example of this improvement is shown in Fig. 7.5. Note that the PDE region and the PME regions can at least be partially resolved into their individual component peaks. The authors also make the point that the increased resolution leads to better estimates of the intracellular pH, since the GPC resonance at 2.97 ppm can be used as an internal reference.

Even without decoupling, it is possible to resolve the ^{31}P–^{31}P splittings in the resonances of ATP.[10] An example is shown in Fig. 7.6. At this level of resolution, the PME and PDE resonances both have shoulders indicating that there are several compounds that contribute to these peaks. These spectra indicate one of the potential advantages of the CSI method. In the CSI methods the spectral resolution is determined by the homogeneity of the field over the voxel rather than over the whole sample.

Quantitation of metabolite levels in the liver of normal adults

In this section we review the values reported for various ratios of ^{31}P metabolites in studies of normal volunteers or control groups for hepatic diseases. Generally these metabolites have been reported relative to β-ATP. For ease of comparison, we have adopted this method for expressing the concentrations of the metabolites present.

Oberhaensli et al[1] employed a field-profiling method with a pulse repetition time of 2 sec. The sensitive volume of liver tissue sampled was about 30 cm^3. The values reported for the ^{31}P metabolites are given in Table 7.2. Luyten et al employed an ISIS technique modified for use with a surface coil at 1.5 T to obtain proton decoupled spectra from 300 cm^3 with a TR of 3 sec.[9] The ratios obtained are given in Table 7.2.

Glazer et al[11] employed a one-dimensional CSI method for selecting sagittal slices of normal liver. These authors placed a 14 cm coil on the lateral thoracic wall. They used a TR of 0.5 sec with 128 phase-encoding steps (field of view 128 cm) and 32 averages per phase encoding view. The values for the metabolite ratios obtained are given in Table 7.2. These authors noted that there was no difference observed in the Pi:β-ATP ratio found in fasting (n = 3) versus non-fasting (n = 9) subjects. Furthermore, an analysis of serial studies carried out in two of the subjects indicated reproducibilities of 3–15 per cent in the metabolite ratios. It is interesting to note that Buchtal et al[6] reached a similar conclusion based on an analysis of one-

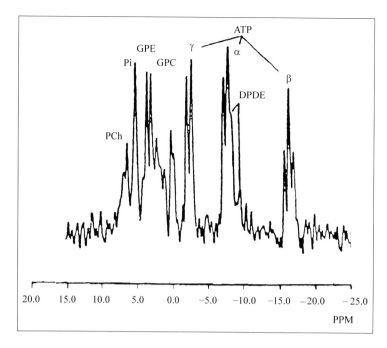

Fig. 7.5 Proton decoupled localized ^{31}P NMR spectrum of the human liver, obtained at 1.5 T, using a 14 cm diameter surface coil. The localization scheme used was ISIS with FM inversion and excitation pulses. The spectrum was obtained in 280 scans with 3 sec repetition time. Proton decoupling has improved resolution in the phosphodiester region. Well resolved resonances are observed of GPE, GPC, and phosphocholine (PCh). Reproduced with permission from Luyten et al.[9]

dimensional CSI data obtained from five consecutive runs on the same subject. The metabolite ratios were found to vary between 5 and 15 per cent.

Meyerhoff et al[7] employed an ISIS method modified for use with surface coils to obtain spectra of 64–120 cm³ of liver using a TR of 1 sec. The peak areas were corrected for saturation effects. The relative values obtained are given in Table 7.2. A procedure for obtaining absolute concentrations of metabolites was presented in this paper and in a report by Matson et al.[13] These results will be discussed in more detail in a following section. The same group published another report[13] using a similar ISIS method to obtain spectra of 53–150 cm³ with a TR of 2 sec. These data are also given in Table 7.2.

Oberhaensli et al extended their original report[1] employing a field profiling method with an 8 cm surface coil at 1.6 T. The sensitive volume of liver tissue sampled was about 30 cm³. The values reported for the ^{31}P metabolites are given in Table 7.2.[14] Dixon et al[15] employed a rotating frame method at 1.9 T to obtain spectra from the liver of normal controls. The metabolite ratios are given in Table 7.2. These ratios are not corrected for either off-resonance distortions or partial saturation.

Cox et al[16] employed either a two-dimensional CSI or a three-dimensional CSI (three spatial dimensions) at 1.6 T to study the liver in normal controls. Spectra were acquired with a 45° pulse and a 1 sec TR. There was a 3.1 ms pre-acquisition delay in this sequence. The sensitive volume sampled was between 8 and 27 cm³. The details of this method have been presented in other reports.[17,18] The same group extended this work to study the reproducibility of the metabolite ratios.[19] These included the variablilities found between successive different examinations, the intrasubject and intersubject variabilities. Values for these variabilities are given in Table 7.3. The

Clinical MRS studies of the liver and spleen

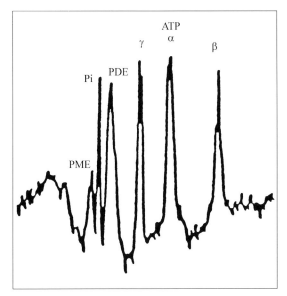

Fig. 7.6 ^{31}P *spectrum of human liver, obtained with phase-encoding experiment to remove contribution from overlying muscle. The 6 cm receive coil was placed in the midaxillary line, over ribs 6–8 on the right. 45 pulse; repetition rate – 1 sec; number of data collections – 1024; number of slices (only one shown), 16; total experiment time – 17.1 minutes line broadening, 2.5 Hz; cosine filter; 'slice' width – 16 mm. Reproduced with permission from Cox et al.*[10]

interexamination and intersubject variabilities were similar, ranging from about 15 per cent for the Pi:ATP ratio to 25 per cent for the PDE:ATP ratio. The values for the intersubject variabilities were somewhat higher.

Meyerhoff et al[20] applied the modified ISIS method to study a group of normal controls. This work was an extension of their previous report.[12] Outer volume suppression was employed to minimize contamination of the signal. The metabolite ratios determined in this study are given in Table 7.2.

Munkata et al[21] employed a one-dimensional CSI sequence with a 90° flip angle and TR of 0.5 sec to obtain spectra from normal liver at 1.5 T. The spectra were analyzed in the frequency domain using a Levenburg–Marquardt algorithm. The metabolite ratios are given in Table 7.2. These authors also reported estimates of the percent changes in the areas of the ^{31}P peaks present in normal liver at TR values of 5 sec and 0.5 sec. These were: PME 2.2; Pi 1.2; PDE 2.6; and β-ATP 1.3. These values are in good agreement with the values for the SHRs of these peaks reported by Cox et al.[8]

From the discussion presented above and the data shown in Table 7.2 it is clear that ^{31}P spectra can be obtained from normal liver tissue using a variety of localization methods. While there have been a relatively significant number of normal subjects studied, there are substantial variations in the metabolite ratios that have been reported. These variations are much larger than the intersubject variations reported by Cox et al[19] given in Table 7.3. The range of reported ratios are: PME:ATP 0.26–1.07; Pi:ATP 0.36–1.1; and PDE:ATP 0.62–2.65.

It is possible that some of the variations are dependent on the particular localization method employed. The choice of parameters employed in the localization method may cause differential distortions in the levels of particular metabolites. For example, consider PDE, whose T_1 is long while its effective T_2 may be short. In localization schemes with delayed acquisition, the relative level of resonances with short T_2 values will be preferentially reduced. Additional distortions may arise when the repetition time is short relative to the T_1 values of the metabolites. Thus, in localization methods with delayed acquisition carried out at a short TR (relative to T_1), it is possible that the level of PDE will be much less than its true value. Virtually all of the data presented in Table 7.2 were collected under conditions where

Table 7.2 Values reported for the ratios of the P-31 metabolites present in normal liver. Values given are mean values with the standard deviations given in parentheses where reported.

n	PME/β:ATP	Pi/β:ATP	PDE/β:ATP	pH	Method	Reference
4	0.61 (0.09)	0.92 (0.07)	2.3 (0.1)	7.19 (0.09)	field profile	1
1	1.07 (0.17)	0.78 (0.12)	a	7.06 (0.006)	modified ISIS	9
12	0.55 (0.06)	0.38 (0.04)	0.62 (0.08)	7.25 (0.04)	1D CSI	11
8	0.4 (0.02)	1.1 (0.3)	2.6 (1.1)	7.44 (0.2)	modified ISIS	7
6	0.45	1.05	2.7	7.36 (0.23)	modified ISIS	13
16	0.44	0.72	1.8		field profile	14
25	0.37 (0.10)	0.64 (0.15)	1.14 (0.33)	7.2 (0.1)	rotating frame	15
28	0.26		1.10		2D and 3D CSI[b]	16
28	0.27 (0.08)	0.36 (0.08)	1.16 (0.3)	7.23 (0.14)	2D and 3D CSI[b]	19
13	0.4 (0.2)	1.1 (0.4)	2.65 (1.9)	7.4 (0.2)	modified ISIS	20
7	0.32	0.55	2.33	7.29	1D CSI	21

a. Values were reported separately for GPE 0.56 (0.10) and GPC 0.29 (0.08)
b. The term 3D CSI refers to three spatial encoding dimensions

Table 7.3 Estimates of the inter-examination, intrasubject and intersubject variabilities in the metabolite ratios determined in normal liver.[19]

Group	PME:ATP ratio	Pi:ATP ratio	PDE:ATP ratio
Intra-exam	15%	14%	24%
Intrasubject	16%	17%	21%
Intersubject	35%	22%	25%

See Meyerhoff et al[20] for details of the statistical analysis

partial saturation of the resonances occurred. There are clearly many other effects that must be considered in comparing spectra obtained using different localization methods. Even without detailed considerations of these effects it is clear that, whatever the reasons, there is no consensus in the literature regarding the 'true' metabolite ratios present in normal liver.

Meyerhoff et al[7,12,20] have reported a method by which spectra obtained with the ISIS method could be analyzed in terms of the absolute concentration of the metabolites present expressed as mmol per kg wet weight of tissue. The details of this method have been reported by Matson et al.[13] The values determined by this approach are presented in Table 7.4 along with the concentrations of the metabolites determined by freeze clamping.[3,4] Although an error analysis of the MRS-based method together with a discussion of other possible sources of differences was presented in this report, it is still somewhat troubling that the absolute concentrations determined by MRS do not agree with values determined biochemically.

Table 7.4 Absolute concentrations of the metabolites present in the spectrum of normal liver (taken from Meyerhoff et al[20]).

Metabolite	Concentration mmol/kg wet weight	Freeze clamping
PME	0.92 (0.41)	<0.3[a]
Pi	2.12 (0.32)	5.13 (1.33)[b]
PDE	5.36 (1.43)	
γ-ATP	2.02 (0.13)	
α-ATP	2.29 (0.31)	2.50 (0.60)[a]
β-ATP	2.16 (0.75)	

a. See Bode et al[3]
b. See Hultman et al[4]

^{31}P Studies of normal liver of pediatric subjects

Iles et al[22] employed a surface receiver coil in a 1.6 T MR scanner to study the relationship of nutrition on hepatic metabolism as well as the influence of age on three neonates and one infant, all of whom had neonatal intracranial problems but normal liver function. A proton image was used for localization, and line width of the water signal was 40–70 Hz after shimming. Two-dimensional CSI spectra were obtained from the whole sensitive volume of the receiver coil. An analysis of the localized ^{31}P spectra revealed that the relative intensities of the PME and PDE resonances in each subject were different when compared to the mean adult values. The spectra from the neonates showed increased PME:ATP ratios (1±0.4) and decreased PDE:ATP ratios (0.4±0.1) when compared with adult means (0.2±1 and 1.3±0.2, respectively). The chemical shift of the PME resonance for neonates was found to be higher (6.8±1 ppm) than for the adults (6.4±0.2 ppm). Since this value is close to that found in neonatal brain, the authors suggest that phosphorylethanolamine (PE) is the major component of the peak. The chemical shift of PDE for the neonates was 2.6±0.2 ppm as compared to 3.0±0.1 ppm for the adult population.

The peak area ratio, PME:ATP, for the infant was slightly greater than the neonates, but much greater than the adult value. The mean peak area ratio, PDE:ATP, for the infant was slightly greater than the neonates, but half the adult value. The authors suggested that the rapid growth characteristic of regenerating livers might have similarities with the neonatal liver.

^{31}P studies of metabolic manipulations of normal adults

One of the earliest applications of ^{31}P MRS involved monitoring the ability of the liver to metabolize fructose. In the liver, fructose is phosphorylated by fructokinase (consuming ATP) to fructose-1-phosphate. Fructose-1-phosphate is then cleaved by aldolase B into two three-carbon fragments, which feed into glycolysis. Radda and coworkers[1,23,24] recognized that the oral ingestion of 50 g of fructose by normal subjects should produce little metabolic change in the ^{31}P spectrum of their livers. In contrast, the infusion of a large dose of fructose (200 mg per kg body weight) can produce spectral alterations in normals. Since the activity of aldolase B in the liver of normals is lower than that of fructokinase, the levels of fructose-1-phosphate will increase with a concomitant decrease in the level of Pi after administration of fructose. Since the ^{31}P resonance of fructose-1-phosphate occurs in the PME region, all these changes can be monitored by ^{31}P MRS.

Oberhaensli et al[24] employed ^{31}P MRS to monitor fructose metabolism *in vivo* in seven healthy subjects. Subjects were examined in a 1.6 T whole-body magnet using surface coils for data acquisition. The region of the liver from which MR signals were collected was selected by magnetic-field profiling. During intravenous injection of fructose, the levels of sugar phosphates, PME, ATP and Pi changed. During the first 5 minutes after a bolus injection of 250 mg fructose per kg body weight, the concentration of PME increased sevenfold whereas Pi and ATP decreased three- to fourfold. The metabolism of sugar phosphates was complete within 20 minutes and could be followed by ^{31}P MR with a time resolution of 5 minutes. On this basis the authors suggested that ^{31}P MRS may provide a means for the non-invasive functional assessment of the liver. Segebarth et al[25] showed that similar measurements could be made on a clinical MR scanner with a temporal resolution of about 2 minutes.

Sakuma et al[26] studied five normal volunteers after intravenous infusion of 20 per cent fructose at a dose of 500 mg per kg body weight. ^{31}P spectra were recorded every 5 minutes. The level of PME was found to increase to 338±76 per cent of control levels at 15–20 minutes post infusion. The level of Pi was depleted at 15–20 minutes then rebounded to 260±67 per cent of its initial value. No metabolic interpretation was offered for this interesting rebound effect. The level of β-ATP dropped to 50 per cent of its initial level and then recovered gradually.

Terrier et al[27] determined the degree of alterations in ^{31}P metabolites at different levels of intravenous injection of fructose in order to arrive at a dose–response relationship in normal subjects. Localization was achieved through the use of a sectorial loop-gap resonator described by Grist et al.[28] Four normal subjects were given two to four doses of fructose (62.5, 125, 250, 375, or 400 mg per kg body weight) on different days. Serial ^{31}P MR spectra were acquired, with a time resolution of 2.5 minutes. The results observed were in agreement with those reported by Oberhaensli et al.[24] The maximum relative alterations observed in PME, Pi, and ATP were fit to linear relationships of the form

$$\Delta \text{Area} = b + mD \qquad [2]$$

where D is the dose of fructose given IV in mg/kg body weight. The values determined for the slope and intercept are given in Table 7.5. The strongest correlation was found between the ratio of PME:Pi and the dose. The authors noted that there was a rebound effect observed on the resonance of Pi after fructose administration. The initial drop was followed by an increase of an average of more than 60 per cent of the control value after 15–20 minutes. The magnitude of this effect was not dose dependent.

Masson et al[29] developed a protocol that made a clinically relevant assessment of the metabolic function of liver (gluconeogenic capacity) after constant infusion of 6–8 mM of fructose in the blood stream using localized ^{31}P spectroscopy and other biochemical analyses. ^{31}P spectra from the liver were collected in a 1.5 T scanner with 16 cm surface coil using the fast rotating gradient spectroscopy (FROGS)[30] pulse sequence. During fructose infusion, the PME:PDE ratio increased to maximum value 206±39 per cent of control value and recovered to control levels within 15 minutes after the discontinuation of fructose infusion. However, Pi and β-ATP:PDE decreased to 55±4

Table 7.5 The dose response of liver metabolites to fructose loading in normal subjects.[27]

Metabolite	Slope	Intercept	r	SE (slope)
PME	175	7	0.72	53
Pi	−100	−21	0.78	25
γ-ATP	−79	−26	0.64	30
α-ATP	−45	−10	0.58	20
β-ATP	−70	−13	0.73	22
PME:Pi ratio	22	1	0.81	5

per cent and 71± per cent of the initial values, respectively. When infusion stopped, Pi overshot its initial level by 234±5 per cent and ATP recovered to a value 88±11 per cent of the control values. Quasi-steady state infusion of fructose in the blood stream induced a significant increase in the area of phosphomonoesters (PME) on the volunteers (206%).

The rate of clearance of fructose from the liver was found to be 0.53 mg per gram liver per minute. The authors found that the time course of changes in metabolite levels could be modeled using first order kinetics. Interestingly, the rate of increase of PME during infusion was the same as the rate of recovery of PME after infusion was halted. A similar inverse relationship was found for ATP, i.e. the rate of depletion of ATP during infusion was the same as the rate for ATP recovery. In contrast, the rate of recovery of Pi was about three times higher than its rate of depletion during infusion with fructose.

The authors pointed out that fructose accumulation in the hepatocytes depends on a number of factors, including hepatic blood flow and transport processes from the vascular space into the cells. Thus, the reduction in concentrations in PME, Pi, and ATP in patients with hepatitis or liver cirrhosis could possibly be explained by reduced blood flow or prolonged diffusion time rather than by changes in the metabolic capacity of hepatocytes. The authors also suggested that the constant fructose infusion technique may be useful for standardization and for the evaluation of hepatic function in patients.

Dagniele et al[31] applied ^{31}P MRS to study changes in hepatic gluconeogenesis in normal human volunteers and perfused rat livers after intravenous infusion of l-alanine. The aim of this investigation was to determine the dose-dependent relationship between l-alanine infusion and metabolic response. A two-dimensional CSI pulse sequence was used to obtain localized ^{31}P spectra. Peak areas for PME, Pi, PDE, and β-ATP were measured using an integration program. There were several important features of PME:ATP, Pi:ATP, and PDE:ATP after infusion of 2.80 mmol of l-alanine per kg body weight. Maximal changes in the ratios of PME:ATP, PDE:ATP and Pi:ATP were 98 per cent, 24 per cent, and −33 per cent of the basal values, respectively. The average time point of these maximal changes in metabolite levels was approximately 45 minutes after infusion for PME:ATP and Pi:ATP and 16 minutes after infusion for PDE:ATP.

An analogous experiment carried out in animals showed that increases in 3-phosphoglycerate and phosphenol-pyruvate concentrations were responsible for the increase in PME and PDE resonances. This observation indicated that no change in phosphorylation status occurred after l-alanine infusion. This observation in is contrast with earlier MRS studies done with fructose infusion. Intravenous infusion of 2.77 mmol of fructose per kg of body weight induced approximate reduction of 80 per cent in Pi and 50 per cent in ATP within 10 minutes of infusion followed by a rapid increase in Pi to almost twice the control concentration. In contrast, l-alanine infusion neither affected hepatic ATP nor caused any rise in Pi above baseline levels, and only a moderate reduction in Pi was detected. The authors suggested that alanine infusion study may serve as a useful tool for the investigation of gluconeogenesis and hepatic biochemical pathology *in vivo*.

The same group investigated the effects of omega-3-fatty acid on the phospholipid metabolism *in vivo* using ^{31}P MRS.[32] A Picker 1.6 T spectrometer with a 15 cm surface coil was used for this investigation. The coil was placed lateral to the liver in the midaxillary line. Five male adult volunteers, aged 30–45 yrs, were provided with a normal, non-slimming diet with one meal of fish per week. A CSI pulse sequences pulse was used to obtain localized spectra in the form of a series of 30 mm thick parasagittal planes. The areas of PME, PDE, Pi, and β-ATP were measured using a computer integration program. Estimates for T_1 saturation factors were calculated by taking peak heights at a TR 5 sec divided by the peak height at a TR of 0.5 sec.

A high dosage of fish oil supplement led to an increase in the ratio of PDE:ATP in healthy subjects. The ratio of Pi:ATP did not change, indicating that the increase in the PDE:ATP ratio was a result of higher PDE levels in the human liver. Similar studies carried out in rats also showed increases in PDE when the rats were fed with high fish oil supplements. Extracts of liver tissue obtained from rats fed with fish oil showed elevated GPC and GPE concentrations. This suggested that these compounds were responsible for elevated PDE in humans. This study indicates the potential utility of applying ^{31}P MRS to monitor the effects of alterations in nutrition on liver metabolism.

^{13}C MRS studies of normal liver

Jue et al[33] reported the first natural abundance ^{13}C NMR spectra of normal liver in humans in a 2.1 T, 1 meter bore NMR spectrometer using a 10 cm surface coil placed above the right lobe of the liver. The TR was 0.4 sec and 4096 scans were acquired in 27 minutes. The ^{13}C spectrum of normal human liver showed prominent lipid and adipose tissue signals at 30 and 130 ppm and C2–C6 glycogen signals overlapped by the triglycerides between 90–120 ppm. The C1 glycogen signal at 101 ppm showed a distinctive splitting of 170 Hz (Fig. 7.7). The signal-to-noise ratio of the liver glycogen signals was approximately 15–1. The authors pointed out that with proton decoupling this signal-to-noise ratio of the glycogen resonance could be increased to about 40–1. The results indicated the potential opportunities for ^{13}C NMR studies of glucose and glycogen metabolism in human liver without surgical intervention.

Having shown that the ^{13}C resonance of glycogen is observable in normal liver, the same group applied ^{13}C NMR to study the rate of glycogen repletion in humans.[34] As before, a 2.1 T, 1 meter bore NMR spectrometer was employed with a concentric ^{1}H/^{13}C surface coil (^{13}C and ^{1}H coils were 8 cm and 11 cm in diameter). The right lobe of the subject was placed on the top of

Clinical MRS studies of the liver and spleen

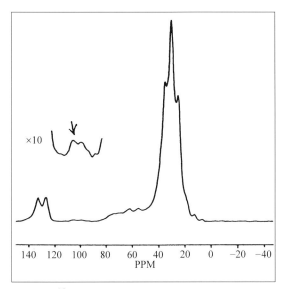

Fig. 7.7 ^{13}C *NMR spectrum of human liver with a 10 cm surface coil placed on the abdomen above the right lobe of the liver. The signals were internally referred to the center of the olefinic lipid peaks as 130 ppm, based on the decoupled* ^{13}C *spectra of rat liver. A 40 Hz apodization was used. Clearly, in the 10-times vertically-expanded spectrum, the coupled C1 glycogen signal with* J_{CH} *of 170 Hz is observed at 170 ppm. Reproduced with permission from Jue et al.*[33]

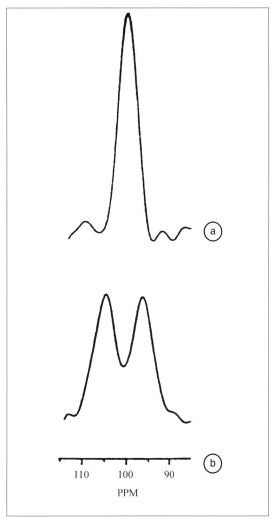

Fig. 7.8 Coupled and decoupled ^{13}C *NMR spectra of C1 hepatic glycogen. In spectrum a, the coupled glycogen C1 signal appears at 100.5 ppm. In spectrum b, with 10 W of 1H irradiation, 30 per cent duty cycle, the* ^{13}C *glycogen signal is decoupled. Reproduced with permission from Jue et al.*[34]

the surface coil. A 2 cm sphere of water served as a calibration marker. A binomial pulse sequence was used to suppress the undesired lipid signals and to excite the C1 glycogen signal at 100.5 ppm. The TR value was optimized for the C1 glycogen signal that has a T_1 of 240 msec and a T_2 of 30 msec, respectively. Thirty thousand scans were collected in 13 minutes. A representative spectrum is shown in Fig. 7.8.

For the repletion study, a male subject fasted for 30 hours and was given a mixed diet after the initial basal spectrum was acquired. The most prominent peaks were from lipids, triglycerol moieties, and olefinic carbon signals. The resonances from C1 glycogen at 100.5 ppm showed a spin–spin coupling to covalent bonded ^1H nucleus of 165 ± 5 Hz. In order to verify the reproducibility of the procedure, the experiment was repeated. The intense signal from the

117

subcutaneous lipid moiety made the quantification difficult. The resolution of the signal of the C1 glycogen at 100.5 ppm was much improved upon ^{1}H decoupling with 10 watt radio frequency power. The hepatic glycogen peaks were completely resolved and line width was similar to the coupled spectra.

The concentric design of the coil saved power loss at the ^{1}H frequency and yielded sufficient sensitivity and resolution to follow glycogen acquisition during the procedure. Moreover, an optimization procedure in data acquisition was used to increase both resolution and the signal-to-noise ratio of the glycogen peak. Because of the FDA power deposition guidelines, the spectra were acquired with no nuclear Overhauser enhancement (nOe)-mediated enhancement. The study demonstrated that the hepatic glycogen concentration increased by a factor of 2.8 over the basal level within 2 hours of refeeding and by a factor 3.8 within 4 hours. From the NMR studies, it was observed that the rate of glycogen repletion was 0.2 mmol per minute per gram of liver. The results obtained by this technique were within 3–5 per cent of the values obtained by needle biopsy technique. The non-invasive nature of the NMR study clearly offers major advantages over needle biopsy for repeated measurements. This study demonstrates the feasibility of employing ^{13}C NMR to study carbohydrate metabolism by investigating the cellular regulation of hepatic glycogen metabolism in humans.

Rothman et al[35] provided an example of this approach to study glycogen metabolism in normal liver. This group employed non-invasive ^{13}C spectroscopy to quantify hepatic glycogenolysis and gluconeogenesis during fasting. The rate of gluconeogenesis was estimated by monitoring the incorporation of radioactively labeled glucogenic precursors into plasma glucose.

Seven healthy volunteers (aged 20–26 years) were fed a high carbohydrate diet (40–45 kCal per kg per day) for 3 days prior to fasting. On day 3, they ingested a liquid meal of 650 kCal within a short period of time. The subjects fasted for 68 hours and were allowed to drink water without restriction. The ^{13}C spectral measurements of glycogen concentration in the liver were done at regular time intervals. The volume of the liver was measured with magnetic resonance imaging at the start and end of fasting in all the subjects. After 67 hours of fasting it was observed that the liver volume was reduced by 23±4 per cent.

The ^{13}C spectra of the C1 position of liver glycogen showed that 4 hours after ingestion of the standard meal, the average concentration of liver glycogen was 396±29 μmol/ml. The glycogen concentration decreased linearly after 22 hours fasting and the average values of glycogen concentration after 40 hours and 64 hours were 66±9 μmol/ml and 42±9 μmol/ml (Fig. 7.9). After 68 hours, when the subjects were refed, the liver glycogen increased to 188±19 μmol/ml within 3 hours.

The rate of glycogenolysis was found to be relatively constant over the first 22 hours of the fast (4.0±1.2 μmol per kg per min for 4–13 hours and 4.3±0.6 μmol per kg per min for 4–22 hours). At 22 hours the rate of glycogenolysis can account for 36% of glucose production. After 44 hours of fasting all of the glucose is being produced by gluconeogenesis. The rate of glucose production was found to be 12.2±0.9 μmol per kg per minute. One of the more interesting findings of this study was the fact that hepatic gluconeogenesis is always occurring at a significant rate in normal subjects.

Beckmann et al[36] showed that the natural abundance ^{13}C spectrum could be obtained from the liver of normal volunteers. This study was carried out at 1.5 T using a Waltz-8 decoupling

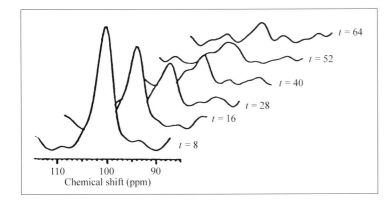

Fig. 7.9 Staggered plot of the ^{13}C NMR spectra of the C1 position of hepatic glycogen obtained at the indicated hours from a subject during 64 hours of fasting. Reproduced with permission from Rothman et al.[35]

method. Under normal nutritional conditions, the ^{13}C spectrum obtained in about 30 minutes from normal volunteers (n = 6) showed no resonance for the C1 carbon of glycogen. When three volunteers were fed a special diet that was designed to increase glycogen, a peak at 100.5 ppm was observed.

This same group later extended their previous study to follow the formation of glycogen under two different conditions: intravenous infusion of glucose labeled with ^{13}C at the 1 position at 99% enrichment; and bolus oral intake of glucose labeled with ^{13}C at the 1 position at 6.6% enrichment.[37] The intravenous infusion was carried out in one volunteer under hyperglycemic and hyperinsulinemic clamp. The level of glucose in the plasma was determined by taking samples of blood from this patient every 10 minutes during infusion. After infusion with 99% enriched glucose was started, spectra were obtained every 8.5 minutes. The initial glucose concentration in the plasma was about 5 mmol. After infusion with the clamp this concentration increased to about 10 mmol. Resonances from the C1 of α- and β-glucose were observed after 10 minutes. These signals arise primarily from glucose in the stomach and gut. The basal concentration of glycogen was about 160 mmol. The concentration of glycogen increased by more than a factor of three 60 minutes after infusion.

Three normal volunteers received an oral dose of 220 g of natural abundance glucose. The initial concentration of glycogen in this group was about 240 mmol. The concentration of glycogen increased by about a factor of two. Four normal volunteers received an oral dose of 150 g of unlabeled glucose. Their basal glycogen concentration was also about 240 mmol. The level of glycogen increased by about a factor of 1.8 after ingestion of glucose. Five volunteers received 50 g of unlabeled glucose and 3 g of 99% labeled glucose (C1). The level of glycogen increased by about a factor of 2.5 in this group. Five volunteers received a dose of 53 g of 6.6% labeled glucose. The level of glycogen increased by about a factor of 2.3 after 60–90 minutes post ingestion. The average basal level of glycogen found in all of the volunteers (n = 18) was 229±34 mmol.

These studies illustrate a number of important points. It is possible to carry out the ^{13}C studies on a commercial MR scanner operating at 1.5 T. The fact that useful spectra can be obtained by employing relatively small amounts of either 99% labeled glucose in combination with unlabeled material or using 6.6% labeled glucose means that the overall cost of the study can be reduced considerably.

Proton MRS studies of liver pathology

Proton image-guided localized MR spectroscopy was used to evaluate the degree of fatty infiltration of the liver in patients with diffuse liver steatosis.[38] The MR-determined fraction was correlated with the ratio between liver and spleen density determined by CT and histological score. A 1.5 T scanner was used for imaging and spectroscopy acquisition. Twenty-six patients who had been histologically diagnosed with diffuse liver steatosis underwent ultrasound-guided biopsies and CT examinations. MR examinations were done 8 hours after fasting. Proton MRS signals of water and fat were determined using a Carr–Purcell double-spin-echo volume selection sequence. A recycle delay (TR) of 3 sec was applied to avoid T_1 effect on the results. The experiment consisted of 6–8 localized spectra with different echo times (50–200 ms). The first series was collected with the excitation frequency centered water signal and the second on the fat signal. The intensities were evaluated by integration of the phase corrected spectra after baseline correction. The water and fat signals were integrated in the region between 3–7.8 ppm and 0–3 ppm, respectively.

The correlation between the hepatic fat content as evaluated by MR or CT and a histologically based fat infiltration scale were $R = 0.68$, $p < 0.001$ and $R = 0.77$, $p < 0.001$ respectively. However, the fat content examined by the CT investigation was expressed as the ratio between the liver and spleen density values whereas, in the case of MR, the ratio between fat and fat-plus-water signal areas was used. The hepatic fat assessed by CT and MR showed a good correlation ($R = 0.83$, $p < 0.001$).

^{31}P MRS studies of liver disease

^{31}P MRS of hepatitis and cirrhosis

Radda et al[23] showed that there were observable changes in the spectrum of patients with viral hepatitis (hepatitis B) as compared with normal controls (Fig. 7.10). In six patients, PME increased and PDE decreased slightly. Also when the patients recovered from this condition, the ^{31}P spectrum returned to normal.

Oberhaensli et al[39] studied 24 patients with various liver diseases such as acute viral hepatitis, alcoholic hepatitis, primary biliary cirrhosis, liver

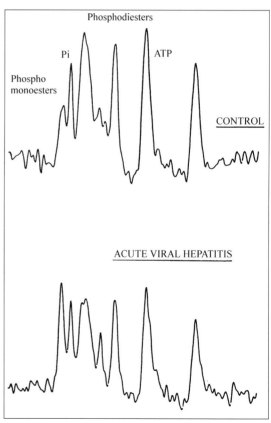

Fig. 7.10. ^{31}P *NMR spectrum of the liver of a patient with viral hepatitis B compared to control subject. Reproduced with permission from Radda et al.*[23]

Table 7.6 Metabolite ratios of patients with a variety of liver diseases taken from Meyerhoff et al.[12]

Disease	n	PME:ATP ratio	Pi:ATP ratio	PDE:ATP ratio	PME:PDE ratio
controls	16	0.44	0.72	1.8	0.24
viral hepatitis	5	0.76	0.52	1.0	0.76
alcoholic hepatitis	4	0.72	0.60	1.9	0.38
primary biliary cirrhosis	3	0.56	0.28	1.4	0.40

tumor, iron overload, and overload of paracetamol. A 1.9 T scanner with an 8 cm surface coil was used to examine patients and age-matched normal controls using field profiling techniques. All of the patients were studied in a fasted state. The metabolite ratios obtained are given in Table 7.6. All of these patients had higher PME:ATP ratios than normal controls. Note also that none of the patients had elevated Pi:ATP ratios. The increases in PME were interpreted as a sign of rapidly regenerating liver tissue in patients with viral hepatitis. These patients also had lower PDE levels than normal. These results indicate that ^{31}P MRS may give insight into the different pathophysiological processes involved in the diseases studied.

Cox et al[19] employed CSI methods to study patients with a variety of liver disease. Twenty-five of these patients had diffuse liver disease. In 14 of these cases an elevation was observed in the PME:PDE ratio. The ^{31}P spectra obtained from patients with well-compensated cirrhosis were very similar to those obtained on normal controls. The authors found a linear correlation between the increase in the PME:PDE ratio and the reduction in plasma albumin concentrations in patients with diffuse liver diseases.

Meyerhoff et al[12] applied ^{31}P MRS to monitor metabolic changes associated with alcoholic liver disease and cirrhosis. A proton image-guided localization technique (ISIS) was used to acquire ^{31}P spectra selectively from the volume of interest within the liver. A 2.0 T (34.78 MHz for ^{31}P) was used for this investigation. A surface coil was placed under the right side of the subject. The patients were divided into three groups: alcoholic hepatitis with fibrosis, alcoholic cirrhosis, and viral hepatitis.

A localized spectrum from the liver of an alcoholic cirrhosis and from the liver of a healthy subject are shown in Fig. 7.11. Both spectra were acquired and processed in a similar manner. The ratios of the metabolites obtained from the three groups are given in Table 7.7. The results obtained on patients with viral hepatitis are in agreement with those reported by Oberhaensli et al;[39] namely, the PME:ATP ratio increased and the PDE:ATP ratio decreased. In contrast, Meyerhoff et al[12] found little difference in metabolite ratios between patients with alcoholic hepatitis and controls. These authors found that these ratios were not altered in patients with alcoholic cirrhosis. The hepatic pH was 7.44 for control subjects and viral hepatitis patients, 7.26 for patients with cirrhosis, and 7.87 for patients with alcoholic hepatitis. Higher pH in an alcoholic hepatitis might be due to the accumulation of HCO_3^-, caused by impaired urea synthesis. Low pH in cirrhosis may be caused by lower oxidative metabolism leading to lactic acidosis.

The absolute concentrations of the metabolites present, together with the percentage difference

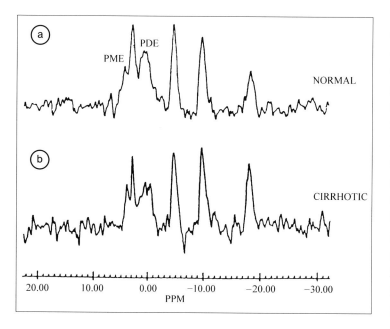

Fig. 7.11 ^{31}P ISIS spectra (2 T, 34.78 MHz) obtained from the liver of a healthy volunteer (a) and a patient with cirrhosis (b). A nominal volume of interest (VOI) of 100 ml was used in both cases. Acquisition values: number of averages – 640 (normal) and 1200 (cirrhotic); TR – 1000 msec; data size – 512 words; spectral width – 3000 Hz; 90 pulse length at VOI – 60 μsec. Processing parameters: convolution difference with 300-Hz exponential line broadening and a scaling factor 0.8; further 10-Hz exponential line broadening zero fill to a data size of 1.024 words; and Fourier transformation. Reproduced with permission from Meyerhoff et al.[12]

from control values, are also given in Table 7.7. Note that the calculation of absolute concentrations reveals significant difference in the three patient populations. There were significantly lower levels of almost all of the metabolites in the patients with alcoholic cirrhosis and alcoholic hepatitis. Patients with viral hepatitis showed a greater than 100% increase in the concentration of PME. These results clearly demonstrate that the analysis of ^{31}P spectra in terms of absolute concentrations provide a powerful method for the discrimination of the three patients groups.

A one-dimensional CSI ^{31}P MRS study of 14 patients with liver cirrhosis of differing severity was carried out using a single turn transmit receiver coil in a 1.5 T GE Signa Scanner.[40] Seven normal volunteers without liver disease acted as controls. Conventional proton gradient SE images with flip angle 30°, TE 30 msec, TR 33 msec, slice thickness 10 mm, and slice spaces 5 mm for the position of the liver were collected. Shimming was done with water proton signal using a body coil as a transmitter and surface coil as a receiver. The data were processed and fitted using Marquardt–Levenburg fitting function with Lorentzian line shapes. Peak areas of phosphorus metabolites were expressed as a percentage of the total mobile phosphorus in ^{31}P MR spectra. Data analyses of ^{31}P spectra showed that the PME peak in cirrhotic liver increased significantly according to the severity of the disease. Furthermore, there was a significant negative linear correlation between PME and hepatic metabolism measured by aminopyrine breath test (AB test). It was also confirmed that saturation effect (T_1 relaxation) had no contribution to the enhancement of PME

Table 7.7 Metabolite ratios and absolute levels of metabolites in patients with a variety of liver diseases, taken from Meyerhoff et al.[12] Standard deviations are in parentheses.

Ratios

Disease	n	PME:ATP ratio	Pi:ATP ratio	PDE:ATP ratio	PME:PDE ratio
controls	8	0.4 (0.2)	1.1 (0.3)	2.6 (1.1)	0.15
viral hepatitis	3	0.8 (0.3)	0.9 (0.3)	2.1 (0.9)	0.38
alcoholic cirrhosis	8	0.5 (0.4)	0.7 (0.3)	2.3 (1.5)	0.22
alcoholic hepatitis	7	0.4 (0.3)	1.1 (0.6)	2.2 (1.0)	0.18

Absolute concentrations (mmol/kg wet weight)

Disease	n	PME	Pi	PDE	γ-ATP	α-ATP
controls	8	0.8 (0.4)	2.2 (0.4)	5.3 (1.9)	2.0 (0.1)	2.7 (0.3)
viral hepatitis	3	1.9 (0.6)	2.1 (0.5)	5.4 (2.0)	2.5 (0.2)	3.1 (0.4)
alcoholic cirrhosis	8	0.7 (0.5)	1.1 (0.3)	3.3 (1.8)	1.5 (0.2)	1.8 (0.4)
alcoholic hepatitis	7	0.6 (0.3)	1.4 (0.4)	2.9 (1.1)	1.3 (0.2)	1.7 (0.2)

Percent change of absolute concentrations from control levels

Disease	n	PME	Pi	PDE	γ-ATP	α-ATP
viral hepatitis	3	138*	−5	2	25*	15
alcoholic cirrhosis	8	−13	−50*	−37*	−25*	−33*
alcoholic hepatitis	7	−25*	−37*	−46*	−36*	−36*

*$p<0.5$

peak in cirrhotic liver. The increase in PME peak in a cirrhotic liver was attributed to the regeneration of hepatocytes in cirrhotic livers. This study highlights the clinical potentials of ^{31}P spectroscopy in assessing the severity of liver cirrhosis.

^{31}P MRS of metabolic disorders, liver poisoning, and the metabolic manipulation of patients with liver disease

Oberhaensli et al[41] carried out a study of patients with hereditary fructose intolerance (n = 5) and subjects who were heterozygotes for hereditary fructose intolerance (n = 9). These heterozygotes were either parents of patients with hereditary fructose intolerance (n = 6) or putative heterozygotes (n = 3). Three of the heterozygotes had recurrent attacks of gout. All of the subjects were examined by ^{31}P MRS after an overnight fast. The patients with hereditary fructose intolerance were presented with a variety of fructose loads. Two patients received an oral dose of 1.5 g of fructose. The other patients were studied after being instructed to eat a fructose-free meal at a local restaurant. One patient was examined after eating a prepared fructose-free meal in the hospital cafeteria. The levels of PME increased by about 50 per cent and the levels of Pi fell about 50 per cent after 20 minutes in the two patients who received oral fructose. One of the two patients who ate in a local restaurant showed increased levels of postprandial

PME, indicating that he had not eaten a fructose-free meal. The one patient who ate a fructose-free meal showed no change in ^{31}P metabolite levels postprandial.

All of the heterozygotes received an oral dose of 50 g of fructose, and ^{31}P spectra were obtained for 60 minutes. Note that previous MRS studies have shown that oral fructose produces minor perturbations in normal controls. In the heterozygotes, the level of PME increased to about 150 per cent of basal value after 20 minutes. This increase was accompanied by about a 50 per cent decrease in the level of Pi. The level of plasma urea was found to increase in both normal controls and heterozygotes, with the levels being higher in the heterozygotes. Moreover, there was a correlation between decreases in Pi and increases in plasma urea ($r = -0.82$). These authors suggested that there might be a predisposition in heterozygotes for gout based on the fructose induced increase in urea. They also pointed out the potential role of MRS in detecting heterozygotes for hereditary fructose intolerance.

The same group[42,43] extended the previous study by employing ^{31}P MRS after an oral dose of 50 g fructose to monitor hepatic metabolism in a group of 11 volunteers with familial gout. Two of these 11 patients showed the ^{31}P spectral pattern observed for subjects who were heterozygotes for hereditary fructose intolerance. Several family members of each of these patients were subsequently studied. The son and daughter of an affected father showed abnormal fructose metabolism. The four sons of an affected mother all showed abnormal fructose metabolism. The genetic and biochemical rationale for interpreting these findings were discussed in detail.

Dufour et al[44] applied ^{31}P MRS to assess hepatic function, by the galactose-elimination capacity test, of six healthy subjects and nine patients with non-alcoholic cirrhosis. Liver spectra were acquired in a 1.5 T GE Signa scanner with 9 cm double tuned surface coil. The shimming was done with tissue water proton signal. The radio frequency pulse amplitude was adjusted to minimize the phosphocreatine (PCr) peak in a non-localized experiment, and a one-dimensional CSI sequence was used for this investigation. The comparison of basal spectra between healthy and cirrhotic subjects shows that the Pi, PME and ATP peaks had similar relative areas. The contribution of PDE signal was significantly smaller in cirrhotic patients than in the healthy volunteers (33 ± 5 per cent and 38 ± 5 per cent respectively). This result is in agreement with the report of Meyerhoff et al.[12] In a time course study, post-fructose administration showed that the increases of PME areas in normal volunteers were more prominent than in patients with liver disease (20 ± 5 per cent compared to 9 ± 5 per cent). The basal value of Pi for normal control subject changed from 14 per cent to 157 per cent, whereas in the case of cirrhotic patients it rose from initial value of 60 per cent to 160 per cent after fructose administration. The increment of Pi was significantly different. The dynamic ^{31}P MRS showed liver metabolic abnormalities upon fructose load in cirrhotic patients. The utilization of Pi and PME formation were decreased in non-alcoholic patients. All relevant information could be obtained within 30 minutes.

All investigations were carried out on the subjects 'in fed state', and not 'in fasted state' as reported in the literature. The investigators claimed that their procedure provided better understanding of the PME response because glucose release through fructose-1-phosphate degradation was inhibited. The diminished metabolism of fructose in cirrhotic livers could be due to the reduced liver uptake, to impaired hepatic phosphorylation, or to both. The investigators, however, could not differentiate the actual cause of fructose metabolism that might be due to reduced cell number, to decreased enzyme substrate, or to both. Further, the procedure might sometimes be risky for patients with severe lactic acidosis or subjects with hereditary fructose intolerance.

Galactosemia is an autosomal-recessive disorder caused by a deficiency of the enzyme UDP glucose (α-d-galactose-1-phosphate uridylyltransferase). Two galactosemic patients were studied in a 1.9 T machine operating at 32.7 MHz for ^{31}P MRS.[44] An 8 cm surface coil was positioned over the liver. The first patient was given 20 mg per kg and the second 10 mg per kg galactose orally, and ^{31}P spectra were collected at different intervals.

The ^{31}P spectrum of the first patient taken after ingestion of galactose (20 mg/kg) displayed a peak at 5.2 ppm, which includes Pi and galactose-1-phosphate. The level of this peak continued to rise even after 190 minutes. In the second patient, there was no significant change observed in the spectrum obtained 30 minutes after ingestion of 10 mg per kg. No higher dosage was possible.

The increase in peak intensity at 5.2 ppm in the first patient was due to the accumulation of galactose-1-phosphate in the liver. Large oral galactose ingestion produced significant changes in the ^{31}P spectrum of the liver, owing to the accumulation of galactose-1-phosphate, with no characteristic spectral changes observed in the brain within the same time interval. This confirms that the liver is the initial site of galactose metabolism.

Dixon et al[45] examined 18 patients with acetaminophen poisoning using ^{31}P MRS at a 1.9 T, 60 cm bore operating at 32 MHz. A double tuned coil was used for this investigation. Liver metabolite levels were measured in the liver with a modified phase-modulated rotating-frame imaging technique. The coil was placed on right side of the liver. The coil homogeneity was optimized with water proton signal from liver tissue. A pyrophosphate marker at the center of the receiver coil served as an indicator for concentration, depth, and frequency. An automated or manual triangulation method was used for peak area calculations.

Phosphorus spectra of acetaminophen damage of the liver showed a decrease in ATP concentration and correlated with blood clotting time, International Normalized Ratio (INR). The concentrations of PME, PDE, and Pi were also diminished with the severity of the poisoning. Statistical (Sparkmann nonparametric rank correlation coefficients) analysis showed that the decrease of ATP, PDE, and Pi significantly correlated only with INR. These results are shown graphically in Fig. 7.12.

This study showed that ^{31}P MRS was a useful technique in estimating high-energy phosphate metabolites in patients with hepatocellular damage. ^{31}P MRS may provide a non-invasive indicator of the severity of liver damage. The metabolic information derived from the spectra may also provide a means for predicting recovery from liver poisoning.

^{31}P MRS of hepatic tumors

Cox et al[17,18] reported the results of a four-dimensional CSI study of a patient with carcinoid metastases in the liver. Elevated phosphomonoester and decreased phosphodiester concentrations relative to ATP were observed. The regions of abnormality corresponded to regions containing metastases identified with plain radiograph, CT, and MRI.

Glazer et al[11] employed one-dimensional CSI to study the metabolism of hepatic tumors. One-dimensional CSI data obtained from ^{31}P MRS can distinguish between the normal liver and the

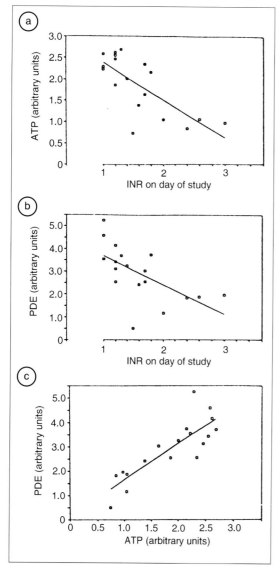

Fig. 7.12 ^{31}P *MRS-determined ATP concentration vs INR (measured on the day of the MRS study). SRCC− −0.72; p < 0.005. (b)* ^{31}P *MRS-determined PDE concentrations vs INR (measured on the day of the study). SRCC− −0.70; p < 0.005. (c) PDE vs ATP concentrations. SRCC−0.84; p<0.001. Reproduced with permission from Dixon et al.*[45]

malignant liver. Malignant hepatic neoplasm differed from normal parenchyma by having a higher concentration of PME. This elevation in PME levels in cancer patients was attributed to increases in PC and PE caused by increased membrane turnover within the tumor. The mean pH value of the tumor was higher by 0.3 units when compared with the mean pH of normal liver.

The investigators also reported ^{31}P MRS results obtained on patients with cavernous haemangiomas. The results of the ^{31}P MRS indicated that these benign tumors had very low ratios of PME:β-ATP. The spectra obtained from the two benign lesions had poor signal-to-noise ratios as compared to those of hepatic tumors. This finding may reflect the relatively low cellularity of these lesions.

Francis et al[46] performed ^{31}P one-dimensional CSI on 37 patients with various malignant neoplasms (30 metastatic lesions and 7 hepatocellular carcinomas). Tumors were grouped according to the percentage of the analyzed section that was occupied by tumor: less than 50 per cent (group A) or more than 50 per cent (group B). In group B, all of the PME:β-ATP ratios were significantly higher than normal (p<0.001).

Oberhaensli et al[14] investigated 24 patients with liver disease using ^{31}P MRS. Two of these patients had hepatic tumors. Liver tumors had raised PME and also showed evidence for altered spin-lattice relaxation of the phosphorus nucleus in the various metabolites.

Dixon et al[47] applied rotating frame ^{31}P MRS to study the 22 patients with either Hodgkin's disease or non-Hodgkin's lymphoma. A 1.9 T, 60 cm bore magnet operating at 32.2 MHz for phosphorus was used in this investigation. The receiver surface coil was isolated and positioned towards the transmitter coil, and the liver of the patient rested on the center of the receiver coil. The field homogeneity was adjusted with liver tissue water proton signal. Each individual study required 30 minutes.

The patients were divided into three groups depending upon the hepatic involvement. Group I (n = 8) had no hepatic infiltration and the radiological and clinical tests were normal. Group III (n = 5) consisted of patients with hepatic disease that had elevated levels of alkaline phosphatase and γ-GTP. Two patients in group II (n = 9) had liver lymphoma and three had diffuse infiltration. Nine patients in group II had enlarged livers with abnormal liver function tests but the CT scans were normal. Two of these nine patients had normal liver biopsy and histology. The MR spectra from group III patients had elevated PME signal when compared to Pi and ATP. Three out of eight patients of group I had elevated PME:Pi and PME:ATP ratios. Two patients of group II had PME:ATP ratios outside two standard deviations of the control mean but their PME:Pi ratios were within the normal range. The remaining four patients of group II had similar abnormalities as seen in group III. Other metabolite ratios (Pi:ATP and PDE:ATP) were normal in 22 patients. The pH, as estimated from the chemical shift of Pi peak, was normal (pH = 7.2±0.1).

Cox et al[19] applied ^{31}P MRS with two-dimensional and three-dimensional CSI localization to 49 adults with different liver diseases. Seventeen of these patients had hepatic malignancies. Elevated PME:ATP ratios and reduction in PDE:ATP ratios were observed both from primary and secondary tumors compared with normal controls.

The same group[16] applied localized ^{31}P MRS *in vivo* using CSI techniques in 32 patients with hepatic malignancies on a 1.6 T MR scanner. A double-tuned surface receiver coil (60 or 150 mm) was placed lateral to the liver in the mid-axillary line. Tissue water proton signal was used for shimming. The *in vivo* result showed that there was a significant increase in the mean PME:ATP ratio and decrease in the mean PDE:ATP ratio in all of the patients compared to normal volunteers. The *in vitro* results showed spectral changes in PE, PC, GPE, and GPC in tumor tissues compared to normal. An increase in PE and PC concentrations and a decrease in GPC concentration were observed. However, the spectral pattern was not specific to the tumor studied. Biopsy-related ischemia may be a limitation of tissue characterization. In almost all the patients, localized *in vivo* spectra from liver lesions showed a significant increase in mean PME–PDE ratio, which was also confirmed with aqueous tissue extraction study after laparotomy.

Meyerhoff et al[20] employed a modified ISIS method to examine five patients with primary or metastatic liver tumors. Four of these cases showed elevated PME:ATP ratios. The authors found low levels of ATP in all of the tumors studied. Low levels of Pi were found in three of the tumors. These results were obtained using the method for obtaining absolute concentrations of metabolites from an analysis of ^{31}P spectra previously described by this group.

We have studied 14 patients (ages 33–76, 10 males and 4 females), with metastatic colorectal carcinoma to the liver using image-guided ^{31}P MRS. All the patients had biopsy-proven metastatic colorectal carcinoma and were studied prior to the start of infusion chemotherapy. The images were used to guide placement of a 12.5 cm transmit and receive surface coil tuned to both proton and phosphorus frequencies. Spectra were obtained by locating the center of the metastases on the images and without moving the patient, taping onto the patient a 12.5 cm round doubly tuned surface coil over the lesion and employing depth resolved surface coil spectroscopy (DRESS) localization for ^{31}P.[48] Anterior metastases were utilized where possible and a coronal DRESS slice was used, although the largest metastases were generally chosen. The slice was chosen with a view to avoiding anterior wall that contained musculature within the limitations of the signal-to-noise drop-off of the coil (the center of the slice was usually at least 5 cm from the coil).

The average depths of the lesions were 5 ± 1 cm from the plane of the coil. Using the Biot–Savart law, we estimate the B_1 field generated at a depth of 5 cm from the center of a 6.25 cm radius coil to be equivalent to a flip angle of less than 25° when the B_1 field is equivalent to a 90° flip angle near the surface of the coil. With the T_1 values given in Table 7.1, using this flip angle and the repetition time employed in this study (1 sec) ensures that the spectra are acquired under fully relaxed conditions.

Typical spectral processing consisted of 10 Hz line broadening, zero- and first-order phasing and deconvolution for baseline correction. The spectral peak areas were determined by finding the best fit to a series of Lorentzian lines. The average ratios for PME to β-NTP, Pi to β-NTP, PDE to β-NTP, PME:PDE, α-NTP to β-NTP, and total low energy phosphates (PME+Pi+PDE) to β-NTP were calculated for normal liver, all metastatic lesions, and all metastatic lesions greater than 6 cm in maximum diameter. Mean ratios for all metastases, metastases 6 cm or greater and normal liver were compared using the unpaired Students' T-test.

Case 1 (Fig. 7.13) is a 43-year-old white male who had resection of a rectal carcinoma in June 1985 and was treated with local radiation therapy and systemic chemotherapy. Two years later, abdominal CT scan demonstrated massive hepatic metastases and he was admitted for placement of an hepatic artery infusion pump. Prior to placement of the pump, he had abdominal MRI, which demonstrated a large left lobe metastasis, which was relatively homogeneous and hyperintense on the T_2-weighted image. The ^{31}P spectroscopy showed a markedly elevated PME:PDE ratio.

Case 2 (Fig. 7.14) is a 72-year-old white female who presented with a cecal carcinoma and diffuse hepatic metastases. She underwent MRI which showed diffuse left and right hepatic metastases that had hyperintense rims with hypointense centers on T_2-weighted images. She had ^{31}P MRS following colon resection and placement of an hepatic artery infusion pump, but prior to the start of the chemotherapy. The PME:PDE ratio was again significantly elevated, even though the MRI signal characteristics of the lesion were different.

Case 3 (Fig. 7.15) is a 62-year-old white male who underwent AP resection for a Duke's B rectal carcinoma. He presented with a massive right hepatic lobe metastasis containing central high signal, presumably liquefactive, necrosis surrounded by a relatively hypointense rim on T_2 weighted images. He had placement of an hepatic artery infusion pump and then had MRI and ^{31}P MRS prior to the start of chemotherapy. In the center of the large necrotic tumor, no phosphorus metabolites could be identified, but the tumor itself demonstrated elevation of the PME:PDE ratio, although it was not as high as the previous two cases.

Many of the patients who were studied had an hepatic artery infusion pump placed in the subcutaneous tissues of the left lower quadrant for the delivery of chemotherapy. This created some signal loss locally on the images but did not affect the spectroscopic exam or the magnetic homogeneity in the region of the liver. The imaging and spectroscopy also did not affect the function of the infusion pump.

Clinical MRS studies of the liver and spleen

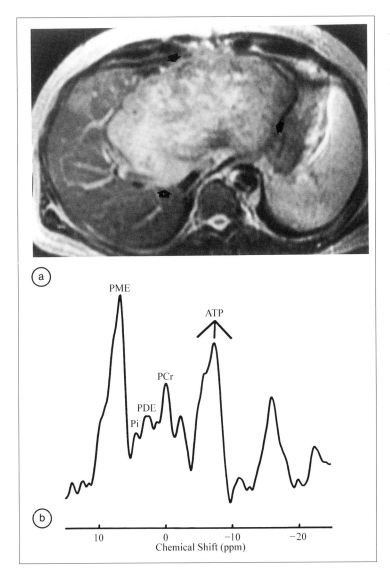

Fig. 7.13 (a) A long TR/long TE (2500/80) transverse MR image demonstrates a large hyperintense mass (arrows) occupying the entire left lobe and the anterior segment of the right lobe of the liver. This image was obtained prior to the beginning of chemotherapy.
(b) The ^{31}P spectrum obtained with a 3 cm coronal DRESS slice from the lesion shown in the figure. The PME:PDE ratio is markedly elevated.

The signal intensity pattern of the hepatic metastases on the long echo time images included relatively homogeneously hyperintense lesions, lesions with hyperintense rims and low signal centers, and lesions with hypointense rims and high signal centers. There was no correlation between the overall signal intensity of the lesion and the PME:PDE ratio. There was a slight tendency toward higher PME:PDE ratios in those metastases that were hyperintense but not high in signal because of liquefactive necrosis. There also was no correlation ($R = 0.3$) between the size of the colorectal metastases and the PME:PDE ratio.

There was a trend toward increased PME and decreased PDE in hepatic colorectal metastases, but the other calculated ratios showed no significant

129

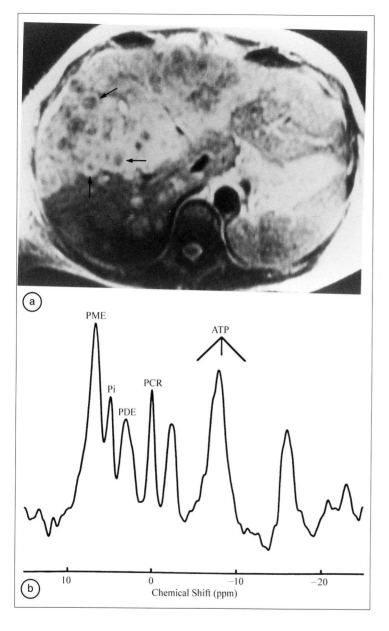

Fig. 7.14 *(a) An axial long TR/ long TE (2500/80) MR image shows multiple confluent hyperintense metastatic lesions involving all four segments of the liver. Many of the metastases have hyperintense rims and low signal intensity centers (arrows). (b) The ^{31}P spectrum obtained with a 3 cm coronal DRESS slice from the lesion shown in the figure also has a markedly elevated PME:PDE ratio relative to normal liver. (The spectra from Figs 7.13 and 7.14 are very similar, despite the difference in their MRI appearances.)*

difference from normal liver. Differences in the PME:PDE ratios for hepatic colorectal metastases versus normal liver were statistically significant when all masses are included ($p<0.05$), but become even more significant when only the large metastases are included ($p<0.02$) (Fig. 7.16). This may be due to the inclusion of more tumor tissue and less normal liver in the spectroscopic volume.

Glazer et al[11] have shown that cavernous haemangiomas have low signal-to-noise in their ^{31}P spectra. However, Case 3 shown in Fig. 7.15

Clinical MRS studies of the liver and spleen

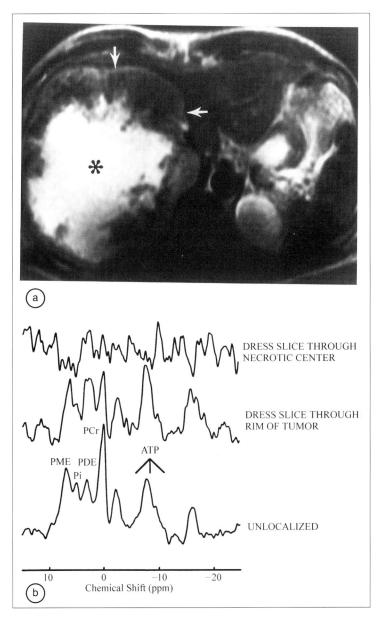

Fig. 7.15 (a) An axial long TR/long TE (2500/80) MR image demonstrates a large necrotic hepatic metastases in the right lobe of the liver. There is a thick rind of low signal tumor (arrows) surrounding the signal necrotic center (asterisk). (b) Unlocalized ^{31}P spectrum and two separate ^{31}P spectra using DRESS localization in the sagittal plane. The middle spectrum is through the rind of tumor and the top spectrum is through the necrotic portion of the tumor There are no resolvable phosphorus metabolites in the necrotic portion of this tumor.

demonstrates that liquefactive necrosis does not contain significant amounts of phosphorus metabolites. Thus it may be difficult to distinguish haemangiomas from large tumors with liquefactive cores.

The finding of an increased PME:PDE ratio is in agreement with many of the other reports in the literature. This finding is probably not specific for tumors alone since it has been observed in other pathologies. Even if the elevation of the PME:PDE ratio is not characteristic of a specific liver lesion, it still represents a marker to differentiate abnormal tissue from normal liver tissue.

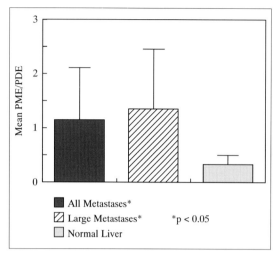

Fig. 7.16 Comparison of the mean of PME:PDE ratios in colorectal metastases and normal liver.

^{31}P MRS studies of tumor response to therapy

Maris et al[49] showed that there were metabolic alterations observable in the ^{31}P MRS spectrum of a neuroblastoma treated with a combination of radiation therapy and chemotherapy. These authors used field profiling MRS methods to study two pediatric patients with metastatic disease of the liver from primary neuroblastoma. One patient with progressing Stage IV disease was studied prior to treatment and was also followed during treatment. The spectra are shown in Fig. 7.17. Spectra obtained prior to treatment (a–c of Fig. 7.17) showed elevated PME:ATP ratios, which increased from 1.3 to 2.3 (normal liver had a ratio of 0.3) as the liver mass increased about a factor of two (Spectrum a; five times normal mass: Spectrum c; ten times normal mass). Spectra a and b were obtained after unsuccessful radiation therapy. Additional radiation therapy and chemotherapy was employed, and this caused the tumor to go into remission. The primary adrenal tumor was resected. Spectra d–f in Fig. 7.17 were obtained after the tumor was treated. Note

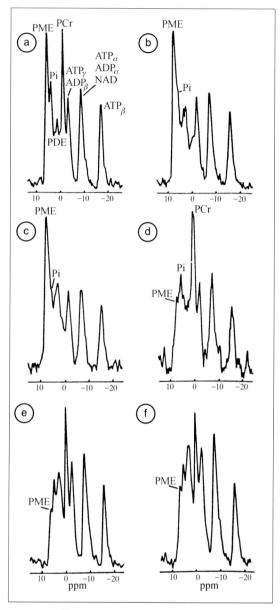

Fig. 7.17 ^{31}P NMR spectra in case 2 as a function of time. (a) shows the spectrum in the first study (13 October, 1984), (b and c) shows the spectra 4 and 8 weeks later, (d) shows the spectrum 16 weeks after the initial response to treatment, and (e) and (f) show the spectra 24 weeks after the initial response to treatment at the edge of the liver (e) and anterolateral to the edge (f). Reproduced with permission from Maris et al.[49]

the decrease in PME:ATP ratio. This ratio decreased earlier than any volume changes were observed in the liver suggesting that the PME:ATP ratio may be an early metabolic predictor of clinical response. These authors also obtained a high-resolution NMR spectrum from a perchloric acid extract of a biopsy sample of the other patient (Stage IV-S disease). This spectrum showed that the PME peak observed was composed of phosphoethanolamine and phosphocholine.

Dixon et al[47] studied 11 patients with hepatic lymphoma who were being treated with systemic chemotherapy. Seven of these patients had elevated PME:ATP ratios before treatment. All of these subjects showed decreases in their PME:ATP ratio after treatment. Some of the ratios returned to normal values. The four cases with normal PME:ATP levels before treatment exhibited no changes in their ^{31}P spectra after treatment. All these results are shown graphically in Fig. 7.18. These authors identified phosphoethanolamine as the major constituent of the PME peak from a high-resolution spectrum of a perchloric acid extract of an involved lymph node.

Cox et al[19] studied four patients who had hepatic malignancies. Localized spectra were obtained before treatment and after treatment. The PME:PDE ratio was elevated in all of these cases before treatment. In three of these subjects, this ratio decreased towards more normal levels after treatment. These three subjects were all judged to be responding to therapy. Two of the four patients underwent arterial embolization of the tumor circulation. The Pi:ATP ratio increased in the patient in whom this procedure was performed successfully. In the other case, in which embolization was unsuccessful, there was a reduction in the PME:PDE ratio, but no change observed in the Pi:ATP ratio.

Meyerhoff et al[20] employed localized ^{31}P MRS to characterize and quantify the phosphorus metabolites in hepatic metastases (n = 2) and primary cancers of the liver of different origin (n = 3) in five cancer patients (aged 51–63 years). MRS was used to monitor the acute and long-term

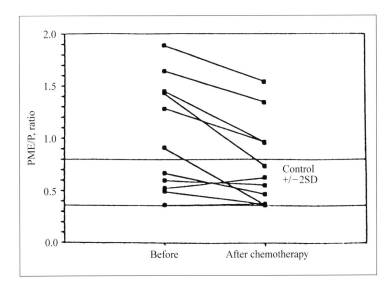

Fig. 7.18 PME–Pi ratio before and after chemotherapy. Reproduced with permission from Dixon et al.[47]

effects of hepatic chemoembolization on tissue metabolites. A 2 T MRI and MRS scanner was used in this investigation. All ^{31}P spectra were acquired using a modified ISIS pulse sequence. Broad spectral components were removed using a convolution difference procedure. All the hepatic lesions had elevated PME:ATP ratios, elevated PME levels, and reduced ATP and Pi levels when compared to normal controls. At 3–4 weeks after chemoembolization, ATP, PME, and PDE concentrations had decreased and Pi concentrations were slightly elevated. Seventeen weeks after chemoembolization, the metabolite concentrations were close to prechemoembolization values. The ratio of PME:ATP values decreased with time. All these changes occurred before any changes in tumor volume were observed on imaging. These studies suggest that ^{31}P MRS can be employed to distinguish patients who respond to chemoembolization therapy from non-responders.

^{13}C MRS studies of metabolic disorders

As part of their initial study of ^{13}C MRS of the liver, Beckmann et al[36] reported the natural abundance spectrum of a patient with type IIIA glycogen storage disease. The level of glycogen was found to be two or three times higher than in normal controls.

Magnusson et al[50] employed ^{13}C MRS to quantitate the glycogen concentration in the livers of patients with type II diabetes mellitus. Five healthy subjects and seven patients with diabetes mellitus were included in this investigation. Three days before the investigation, all the subjects were given a high carbohydrate diet (60 per cent) with an energy content was 30–33 kCal per kg of body weight. On the day of the investigation, subjects were given total 650 kCal in 640 ml. A sequence of ^{13}C NMR measurements of liver glycogen concentration were started after 23 hours of fasting at different time intervals. Liver volumes after 14.5 hours fast were measured using MRI. These volumes were 1.42±0.1 liter and 1.19±0.13 liter in the diabetic and control subjects, respectively. The subject was cannulated, a base line spectrum was collected, continuous infusion of [6-^{3}H] glucose was started, and blood samples were collected at different intervals. Plasma glucose, insulin, glucagon and cortisol concentrations were measured.

Plasma glucose concentration was significantly elevated throughout the study in the diabetic patients compared with the controls. Figure 7.19 shows a typical decoupled ^{13}C NMR spectra (100.4 ppm) of liver glycogen for a diabetic and a control subject 4 hours after ingesting the liquid meal. The concentration of glycogen was lower in the diabetic patients compared to the control subjects (131±20 mmol/liter of liver compared to 282±60 mmol/liter of liver, $p<0.05$).

There was a significant decline in liver glycogen concentration in both groups during the fast. Further, the rate of glycogen breakdown was higher in the control subjects than in the diabetic subjects (10.5±2.0 mmol/liter of liver per hour compared to 4.6±0.9 mmol/liter of liver per hour, $p<0.05$). The mean rate of net hepatic glycogenolysis was lower in the diabetics than in the control subjects (1.3±0.2 µmol/kg body weight per hour compared to 2.8±0.7 µmol/kg body weight per hour, $p<0.05$). Total glucose production was measured after 22 hours of fasting, following 6 hours of [6-^{3}H] glucose infusion. The mean rate of total glucose production increased by 25±13 per cent in the diabetic compared to the control subjects. The hepatic glycogenolysis was decreased and the total glucose production rate was enhanced in the diabetic subjects. Consequently, the rate of gluconeogenesis was

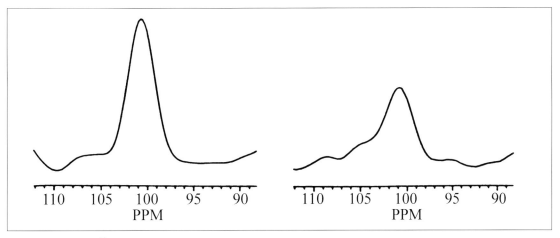

Fig. 7.19 Typical ^{13}C NMR spectrum of the C1 position of liver glycogen from one control subject (left panel) and one type II diabetic patient (right panel) 4 hours after the liquid meal. Reproduced with permission from Magnusson et al.[50]

found to be increased in the diabetic as compared to the control subjects (9.8±0.7 μmol/kg body weight per minute compared to 6.1±0.5 μmol/kg body weight per minute, p<0.01). This study illustrates the utility of ^{13}C MRS in studying metabolic disorders.

SPLEEN

Smith et al[51] reported the first *in vivo* phosphorus spectra obtained from the human spleen both in normals and in disease. The spleen is of secondary lymphoid origin and is often involved in malignancies of hemopoietic origin. One-dimensional CSI ^{31}P spectroscopy with a surface coil at 1.5 T was performed on normal volunteers, patients (n = 13) with hematological malignancies, and patients (n = 2) with benign splenomegaly. The patients with malignancies had higher PME:β-ATP ratios than normal controls. A scatter plot of PME:Pi and PME:β-ATP ratios showed that the PME:Pi ratio discriminated most clearly between low grade non-Hodgkin's lymphoma (NHL), myeloproliferative disorder, and normals. The mean PME:Pi ratio was higher for the low grade NHL than myeloproliferative disorders, producing two characteristic pathological groups. The mean pH values for the normal control group and the patient group were not specific and both groups had an alkaline pH.

Four patients with high grade NHL were followed serially during the course of treatment with combination chemotherapy. The most striking alterations observed were increases in the level of PDE at week 4 of the treatment followed by a decrease in the level of PME at week 12 of treatment.

This report indicates the feasibilty of employing ^{31}P MRS to study splenic masses. Clearly, more studies are needed before definitive conclusions can be reached. Nevertheless it is important to note that the detection of splenic malignancies has clinical utility.

SUMMARY

As was pointed out in the Introduction, the first localized ^{31}P MR spectrum of human liver was published almost 10 years ago. In retrospect, this spectrum was of high quality. The question to be answered is: How much progress has been made in the intervening ten years in terms of our collective understanding of liver metabolism and disease? Several groups working independently at different institutions have found good intrasubject and intersubject reproducibility in the metabolite ratios derived from ^{31}P MRS at their own institutions. However, there are still substantial differences in these ratios when results from different institutions are compared. The reasons for these differences should be determined before large-scale interinstitutional clinical studies are undertaken.

The development of tests for liver function based on the liver's ability to handle fructose appears to be an area ready for further study. There is agreement in the literature regarding the magnitudes of spectral changes observed on ^{31}P MRS after fructose challenge in normal controls, patients with impaired liver function, and in subjects with hereditary fructose intolerance.

There is also general agreement on the spectral pattern observed on ^{31}P MRS in hepatic malignancies. These involve elevation of the PME resonance. It is also clear that the ratio of PME:PDE can be employed as an assessment of tumor response to therapy. The availibility of proton decoupling in combination with localization methods that result in good spectral resolution may aid in the assignment of these alterations to changes in specific compounds present. *In vitro* and animal studies have already been helpful in this area. The reports of MRS of lesions in the spleen also indicate a great deal of promise in the clinical setting.

The development of ^{13}C MR methods which permit the detection of glycogen in the liver has wide clinical utility. This methodology has already provided insights into the rates of carbohydrate metabolism in normal subjects. The Yale group headed by Shulman have also shown the utility of these methods in studying carbohydrate metabolism in patients with diabetes. The ability to perform this test at natural abundance makes this methodolgy financially competitive.

REFERENCES

1. Oberhaensli RD, Galloway GJ, Taylor DJ, Bore PJ, Rajagopalan B, Radda GK. First year experience with ^{31}P magnetic resonance studies of human liver. *Magn Reson Imaging* 1986;4:413–16.
2. Bell JD, Cox IJ, Sargentoni J, et al. A ^{31}P and ^{1}H-NMR investigation of normal and abnormal human liver. *Biochim Biophys Acta* 1993;**1225**:71–7.
3. Bode JC, Zelder O, Rumpelt HJ, Wittkamp U. Depletion of liver adenosine phosphates and metabolic effects of intravenous infusion of fructose or sorbitol in man and in the rat. *Eur J Clin Invest* 1973;3:436–41.
4. Hultman E, Nilsson HL, Sahlin K. Adenine nucleotide content of human liver. Normal values and fructose-induced depletion. *Scand J Clin Lab Invest* 1975;35:245–51.
5. Blackledge MJ, Oberhaensli RD, Styles P, Radda GK. Measurement of *in vivo* ^{31}P relaxation rates and spectral editing in human organs using rotating-frame depth selection. *J Magn Reson* 1987;71:331–6.
6. Buchthal SD, Thoma WJ, Tayor JS, Nelson SJ, Brown TR. In vivo T_1 values of phosphorous metabolites in human liver and muscle determined at 1.5 T by chemical shift imaging. *NMR Biomed* 1989;**2**:298–304.
7. Meyerhoff DJ, Karczmer GS, Matson GB, Boska MD, Weiner MW. Non-invasive quantitation of human liver metabolites using image-guided ^{31}P magnetic resonance spectroscopy. *NMR Biomed* 1990;3:17–22.
8. Cox IJ, Coutts GA, Gadian DG, Ghosh P, Sargentoni J, Young IR. Saturation effects in phosphorous-31 magnetic resonance spectra of the human liver. *Magn Reson Med* 1991;**17**:53–61.
9. Luyten PR, Bruntink G, Sloff FM, et al. Broadband decoupling in human ^{31}P NMR spectroscopy. *NMR*

Biomed 1989;1:177–83.
10. Cox IJ, Bryant DJ, Ross BD, et al. Spectral resolution in clinical magnetic resonance spectroscopy. *Magn Reson Med* 1987;**6**:186–90.
11. Glazer GM, Smith SR, Chenevert TL, Martin PA, Stevens AN, Edwards RHT. Image localized ^{31}P magnetic resonance spectroscopy of the human liver. *NMR Biomed* 1989;**1**:184–9.
12. Meyerhoff DJ, Boska MD, Thomas AM, Weiner MW. Alcoholic liver disease: quantitative image-guided ^{31}P MR spectroscopy. *Radiology* 1989;**173**:393–400.
13. Matson GB, Meyerhoff DJ, Lawry TJ, et al. The use of computer simulations for quantitation of ^{31}P ISIS MRS results. *NMR Biomed* 1993;**6**:215–24.
14. Oberhaensli R, Rajagopalan B, Galloway GJ, Taylor DJ, Radda GK. Study of human liver disease with ^{31}P magnetic resonance spectroscopy. *Gut* 1990;**31**:463–7.
15. Dixon RM, Angus PW, Rajagopalan B, Radda GK. Abnormal phosphomonoester signals in ^{31}P MR spectra from patients with hepatic lymphoma. A possible marker of liver infiltration and response to chemotherapy. *Br J Cancer* 1991;**63**:953–8.
16. Cox IJ, Bell JD, Peden CJ, et al. *In vivo* and *in vitro* [31]P magnetic resonance spectroscopy of focal hepatic malignancies. *NMR Biomed* 1992;**5**:114–20.
17. Cox IJ, Calam J, Sargentoni J, Bryant DJ, Illes RA. Four-dimensional phosphorus-31 chemical shift imaging of carcinoid metastases in the liver. *NMR Biomed* 1988;**1**:56–60.
18. Cox IJ, Bryant DJ, Collins AG, et al. Four-diamensional chemical shift MR imaging of phosphorous metabolites of normal and diseased human liver. *J Comput Assist Tomogr* 1988;**12**:369–76.
19. Cox IJ, Menon DK, Sargentoni J, et al. Phosphorus-31 magnetic resonance spectroscopy of the human liver using chemical shift imaging techniques. *J Hepatol* 1992;**14**:265–75.
20. Meyerhoff DJ, Karczmar GS, Valone F, Venook A, Matson GB, Weiner MW. Hepatic cancers and their response to chemoembolization therapy: quantitative image-guided ^{31}P magnetic resonance spectroscopy. *Invest Radiol* 1992;**27**:456–64.
21. Munakata T, Griffiths RD, Martin PA, Jenkin SA, Shields R, Edwards RHT. An *in vivo* ^{31}P MRS study of patients with liver cirrhosis: progress towards a noninvasive assessment of disease severity. *NMR Biomed* 1993;**6**:168–72.
22. Iles RA, Cox IJ, Bell JD, Dubowitz LMS, Cowan F, Bryant DJ. [31]P magnetic resonance spectroscopy of the human paediatric liver. *NMR Biomed* 1990;**3**:90–4.

23. Radda GK, Oberhaensli RD, Taylor DJ. The biochemistry of human diseases as studied by ^{31}P NMR in man and animal models. *Ann NY Acad Sci* 1987;**508**:300–7.
24. Oberhaensli RD, Galloway GJ, Taylor DJ, Bore PJ, Radda GK. Assessment of human liver metabolism by phosphorus-31 magnetic resonance spectroscopy. *Br J Radiol* 1986;**59**:695–9.
25. Segebarth C, Grivegnee AR, Longo R, Luyten PR, den Hollander JA. *In vivo* monitoring of fructose metabolism in the human liver by means of ^{31}P magnetic resonance spectroscopy. *Biochimie* 1991;**73**:105–8.
26. Sakuma H, Itabashi K, Takeda K, et al. Serial ^{31}P MR spectroscopy after fructose infusion in patients with chronic hepatitis, J Magn Reson Imaging 1991;**6**:701–4.
27. Terrier F, Vock P, Cotting J, Ladebeck R, Reichen J, Hentschel D. Effect of intravenous fructose on the ^{31}P MR spectrum of the liver: dose response in healthy volunteers. *Radiology* 1989;**171**:557–63.
28. Grist TM, Jesmanowicz A, Froncisz W, Hyde JS. 1.5 T *in vivo* ^{31}P NMR spectroscopy of human liver using a sectorial resonator. *Magn Reson Med* 1986;**3**:135–9.
29. Masson S, Henrikson O, Stengaard A, Thomsen C, Quistorff B. Hepatic metabolism during constant infusion of fructose; comparative studies with 31P-magnetic resonance spectroscopy in man and rats. *Biochem Biophys* Acta 1994;**1199**:166–74.
30. Sauter R, Mueller S, Weber H. Localization in *in vivo* ^{31}P NMR spectroscopy by combining surface coils and slice-selective saturation. *J Magn Reson* 1987;**75**:167–73.
31. Dagnelie PC, Menon DK, Cox IJ, et al. Effect of L-alanine infusion on ^{31}P nuclear magnetic resonance spectra of normal human liver: towards biochemical pathology *in vivo*. *Clin Sci* 1992;**83**:183–90.
32. Dagnelie PC, Bell JD, Cox IJ, et al. Effects of fish oil on phospholipid metabolism in human and rat liver studied by ^{31}P NMR spectroscopy *in vivo* and *in vitro*. *NMR Biomed* 1993;**6**:157–62.
33. Jue T, Lohman JAB, Ordidge RJ, Shulman RG. Natural abundance ^{13}C NMR spectrum of glycogen in humans. *Magn Reson Med* 1987;**5**:377–9.
34. Jue T, Rothman DL, Tavitian BA, Shulman RG. Natural-abundance ^{13}C NMR study of glycogen repletion in human liver and muscle. *Proc Natl Acad Sci U S A* 1989;**86**:1439–42.
35. Rothman DL, Magnusson I, Katz LD, Shulman RG, Shulman GI. Quantitation of hepatic glycogenolysis and gluconeogenesis in fasting humans with ^{13}C NMR. *Science* 1991;**254**:573–6.

36. Beckmann N, Seelig J, Wick H. Analysis of glycogen storage disease by *in vivo* ^{13}C NMR: comparison of normal volunteers with a patient. *Magn Reson Med* 1990;**16**:150–60.
37. Beckmann N, Fried R, Turkalj I, Seelig J, Keller U, Stalder G. Noninvasive observation of hepatic glycogen formation in man by ^{13}C MRS after oral and intravenous glucose administration. *Magn Reson Med* 1993;**29**:583–90.
38. Longo R, Ricci R, Masutti F, et al. Fatty infiltration of the liver: quantification by ^{1}H localized magnetic resonance spectroscopy and comparison with computed tomography. *Invest Radiol* 1993;**28**:297–302.
39. Oberhaensli RD, Rajagopalan B, Galloway GJ, Taylor DJ, Radda GK. Study of human liver disease with ^{31}P magnetic resonance spectroscopy. *Gut* 1990;**31**:463–7.
40. Munakata T, Griffiths RD, Martin PA, Jenkin SA, Shields R, Edwards RHT. An *in vivo* ^{31}P MRS study of patients with liver cirrhosis: progress towards a noninvasive assessment of disease severity. *NMR Biomed* 1992;**6**:168–72.
41. Oberhaensli RD, Rajagopalan B, Taylor DJ, et al. Study of hereditary fructose intolerance by use of ^{31}P magnetic resonance spectroscopy. *Lancet* 1987;**2**:931–4.
42. Seegmiller JE, Dixon RM, Kemp GJ, et al. Fructose-induced aberration of metabolism in familial gout identified by ^{31}P magnetic resonance spectroscopy. *Proc Natl Acad Sci U S A* 1990;**87**:8326–30.
43. Kalderon B, Dixon RM, Rajagopalan B, et al. A study of galactose intolerance in human and rat liver *in vivo* by ^{31}P magnetic resonance spectroscopy. *Pediatr Res* 1992;**32**:39–44.
44. Dufour JF, Stoupis C, Lazeyras F, Vock P, Terrier F, Reichen J. Alterations in hepatic fructose metabolism in cirrhotic patients demonstrated by dynamic ^{31}Phosphorus spectroscopy. *Hepatology* 1992;**15**:835–41.
45. Dixon RM, Angus PW, Rajagopalan, Radda GK. ^{31}P magnetic resonance spectroscopy detects a functional abnormality in liver metabolism after acetaminophen poisoning. *Hepatology* 1991;**16**:943–8.
46. Francis IR, Chenevert TL, Gubin B, et al. Malignant hepatic tumors: P-31 MR spectroscopy with one-dimensional chemical shift imaging. *Radiology* 1991;**180**:341–4.
47. Dixon RM, Angus PW, Rajagopalan B, Radda GK. Abnormal phosphomonoester signals in ^{31}P MR spectra from patients with hepatic lymphoma. A possible marker of liver infiltration and response to chemotherapy. *Br J Cancer* 1991;**63**:953–8.
48. Bottomley PA, Foster TB, Darrow RD. Depth-resolved surface-coil spectroscopy (DRESS) for *in vivo* ^{1}H, ^{31}P, and ^{13}C NMR. *J Magn Reson* 1984;**59**:338–42.
49. Maris JM, Evans AE, McLaughlin AC, et al. ^{31}P nuclear magnetic resonance spectroscopic investigation of human neuroblastoma *in situ*. *N Engl J Med* 1985;**312**:1500–5.
50. Magnusson I, Rothman DL, Katz LD, Shulman RG, Shulman GI. Increased rate of gluconeogenesis in type II diabetes mellitus. *J Clin Invest* 1992;**90**:1323–7.
51. Smith SR, Martin PA, Davies JM, Edwards RH. Characterization of the spleen by *in vivo* image guided ^{31}P magnetic resonance spectroscopy. *NMR Biomed* 1989;**2**:172–8.

Practical aspects of clinical applications of MRS in the brain

Douglas L Arnold, Paul M Matthews

INTRODUCTION

This chapter will review clinical applications of magnetic resonance spectroscopy (MRS) to neurologic diseases of the brain. Up to this point, only preliminary studies have been performed and possible applications have only been suggested. Although exciting, results are still too limited to determine whether MRS will find a role in usual neurologic practice. This chapter focuses on the evaluation of applications using phosphorus and proton MRS, which are by far the most widely studied nuclei and the ones most likely to enter general clinical use. No attempt is made to be exhaustive: for example, exciting applications of ^{13}C MRS for measuring cerebral metabolic fluxes will not be addressed, as these applications are likely to be confined to specific research programs for the foreseeable future.

NORMAL BRAIN

Importance of localization

Special problems are encountered in brain MRS, which are different from those in studies of muscle (reviewed by Argov and Bank).[1] Localization is a much more significant problem for MRS studies of brain. Abnormalities of the brain are often focal. The fatty dipole of the skull and muscle surrounding the brain contain substantial amounts of the same compounds that can be seen in normal or pathological brain, but in very different relative amounts. Thus, the ability to obtain signals from anatomically defined regions of the brain with relatively little contamination from outside the desired volume is critical. Limitations of real spatial resolution using current volume-selection protocols must be appreciated if spectroscopic data from 'borders' between normal and pathological brain are to be correctly interpreted.

A second fundamental difference is anatomical. While muscle is a relatively homogeneous tissue anatomically, brain is not. Brain volumes studied include varying proportions of neurons, neuronal projections, and glial cells, and the biochemical and physiological characteristics of these tissues differ greatly. Furthermore, from morphological and functional differences, as well as preliminary observations by MRS,[2,3] one may expect biochemical differences between the responses of cells in different regions of the brain. Thus, control data must come from appropriately anatomically correlated volumes. As spectroscopic data is averaged across a complex multicompartment biochemical system, sensitivity to biochemical

139

changes in any single group of cells is reduced and interpretation is complicated.

Phosphorus spectra of human brain

The normal phosphorus MR spectrum of brain has seven major resonances (Fig. 8.1):

- one from each of the three phosphates of ATP;
- one from phosphocreatine (PCr), a high energy buffer compound;
- one from inorganic phosphate (Pi), a product of ATP breakdown; and
- one each from phosphodiesters and phosphomonoesters, which are found mainly in phospholipids of membranes.

These relatively narrow resonances are found superimposed on a broad resonance thought to arise from relatively immobilized (and therefore short T_2) brain phospholipids. Because of the broad underlying resonance and the fact that resonances are overlapping due to intrinsic T_2 and inhomogeneity broadening effects, precise measurement of resonance areas is a particularly complicated problem and a unique solution to the problem of line fitting cannot be found. Fitting of spectra in the time domain may offer the best available approach to this problem.[4]

Intracellular pH measurements based on Pi resonance frequency

The ability to measure intracellular pH (pHi) with phosphorous MRS is a major area of interest, because other approaches for obtaining this value *in vivo* are limited and less reliable. Pi is a weak acid that titrates in the physiological pH range. The acidic form of Pi resonates at 3.3 parts per million (ppm) and the basic form resonates at 5.7 ppm. However, because the protonation–deprotonation reaction is extremely fast, these individual peaks are not seen; rather, a single peak is seen, the position of which moves between these two extremes depending on the relative concentrations of the acidic and basic species. The position of the Pi peak thus describes a titration curve that reports on the pH environment of the phosphate. Since the vast majority of this is in the cell cytoplasm, the position of the Pi peak reports the intracellular, cytoplasmic pH. There is no significant contribution of the low CSF Pi concentration to that measured. However, the measured pH represents an average across all cells in the brain parenchyma, weighted by their relative numbers and cytoplasmic Pi concentrations. Interpretation of changes can therefore be difficult.

The creatine kinase equilibrium

ATP in the cytoplasm is in chemical equilibrium with PCr through the reaction catalyzed by creatine kinase (CK),

$$PCr + ADP + H^+ \rightleftharpoons ATP + Cr$$

Since the equilibrium constant for this reaction lies far to the right (in the direction of ATP formation),

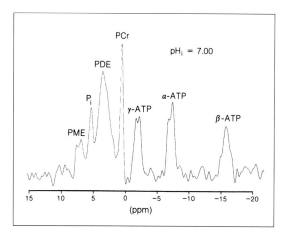

Fig. 8.1. Phosphorus MR spectrum of normal human brain.

demand for energy in excess of the capacity of cells for oxidative ATP synthesis is met initially by a shift in this equilibrium that maintains ATP at the expense of PCr hydrolysis. To the extent that this supply is inadequate, ATP is also synthesized anaerobically with resultant generation of lactic acid.

Considerable emphasis has been placed by us and others on the possibility of using the CK equilibrium constant together with measurement of relative PCr:ATP concentrations, and intracellular pH to estimate free ADP concentrations in muscle. Extension of this approach to brain is problematical, however, because the observed high energy phosphate compounds are not in a single cytoplasm or even a single cell type. Our own feeling is that such estimations are inappropriate in brain.[5]

Localization

Because the T_2 of brain phosphates is relatively short, early studies *in vivo* relied on single voxel subtraction techniques for localization that acquired free induction decays (FIDs), e.g., ISIS.[6] These methods are not entirely satisfactory because the spectra obtained can be contaminated by very significant amounts of signal from outside the volume of interest (VOI). This is especially true when the desired VOI is relatively small (which is most of the time).

More recently, spectroscopic imaging techniques have been implemented. These usually employ spin echoes and are associated with a small but significant loss of signal due to the short values of T_2 of some of the phosphate-containing metabolites, such as ATP. Combined with the relative insensitivity of phosphorus, this means that individual voxels contain in the order of 12–25 cm³, or more. This, in turn, means that the number of phase-encoding profiles is limited, which again tends to lead to contamination of spectra by signals from outside the VOI.

Proton spectra of human brain

Water-suppressed, localized proton MR spectra of normal human brain at 'long' echo times (TE 136 or TE 272 msec) reveal four major resonances (Fig. 8.2):

- one at 3.2 ppm, which arises from tetramethylamines (mainly from choline-containing phospholipids) (Cho);
- one at 3.0 ppm, which arise primarily from creatine (alone or in phosphocreatine) (Cr);
- one at 2.0 ppm, which arises from N-acetyl groups (mainly N-acetylaspartate) (NAA); and

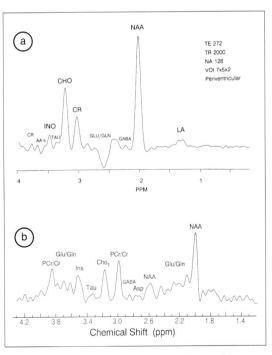

Fig. 8.2. (a) Proton MR spectrum of normal human brain (TE 272 msec). (b) Proton MR spectrum of normal human brain (TE 20 msec).

- one at 1.3 ppm, which arises from the methyl resonance of lactate (LA), or, in certain pathological conditions, methyl groups of lipids.

Lipids and several other compounds (including inositol, GABA, and glutamate) are better observed using much shorter echo times because of short values of T_2 or J-modulation effects.

Multiple lines of evidence suggest that NAA can be used as a neuronal marker as it is found exclusively in neurons and their processes in the mature brain.[7–9] In human brain spectra, *in vivo* NAA is reduced in situations known to be associated with neuronal loss; e.g. neuronal degenerative disorders,[10] stroke,[11,12] and glial tumors.[13] When decreases in the relative NAA signal arise from neuronal or axonal degeneration, irreversible changes are expected. However, we have observed reversible decreases in NAA in a number of conditions, emphasizing that neuronal dysfunction or transient relative volume changes can also lead to decreased NAA. The ability to quantify neuronal loss or damage specifically is one of the most interesting potential applications of MRS in cerebral disorders.

Changes in the resonance intensity of Cho probably result mainly from increases in the steady-state levels of phosphocholine and glycerol-phosphocholine. These choline-containing membrane phospholipids are released during active myelin breakdown. Thus, the resonance intensity of Cho increases in acute demyelinating lesions in humans.[14,15] Chronic, slowly progressive leukodystrophies are associated with normal Cho:Cr resonance intensity ratios,[10] presumably because the loss of myelin is so slow that significant increases of released membrane phospholipid do not accumulate. Certain tumors, e.g. meningiomas, may be associated with high steady state levels of choline-containing phospholipids.

Total Cr concentration is relatively constant throughout the brain and tends to be relatively resistant to change. However, large changes can be seen with destructive pathology such as malignant tumors. Therefore, although in many conditions, it is reasonable to use creatine as an internal standard to normalize resonance intensities of NAA and Cho in order to correct for artifactual variations in signal intensity over space due to magnetic field and radiofrequency inhomogeneity, this must be done with care. Use of an external concentration reference can be reliable if factors such as radiofrequency field inhomogeneity and coil tuning and coupling can be adequately controlled.

Lactic acid (LA) is the endproduct of glycolysis and accumulates when oxidative metabolism cannot meet energy requirements. Elevation of LA in cerebral neoplasms correlates approximately with relative rates of glucose uptake, for example. However, lactate is both intracellular and extracellular, and large amounts may be accumulated *outside* actively anaerobic tissue (e.g. in necrotic tissue or fluid-filled cysts).

In conditions associated with inflammation, LA accumulation may also reflect metabolism of inflammatory cells rather than of brain parenchyma. For example, the prolonged elevation of LA after ischemic infarction may result more from the metabolism of infiltrating macrophages than from ischemic brain.[16]

Proton MRS is currently a more promising field for neurologic applications to the brain than phosphorus MRS. A wider range of compounds can be studied. These include:
- NAA, a marker of neuronal integrity;
- Cho and lipids, markers of myelin breakdown (under some circumstances);
- compounds directly linked with disease (e.g. elevated lactate with infarction[11,12,17–19]); and key neurotransmitters (e.g., GABA[20] and glutamate[21,22]).

It is also theoretically possible to study exogenously derived agents if they are present at high enough concentrations. Thus, for example, the time course of cerebral ethanol concentrations has been directly measured.[23]

However, proton MRS is technically more challenging than phosphorus MRS, despite the higher sensitivity of the MR experiment for protons. Observing metabolites present at millimolar concentrations in a largely aqueous place (approximately 100 molar water protons) demands suppression of the water signal by means of special pulse sequences.

Localization

Proton spectra of brain *in vivo* require effective localization to eliminate signals from fat that is present in the scalp and skull at approximately three orders of magnitude greater concentration than the metabolites of interest in the brain. As noted above, suppression of the signals from water, which is present at almost four orders of magnitude greater concentration than the metabolites of interest, is also necessary. These requirements are most easily achieved by spin echo sequences using relatively long echo times. This takes advantage of the relatively long T_2 of the metabolites mentioned above compared to fat, and avoids problems from eddy currents generated by gradient coils. Spectra are often obtained at either TE 272 or TE 136 msec, because LA, a doublet that phase modulates in spin echo experiments, is respectively in phase and 180° out of phase at these echo times. Very short echo times avoid the problem of phase modulation and allow the observation of additional metabolites with complicated spin systems and short values of T_2, e.g. amino acids and inositol.[12,24–27] However, these short echo times are much more demanding of hardware, the spectra are more susceptible to artifact, and they are more difficult to quantitate because of the appearance of multiple overlapping small peaks that make the baseline less certain (see Fig. 8.2).

APPLICATIONS TO NEUROLOGICAL DISEASE

As neurologists, we see potential clinical roles for MRS in four general areas.

1. MRS has the ability to define pathological changes not associated with structural abnormalities that are significant or easily discernible using conventional clinical imaging.
2. MRS offers the unique potential for easy serial studies of chemical changes.
3. Phosphorus MRS can monitor pHi, providing a new, potentially useful parameter for assessment of clinical pathology.
4. Quantitative measurements of non-radioactive tracer studies of metabolite turnover in the brain may be useful, though their clinical role is far from clear.

MRS in non-structural lesions

Applications of MRS in metabolic disorders, AIDS, postanoxic encephalopathies, and its use in lateralization of temporal lobe seizure disorders illustrate that it can be used to define lesions which are either not associated with structural abnormalities or that are not easily seen on conventional imaging (see below). Of particular importance has been use of relative NAA resonance intensity as a marker of neuronal loss. More generally, the concept of using gross chemical changes in brain parenchyma as an index of pathological changes is becoming better established.

Serial study of chemical changes

MRS ability to study serial chemical changes can be used to define the natural history of a disease process, as, for example, in recent studies of demyelinating disease.[15] Direct monitoring of biochemical responses to therapeutic agents may be performed, as demonstrated in chemotherapy of brain neoplasms.[28,31] This offers the potential for rapid selection of appropriately individualized therapy for patients with development of early biochemical criteria for defining tumors as responsive or not. A more recent prospect (as yet less well established) is the use of MRS for monitoring dynamic changes in biochemical state with functional activation of specific cortical areas.[32,33] Potentially, the nature and time course of change induced by specific tasks might provide a new physiological index for pathology. Where appropriate, MRS offers potential advantages over PET of cost, ease of application, and lack of radiation.

Monitoring of pH with phosphorus MRS

This provides a new, potentially useful parameter for assessment of clinical pathology. As discussed below, changes in local brain pH may be a marker of epileptic foci.[34,35] In ischemic stroke, the time of the shift from acidotic to alkalotic may correlate or be associated with the shift from largely reversible to largely irreversible damage.[36] Tumor pH may be an important factor determining the distribution of chemotherapeutic agents and responses to them.[28]

Quantitative measurements of non-radioactive studies of metabolite turnover

Although the clinical role is far from clear, it is intriguing to consider the possible importance of quantitative measurements of non-radioactive tracer studies of metabolite turnover in the brain. Elegant studies of ^{13}C-glucose transport into brain and its catabolism have demonstrated the feasibility of such measurements.[37,38] With editing techniques, the sensitivity of proton MRS can be used, allowing the higher sensitivity of this nucleus to be exploited for increased spatial resolution.[39] Several potential applications could be identified, e.g. assessment of exogenous glucose uptake as a measure of viability of ischemia brain, determination of GABA turnover rates to assess the potential efficacy of GABA-specific anticonvulsants, use of labelled choline incorporation as an index of remyelination of lesion volumes in MS, and so on. Unfortunately, this type of study remains difficult and expensive and, with the exception of the pioneering work from Yale, is not frequently reported.

Nonetheless, on a more sombre note, despite more than a decade since the initial demonstration of the potential for *in vivo* MRS to elucidate aspects of pathology in neurologic disease, little more than preliminary trials or feasibility studies have appeared in the literature. An important challenge for the next decade will be to assess the sensitivity and specificity of MRS measurements relative to currently employed clinical tests and to define the clinical significance of novel metabolic parameters potentially available from MRS.

Here we will briefly review early studies of potentially important applications in a range of neurologic diseases affecting the brain.

Stroke

Ischemic–anoxic strokes are the commonest disorders of cerebral energy metabolism. Brain cells normally depend heavily on oxidative metabolism to supply their energy requirements. Although the brain can metabolize glucose anaerobically for brief periods in the absence of oxygen, this is done at the expense of accumulation of LA. This LA or the associated acidosis may actually exacerbate the amount of neural damage that occurs during stroke.

Phosphorus spectroscopy of brain-damaged newborn infants was one of the earliest applications of MRS in the brain. This is because morphologic assessment of brain damage early after anoxic insults by conventional imaging is difficult, and infants could be accommodated inside the 30 cm bore magnets that were available before the 1 m bore magnets that are now common. As a result of these studies, normal developmental changes in phosphorus[40] and proton[40,41] spectra have been well described. Phosphorus spectra of brain from infants who have suffered brain damage during delivery demonstrate impairment of the energy state of brain cells that is predictive of outcome after birth asphyxia[42–48] and intraventricular haemorrhage.[49] The energetic compromise is more pronounced in subcortical structures, in keeping with the clinical susceptibility of these structures.[50] Some studies have shown progressive energy failure over several hours and days after birth, emphasizing the ongoing, potentially reversible longer-term components to anoxic injury.[51] More recent proton spectra have allowed some quantitation of lactic acidosis associated with this metabolic compromise. They also illustrate the use of changes in NAA to estimate the extent of neuronal damage.[52]

In adult patients with stroke, the main questions are also to determine to what extent the cerebral damage is reversible and to define agents that can improve the outcome. As in the case of perinatal brain damage, MRS has potentially important information to offer.[53] First, phosphorus MRS can document the extent of focal energy failure and of potentially damaging acidosis.[36,54,55]

Measurements of pHi and high energy phosphates may also be able to define the volume of the brain around irreversibly damaged tissue that is 'at risk' for delayed cell death, but that is still potentially salvageable with appropriate treatment.[56] Proton MRS of stroke shows very high concentrations of lactate consistent with anaerobic metabolism of glucose, as well as loss of NAA, reflecting the associated neuronal damage (Fig. 8.3).[11,12,19] It has been shown that loss of NAA progresses for days after an ischemic stroke, suggesting that the 'window' for salvage of damaged neurons may be greater than previously expected.[57] Persistent increase of lactate adjacent to infarcts may indicate ongoing ischemia.[12,17,39] The fact that MRS can demonstrate these changes offers exciting possibilities for its use in monitoring the efficacy of therapeutic interventions in stroke and as an adjunct in selection and assessment of rehabilitation strategies.

Dementia

Neurodegenerative diseases

MRS may provide a useful index of progression in neurodegenerative disease that is more sensitive than clinical imaging, more quantitative than nuclear medicine techniques such as SPECT, and more practical to apply to large populations than PET. Proton spectra show an accelerated decrease in NAA with age in patients with senile dementia of the Alzheimer's type (SDAT). Peer-reviewed published studies are lacking, but many abstracts on this topic have been presented (see, for example, the *Proceedings of the Society of Magnetic Resonance Medicine*, 1993). Regional decreases in NAA or multifeature analysis combining MRI measurement of hippocampal atrophy with MRS measurement of low NAA:Cr ratios provide more

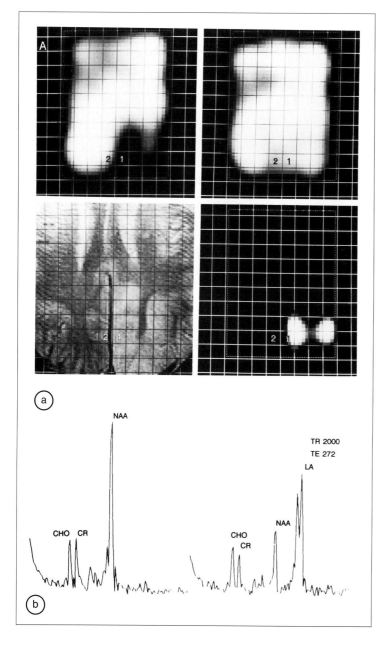

Fig. 8.3. (a). Conventional (lower left) and spectroscopic MR images of an acute cortical infarction in a patient with mitochondrial encephalomyopathy, lactic acid and stroke-like episodes (MELAS). The upper left image is based on the resonance intensity of NAA, the upper right image on the resonance intensity of Cho and the lower right image is based on the resonance intensity of LA. (b). Sample spectra from voxels 1 and 2 indicated on the images. Note the low NAA:Cr and high LA:Cr in the infarcted tissue compared to normal.

specific data that may additionally be helpful in establishing specific etiologies.[58] Phosphorus spectroscopy of brains of patients with SDAT do not show usefully specific changes.[59,60]

Proton MRSI can be used to define focal neurodegenerative pathology, such as that in amyotrophic lateral sclerosis (ALS).[61] This study has demonstrated that changes in NAA are closely linked to focal neuronal loss or dysfunction. MRS appears able to demonstrate directly early premorbid changes in ALS, potentially providing a quantitative index for the evolution of the pathology of this disease that could be used to evaluate response to experimental treatment.

Although only a single report has appeared thus far, the potential for MRS to define unexpected mechanisms of pathology has been illustrated by evidence for impairment of energy metabolism in Huntington's disease. Although the gene defect underlying Huntington's disease has been found, the function of the protein encoded remains obscure. Recently, Jenkins and colleagues[32,62] have shown that there are increases in LA in the occipital lobes and basal ganglia of affected patients, adding significantly to evidence that mitochondrial dysfunction may underlie this enigmatic disorder. LA provides an additional index for monitoring efficacy of therapy, in addition to the decreased NAA and increased choline.

AIDS

AIDS dementia is the major acquired neurodegenerative brain disease. As in Alzheimer's disease and ALS, spectroscopic abnormalities in AIDS dementia occur before changes in conventional MRI. Phosphorus spectra of the brain of patients with AIDS dementia complex show a low concentration of PCr and ATP that cannot be accounted for by atrophy and is consistent with a generalized virus-associated toxic process.[63] Ratios of high to low energy phosphates may also be adversely affected.[64] Proton spectra of the brain of patients with AIDS dementia complex show reduced NAA and increased Cho signals.[65-69]

Multiple sclerosis

Multiple sclerosis (MS) is a disease of the central nervous system that is postulated to result from an immunologically mediated attack on myelin or oligodendrocytes. Post-mortem observations suggest a pathological model of MS that involves inflammation, destruction of myelin, macrophage activation, astrocytic gliosis, and, eventually, loss of axons. It has been difficult to directly test this model because of the limited ante-mortem information on what is happening in tissue affected by MS. MRI shows the lesions very well, but it primarily reports changes in water content and mobility, which reflect only in an indirect way the demyelination and associated tissue damage or loss, and it does not discriminate these from edema.

MRS of acute MS lesions can provide information on the chemical pathology of the disease that was hitherto unavailable (Fig. 8.4). Proton spectra at long echo times of acute lesions reveal that the Cho resonance increases early in the demyelinating process and that this is followed, if the lesion is severe enough, by a decrease in NAA, presumably from secondary axonal damage/loss.[2,14,70-78] The available evidence suggests that these changes result predominantly from changes in NAA concentration rather than relaxation times.[14,15]

MRS of MS has revealed that axonal damage and loss are much more prominent than previously appreciated. Proton spectra at short echo times can demonstrate the neutral lipid released during myelin breakdown.[71,79-82] Phosphorus spectra of MS are also abnormal and show changes in signals from phosphodiesters (PDE) and phosphomonoesters (PME) associated with gliosis.[83,84] Together these measurements may provide an approach for monitoring the pathological evolution of lesions and for pathological classification of individual lesions or overall patterns of progression in demyelinating disease.

Although many treatments directed at the immune system have been tried in MS, proof of efficacy has been elusive. A major reason for this

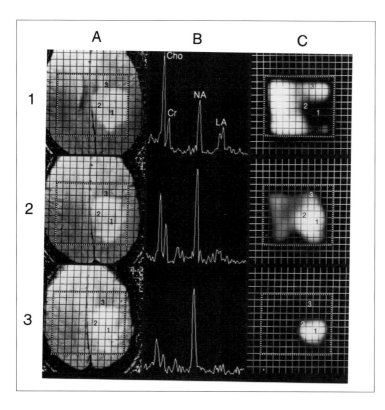

Fig. 8.4. (Left) Three 6mm slices from a conventional MRI through a demyelinating lesion. (Right) Three MR spectroscopic images through the same lesion. Each MR spectroscopic image (MRSI) originates from a slice thickness covered by the three conventional MR images. The top image is based on the signal intensity of NAA, the middle image is based on the signal intensity of Cho and the bottom image is based on the signal intensity of LA. (Center) Sample spectra from individual voxels in the center (1), edge (2), and periphery (3) of the lesion on MRI.

has been difficulty in assessing lesion load and disease progression. Clinical measures reflect lesion load poorly and underestimate disease activity. MRI, despite its power for demonstrating the lesions of MS, also has limitations in estimating the volume of chronic irreversible disease load, in part because of its exquisite sensitivity to changes in water. MRI does not discriminate inflammation and reversible edema from irreversible demyelination and gliosis.[85] MRS may eventually play a major role in the assessment of disease once the sensitivity, specificity, and significance of observed changes have been better defined.

Preliminary data using proton MRS to follow patients with MS has also emphasized the importance of progressive axonal damage and loss as MS evolves,[86] a finding that was not expected on the basis of the standard pathological studies.[87] This conclusion was drawn from observations of progressive decreases in the relative NAA resonance intensity with time in individual lesions, in longitudinal studies of patients with time, and in cross-sectional studies showing that patients with more severe disease had lower NAA than those with mild disease. On the basis of this, we have proposed the NAA:Cr ratio as a useful index of accumulated damage with disease progression.[86]

However, while probably much less sensitive to reversible consequences of acute inflammatory changes, recent studies in our laboratory have clearly demonstrated that NAA decreases are not always irreversible in MS or other pathologies.[79,88,89] As the capacity for neuronal regeneration is limited in the CNS, we interpret this as evidence that NAA changes can reflect reversible neuronal dysfunction and axonal caliber loss, as well as irreversible neuronal loss.

Brain tumors

To date, applications of MRS to the study of brain tumors have generally focused on differential diagnosis. Detection and possible prediction of response to treatment have been studied to a lesser extent, although we believe that assessment of early biochemical response to treatment (including direct assessment of drug delivery) may eventually be an important area for clinical applications.[29,31,90,91]

Even with modern imaging, diagnoses of brain tumors can be problematic. Although x-ray, CT, and MRI detect brain tumors well, diagnosis of the tumor type based on these methods is incorrect in 25–30% of cases. A method that could accurately assign observed lesions to specific diagnostic categories (tumor or non-tumor; malignant or benign) would reduce the need for surgical biopsy and thus reduce patient morbidity and mortality and health care expenses. Proton MRS of brain tumors reveals changes in the relative resonance intensities of chemical compounds normally observed (Cho, Cr, NAA, and LA), as well as new signals from compounds not normally observed in MR spectra obtained at long echo times, such as neutral lipid, alanine, citrate, and other unidentified resonances (Fig. 8.5).[13,92–98] Meningiomas show a characteristic prominent signal from alanine (in the absence of LA) and essentially no signals from NAA (see Fig. 8.5). Metastases display a prominent lipid peak, very high LA signals, and lower NAA levels than glioblastomas and low-grade gliomas. Glioblastomas show more LA and less NAA than low-grade gliomas. Choline levels are variable, as is the presence of a lipid peak. Low-grade gliomas show high levels of Cho, and lower levels of LA and higher levels of NAA than glioblastomas.

Despite the significant differences that exist for individual resonance intensities between different tumor types, attempts at classification based on individual peaks have generally been found to lack the specificity required.[94,96,99] To some extent, this can be explained by inclusion of tumors whose metabolism had been altered by treatment (since treatment is known to alter metabolite ratios[30]) or inclusion in the VOI of fluid filled cysts which contain large amounts of lactate and little else and are generally found in low-grade tumors. However, to a large extent it simply reflects the fact that one is dealing mostly with a limited number of resonances, which are present in most tumors and have a limited dynamic range. More recent studies using spectroscopic imaging, which is more suitable to the heterogeneous nature of the pathology,[100] and artificial intelligence methods for pattern recognition, so as to take into account the information available from all the resonances in the spectrum simultaneously, may be able to improve specificity.[101]

Fluorodeoxyglucose uptake measured by PET was previously proposed as a useful index of tumor malignancy. A number of studies have investigated the correlation between fluorodeoxyglucose metabolism assessed by PET and changes in LA and other metabolites on MR spectroscopic images. The relationship is not necessarily straightforward.[99,102] LA may not always be observed in hypermetabolic lesions on PET, possibly for technical reasons.[103] Maximal LA concentrations can be found in cystic or necrotic regions[104] and not necessarily co-localized with areas of hypermetabolism on PET.

Phosphorus spectra of brain tumors may differ from normal brain[105–108] and can have some value for differential diagnosis of different tumor types[109–112] and grading of gliomas.[13] Phosphorus spectra can reveal metabolic responses

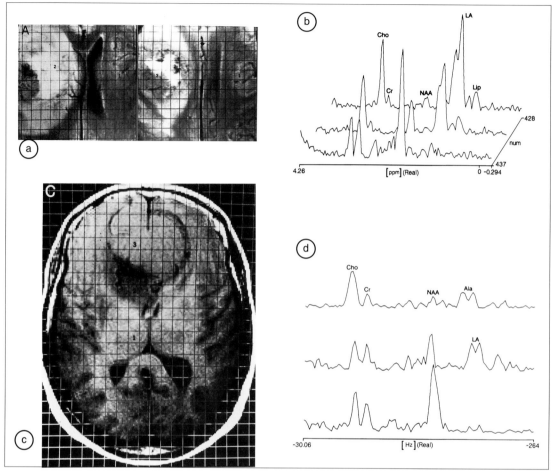

Fig. 8.5. (a). Conventional MRI of a malignant glial tumour (glioblastoma) with the phase-encoding grid for the MR spectroscopic image overlaid on it. (b). Sample MR spectra (bottom to top) from the voxels indicated as 1, 2, and 3. Note the large resonance from Cho, the almost complete absence of Cr and NAA, and the large signals from LA and lipid. (c). Conventional MRI of a meningioma with the phase-encoding grid for the MR spectroscopic image overlaid on it. (d). Sample MR spectra (top to bottom) from the voxels indicated as 3, 2, and 1. Note the alanine peak in the tumour resonates at a slightly different frequency from the LA in adjacent metabolically compromised brain. The tumour has essentially no NAA, very little Cr, and greatly increased Cho.

to therapy within hours, long before they are observable on conventional images.[28] However, the large voxel size required for phosphorus spectroscopy severely limits the proportion of patients in whom phosphorus spectroscopy of tumors is feasible.

Epilepsy

Ictal changes

Epilepsy is a common disorder affecting about 1% of the population. MRS has demonstrated directly that active epileptic foci are associated with potentially damaging bioenergetic abnormalities. Phosphorus spectra from infants actively seizing reveal large decreases in the PCr:Pi resonance intensity ratio, sometimes to very low levels associated with brain damage.[113] One infant whose seizures were stopped with medication in the magnet showed a prompt increase in its PCr:Pi ratio.[113] Proton spectra from rare cases of continuous focal seizure activity show increased lactate and decreased NAA consistent with energy compromise and neuronal damage.[114] Proton spectra have also been obtained before and after electroconvulsive therapy (ECT). These have shown an increased resonance intensity in the lipid region.[115] The significance of this alteration after ECT remains to be determined, but it may reflect changes in membrane lipids. These observations suggest that the brain's metabolic requirements during seizure activity exceed its capacity to provide energy in this situation, and they explain, at least in part, the progressive brain damage that can occur after repeated seizures.

Interictal changes

Most of the time, seizures can be controlled with medication. However, in a significant number of patients, seizures cannot be stopped with drugs, and surgical removal of the part of the brain that initiates or propagates the seizures has to be considered. The part of the brain responsible is usually one of the temporal lobes, but it can be difficult to determine which one. Localization of the seizure focus is therefore a major issue in surgical treatment of epilepsy. Methods for clearly defining the sole or most severely involved region of brain are not yet available. Many techniques are currently employed, including electrical recordings, both from the scalp and from intracerebrally implanted electrodes, MRI, and PET/SPECT scanning. Recent proton MRS observations demonstrate significant reductions in NAA in temporal lobes of such epileptic patients (Fig. 8.6).[116] This reduction in NAA reflects the neuronal damage that is associated with the seizures and appears to correlate very well with the side from which the seizures originate[117–119] as well as with cognitive dysfunction.[120]

Phosphorus spectra may also reveal lateralized abnormalities in patients with temporal lobe epilepsy. Intracellular pH appears to be significantly higher in the temporal lobe on the side from which the seizures originate.[34,121]

Metabolic disorders

MRS has been used to demonstrate chemical pathology in a variety of metabolic disorders. Those resulting from mitochondrial dysfunction were among the earliest to be studied because of the importance of the cellular energy state, which can be determined easily by MRS. The anticipated reduction in the energy state of phosphate-containing metabolites in mitochondrial encephalopathies has been observed by some groups, but not by others, for reasons that are not entirely clear.[5,122–124] Proton spectra of mitochondrial encephalopathies can reveal dramatic evidence of acute and chronic metabolic compromise (in the form of cerebral parenchymal lactic acidosis), evidence of progressive neuronal damage and death (in the form of decreased NAA),[125–127] and evidence of leukoencephalopathy (reflected in low Cho) (Fig. 8.7).[125] The spatial and temporal distribution of metabolic abnormalities can correlate well with clinical symptoms and signs.

A number of metabolic disorders are associated with signals from metabolites that are not normally observed or that cause substantial changes in the amounts of metabolites that are normally

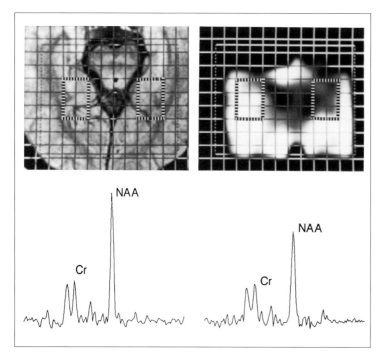

Fig. 8.6. *Conventional MR image (upper left) and MR spectroscopic image (upper right) from a patient with temporal lobe epilepsy. The phase-encoding grid for the spectroscopic image is shown. The spectra shown below the images are averaged from the six voxels outlined in the mesial temporal lobes. The NAA is lower (relative to Cr) on the side from which the seizures originate.*

present in brain. Proton MRS at short echo times is particularly useful in this situation because it can demonstrate the accumulation of abnormal metabolites that are not well seen at longer echo times. Examples include measurement of ketone bodies in diabetes mellitus,[128] measurement of glycine in children with nonketotic hyperglycinemia,[129] and measurement of abnormal lipids in Neiman–Pick disease.[130] Changes in the intensities of resonances normally present can also be of diagnostic and therapeutic interest. Increases in NAA and decreases in Cho resonance intensities have been reported in Canavan's disease.[131–133] Accumulation of glutamine and decreased levels of Cho and myo-inositol can be correlated with development of hepatic encephalopathy.[26,27,134] Increases of glutamine and decreases of cytosolic myo-inositol, as well as decreases in NAA and increases in mobile lipids, have been observed in peroxisomal disorders.[135]

CONCLUSION

Many potentially useful clinical applications of MRS, examples of which have been described above, are under investigation. Progress has been slow compared to MRI for a number of reasons: the resources invested have been much less; many of the early instruments were, in fact, unable to reliably obtain meaningful spectra; and spectroscopy has often been performed outside usual clinical settings. In addition, the chemical data obtained is less readily adapted to standard clinical paradigms in neuropathology and neuroradiology, which are traditionally based in morphologic definitions of disease. Over the past few years, much progress has been made in terms of hardware, software, and pulse sequences for spectroscopy and spectroscopic imaging, and clinical instruments currently becoming available in hospital settings should be capable of adequate spectroscopy by indications for appropriate clinical use in evaluation of at least some of the diseases reviewed here.

REFERENCES

1 Argov Z, Bank WJ. Phosphorus magnetic resonance spectroscopy (31P MRS) in neuromuscular disorders. *Ann Neurol* 1991; 30:90–7.
2 Frahm J, Bruhm H, Gyngell ML, Merboldt KD, Hanicke W, Sauter R. Localized proton NMR spectroscopy in different regions of the human brain *in vivo*. Relaxation times and concentration of cerebral metabolites. *Magn Reson Med* 1989; 11:47–63.
3 Collins DL, Fu L, Pioro E, Evans AC, Arnold DL. *Generation of Normal Average Metabolite Images from MRSI in Stereotaxic Space.* 1993 (unpublished).
4 Knijn A, de Beer R, van Ormondt D. Frequency-selective quantification in the time domain. *J Magn Reson* 1992; 97:444–50.
5 Matthews PM, Arnold DL. Phosphorus magnetic resonance spectroscopy of brain in mitochondrial cytopathies. *Ann Neurol* 1990; 28:839–40 [letter, comment].
6 Ordidge RJ, Mansfield P, Lohman JA. Volume selection using gradients and selective pulses. *Ann N Y Acad Sci* 1987; 508:376–85.
7 Simmons ML, Frondoza CG, Coyle JT. Immunocytochemical localization of N-acetyl-aspartate with monoclonal antibodies. *Neuroscience* 1991; 45:37–45.
8 Moffett JR, Namboodiri MAA, Cangro CB, Neale JH. Immunohistochemical localization of N-acetylaspartate in rat brain. *NeuroReport* 1991; 2:131–4.
9 Birken DL, Oldendorf WH. N-acetyl-L-aspartic acid: a literature review of a compound prominent in 1H-NMR spectroscopic studies of brain. *Neurosci Biobehav Rev* 1989; 13:23–31.
10 van der Knaap MS, van der Grond J, Luyten PR, den Hollander JA, Nauta JJP, Valk J. 1H and 31P magnetic resonance spectroscopy of the brain in degenerative cerebral disorders. *Ann Neurol* 1992;31:202–11.
11 Duijn JH, Matson GB, Maudsley AA, Hugg JW, Weiner MW. Human brain infarction: proton MR spectroscopy. *Radiology* 1992; 183:711–18.
12 Graham GD, Blamire AM, Howseman AM, et al. Proton magnetic resonance spectroscopy of cerebral lactate and other metabolites in stroke patients. *Stroke* 1992; 23:333–40.

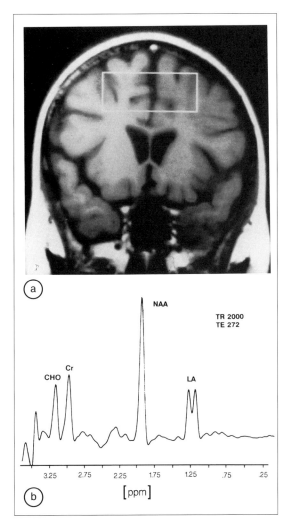

Fig. 8.7. (a). *Conventional MRI showing a large central supraventricular volume of interest for a single voxel proton MRS in a patient with a mitochondrial encephalomyopathy (Kearn–Sayre syndrome).*
(b). *Proton MRS from this volume showing a marked increase in the relative resonance intensity of LA, consistent with chronically impaired oxidative metabolism.*

available, trained personnel. In consequence, we are optimistic that the next decade will bring better clinical evaluation of MRS and establishment of

13 Arnold DL, Shoubridge EA, Villemure JG, Feindel W. Proton and phosphorus magnetic resonance spectroscopy of human astrocytomas in vivo. Preliminary observations on tumor grading. *NMR Biomed* 1990; 3:184–9.

14 Matthews PM, Francis G, Antel J, Arnold DL. Proton magnetic resonance spectroscopy for metabolic characterization of plaques in multiple sclerosis. *Neurology* 1991; 41:1251–6.

15 Arnold DL, Matthews PM, Francis GS, O'Connor J, Antel JP. Proton magnetic resonance spectroscopic imaging for metabolic characterization of demyelinating plaques. *Ann Neurol* 1992; 31:235–41.

16 Petroff OA, Graham GD, Blamire AM, et al. Spectroscopic imaging of stroke in humans: histopathology correlates of spectral changes. *Neurology* 1992; 42:1349–54.

17 Berkelbach van der Sprenkel JW, Luyten PR, van Rijen PC, Tulleken CA, den Hollander JA. Cerebral lactate detected by regional proton magnetic resonance spectroscopy in a patient with cerebral infarction. *Stroke* 1988; 19:1556–60.

18 Bruhn H, Frahm J, Gyngell ML, Merboldt KD, Hanicke W, Sauter R. Cerebral metabolism in man after acute stroke: new observations using localized proton NMR spectroscopy. *Magn Reson Med* 1989; 9:126–31.

19 Henriksen O, Gideon P, Sperling B, Olsen TS, Jorgensen HS, Arlien-Soborg P. Cerebral lactate production and blood flow in acute stroke. *J Magn Reson Imaging* 1992; 2:511–17.

20 Rothman DL, Petroff OAC, Behar KL, Mattson RH. Localized 1H MRS measurements of GABA in human brain. *Proc Soc Magn Reson Med* 1993; 1:129 [abstract].

21 Rothman DL, Novotny EJ, Shulman GI, et al. 1H-[13C] NMR measurements of [4-13C]glutamate turnover in human brain. *Proc Natl Acad Sci U S A* 1992; 89:9603–6.

22 Rothman DL, Hanstock CC, Petroff OA, Novotny EJ, Prichard JW, Shulman RG. Localized 1H NMR spectra of glutamate in the human brain. *Magn Reson Med* 1992; 25:94–106.

23 Hanstock CC, Rothman DL, Shulman RG, Novotny EJ Jr, Petroff OA, Prichard JW. Measurement of ethanol in the human brain using NMR spectroscopy. *J Stud Alcohol* 1990; 51:104–7.

24 Frahm J, Bruhn H, Hanicke W, Merboldt KD, Mursch K, Markakis E. Localized proton NMR spectroscopy of brain tumors using short-echo time STEAM sequences. *J Comput Assist Tomogr* 1991; 15:915–22.

25 Frahm J, Michaelis T, Merboldt KD, et al. Localized NMR spectroscopy in vivo. Progress and problems. *NMR Biomed* 1989; 2:188–95.

26 Ross BD. Biochemical considerations in 1H spectroscopy. Glutamate and glutamine; myo-inositol and related metabolites. *NMR Biomed* 1991; 4:59–63.

27 Kreis R, Farrow N, Ross BD. Localized 1H NMR spectroscopy in patients with chronic hepatic encephalopathy. Analysis of changes in cerebral glutamine, choline and inositols. *NMR Biomed* 1991; 4:109–16.

28 Arnold DL, Shoubridge EA, Emrich J, Feindel W, Villemure JG. Early metabolic changes following chemotherapy of human gliomas in vivo demonstrated by phosphorus magnetic resonance spectroscopy. *Invest Radiol* 1989; 24:958–61.

29 Semmler W, Gademann G, Bachert Baumann P, Zabel HJ, Lorenz WJ, van Kaick G. Monitoring human tumor response to therapy by means of P-31 MR spectroscopy. *Radiology* 1988; 166:533–9.

30 Szigety SK, Allen PS, Huyser-Wierenga D, Urtasun RC. The effect of radiation on normal human CNS as detected by NMR spectroscopy. *Int J Radiat Oncol Biol Phys* 1993; 25:695–701.

31 Preul M, Villemure JG, Langleben A, Bahary J, Shenouda G, Arnold DL. Metabolic response to high dose tamoxifen for recurrent malignant glioma monitored with ^{1}H MR spectroscopic imaging. *Proc Soc Magn Reson Med* 1993; 2:1021 [abstract].

32 Jenkins BG, Koroshetz WJ, Beal MF, Rosen BR. Assessment of energy metabolism defects in Huntington's disease using ^{31}P and ^{1}H localized spectroscopy and functional MRI. Possible therapy with Coenzyme Q_{10}. *Proc Soc Magn Reson Med* 1993; 1:134 [abstract].

33 Xue M, Ng TC, Comair Y, Modic M. Decreased NAA and increased lactate in the activated motor cortex detected with localized spectroscopy guided with functional MRI. *Proc Soc Magn Reson Med* 1993; 1:59 (abstract).

34 Laxer KD, Hubesch B, Sappey-Marinier D, Weiner MW. Increased pH and inorganic phosphate in temporal seizure foci demonstrated by [31P]MRS. *Epilepsia* 1992; 33:618–23.

35 Hugg JW, Matson GB, Twieg DB, Maudsley AA, Sappey-Marinier D, Weiner MW. Phosphorus-31 MR spectroscopic imaging (MRSI) of normal and pathological human brains. *Magn Reson Imaging* 1992; 10:227–43.

36 Welch KM, Levine SR, Helpern JA. Pathophysiological correlates of cerebral ischemia the significance of cellular acid base shifts. *Funct Neurol* 1990; 5:21–31.

37 Gruetter R, Novotny EJ, Boulware SD, et al. Direct measurement of brain glucose concentrations in humans by 13C NMR spectroscopy. *Proc Natl Acad Sci U S A* 1992; **89**:1109–12.

38 Gruetter R, Novotny EJ, Boulware SD, et al. Non-invasive measurements of the cerebral steady-state glucose concentration and transport in humans by 13C nuclear magnetic resonance. *Adv Exp Med Biol* 1993; 331:35–40.

39 Rothman DL, Howseman AM, Graham GD, et al. Localized proton NMR observation of [3-13C]lactate in stroke after [1-13C]glucose infusion. *Magn Reson Med* 1991; **21**:302–7.

40 van der Knaap MS, van der Grond J, van Rijen PC, Faber JA, Valk J, Willemse K. Age-dependent changes in localized proton and phosphorus MR spectroscopy of the brain. *Radiology* 1990; **176**:509–15.

41 Huppi PS, Posse S, Lazeyras F, Burri R, Bossi E, Herschkowitz N. Magnetic resonance in preterm and term newborns: 1H-spectroscopy in developing human brain. *Pediatr Res* 1991; **30**:574–8.

42 Younkin DP, Delivoria-Papadopoulos M, Leonard JC, et al. Unique aspects of human newborn cerebral metabolism evaluated with phosphorus nuclear magnetic resonance spectroscopy. *Ann Neurol* 1984; **16**:581–6.

43 Hope PL, Reynolds EO. Investigation of cerebral energy metabolism in newborn infants by phosphorus nuclear magnetic resonance spectroscopy. *Clin Perinatol* 1985; **12**:261–75.

44 Hamilton PA, Hope PL, Cady EB, Delpy DT, Wyatt JS, Reynolds EO. Impaired energy metabolism in brains of newborn infants with increased cerebral echodensities. *Lancet* 1986; **1**:1242–6.

45 Delpy DT, Cope MC, Cady EB, et al. Cerebral monitoring in newborn infants by magnetic resonance and near infrared spectroscopy. *Scand J Clin Lab Invest Suppl* 1987; **188**:9–17.

46 Azzopardi D, Wyatt JS, Cady EB et al. Prognosis of newborn infants with hypoxic-ischemic brain injury assessed by phosphorus magnetic resonance spectroscopy. *Pediatr Res* 1989; **25**:445–51.

47 Kato T, Tokumaru A, O'uchi T, et al. Assessment of brain death in children by means of P-31 MR spectroscopy: preliminary note. Work in progress. *Radiology* 1991; **179**:95–9.

48 Roth SC, Edwards AD, Cady EB, et al. Relation between cerebral oxidative metabolism following birth asphyxia and neurodevelopmental outcome and brain growth at one year. *Dev Med Child Neurol* 1992; **34**:285–95.

49 Younkin D, Medoff-Cooper B, Guillet R, Sinwell T, Chance B, Delivoria-Papadopoulos M. In vivo 31P nuclear magnetic resonance measurement of chronic changes in cerebral metabolites following neonatal intraventricular hemorrhage. *Pediatrics* 1988; **82**: 331–6.

50 Moorcraft J, Bolas NM, Ives NK, et al. Global and depth resolved phosphorus magnetic resonance spectroscopy to predict outcome after birth asphyxia. *Arch Dis Child* 1991; **66**:1119–23.

51 Reynolds EO, McCormick DC, Roth SC, Edwards AD, Wyatt JS. New non-invasive methods for the investigation of cerebral oxidative metabolism and haemodynamics in newborn infants. *Ann Med* 1991; **23**:681–6.

52 Peden CJ, Cowan FM, Bryant DJ, et al. Proton MR spectroscopy of the brain in infants. *J Comput Assist Tomogr* 1990; **14**:886–94.

53 Welch KM, Levine SR, Martin G, Ordidge R, Vande Linde AM, Helpern JA. Magnetic resonance spectroscopy in cerebral ischemia. *Neurol Clin* 1992; **10**:1–29.

54 Levine SR, Welch KM, Helpern JA, et al. Prolonged deterioration of ischemic brain energy metabolism and acidosis associated with hyperglycemia: human cerebral infarction studied by serial 31P NMR spectroscopy. *Ann Neurol* 1988; **23**:416–18.

55 Levine SR, Helpern JA, Welch KM, et al. Human focal cerebral ischemia: evaluation of brain pH and energy metabolism with P-31 NMR spectroscopy. *Radiology* 1992; **185**:537–44.

56 Prichard JW. The ischemic penumbra in stroke: prospects for analysis by nuclear magnetic resonance spectroscopy. [Review]. *Research Publications – Association for Research in Nervous and Mental Disease* 1993; **71**:153–74.

57 Fenstermacher MJ, Narayana PA. Serial proton magnetic resonance spectroscopy of ischemic brain injury in humans. *Invest Radiol* 1990; **25**: 1034–9.

58 Carr CA, Guimaraes AR, Growdon JH, Gonzalez RG. Combining proton MRS and MRI morphometry increases accuracy in the diagnosis of Alzheimer's disease. *Proc Soc Magn Reson Med* 1993; **1**:233 [abstract].

59 Pettegrew JW, Withers G, Panchalingam K, Post JF. 31P nuclear magnetic resonance (NMR) spectroscopy of brain in aging and Alzheimer's disease. *J Neural Transm Suppl* 1987; **24**:261–8.

60 Bottomley PA, Cousins JP, Pendrey DL, et al. Alzheimer dementia: quantification of energy metabolism and mobile phosphoesters with P-31 NMR spectroscopy. *Radiology* 1992; **183**:695–9.

61 Pioro E, Preul MC, Antel JP, Arnold DL. 1H magnetic resonance spectroscopy demonstrates decreased N-acetylaspartate in the cerebrum of patients with amyotrophic lateral sclerosis. *Neurology* 1993; **43**:A257 [abstract].

62 Jenkins BG, Koroshetz WJ, Beal MF, Rosen BR. Evidence for an energy metabolism defect in Huntington's disease using localized proton spectroscopy. *Proc Soc Magn Reson Med* 1992; **1**:755 [abstract].

63 Bottomley PA, Hardy CJ, Cousins JP, Armstrong M, Wagle WA. AIDS dementia complex: brain high-energy phosphate metabolite deficits. *Radiology* 1990; **176**:407–11.

64 Deicken RF, Hubesch B, Jensen PC, et al. Alterations in brain phosphate metabolite concentrations in patients with human immunodeficiency virus infection. *Arch Neurol* 1991; **48**:203–9.

65 Menon DK, Baudouin CJ, Tomlinson D, Hoyle C. Proton MR spectroscopy and imaging of the brain in AIDS: evidence of neuronal loss in regions that appear normal with imaging. *J Comput Assist Tomogr* 1990; **14**:882–5.

66 Menon DK, Ainsworth JG, Cox IJ, et al. Proton MR spectroscopy of the brain in AIDS dementia complex. *J Comput Assist Tomogr* 1992; **16**:538–42.

67 Chong WK, Sweeney B, Wilkinson ID, et al. Proton spectroscopy of the brain in HIV infection: correlation with clinical, immunologic, and MR imaging findings. *Radiology* 1993; **188**:119–24.

68 Meyerhoff DJ, MacKay S, Bachman L, et al. Reduced brain N-acetylaspartate neuronal loss in cognitively impaired human immunodeficiency virus-seropositive individuals: *in vivo* 1H magnetic resonance spectroscopic imaging. *Neurology* 1993; **43**:509–15.

69 Jarvik JG, Lenkinski RE, Grossman RI, Gomori JM, Schnall MD, Frank I. Proton MR spectroscopy of HIV-infected patients: characterization of abnormalities with imaging and clinical correlation. *Radiology* 1993; **186**:739–44.

70 Arnold DL, Matthews PM, Francis G, Antel J. Proton magnetic resonance spectroscopy of human brain in vivo in the evaluation of multiple sclerosis: assessment of the load of disease. *Magn Reson Med* 1990; **14**;154–9.

71 Larsson HB, Christiansen P, Jensen M, Frederiksen J, Heltberg A, Olesen J, Henriksen O. Localized in vivo proton spectroscopy in the brain of patients with multiple sclerosis. *Magn Reson Med* 1991; **22**:23–31.

72 Richards TL. Proton MR spectroscopy in multiple sclerosis: value in establishing diagnosis, monitoring progression, and evaluating therapy. *AJR Am J Roentgenol* 1991; **157**:1073–8.

73 Van Hecke P, Marchal G, Johannik K, Demaerel P, Wilms G, Carton H, Baert AL. Human brain proton localized NMR spectroscopy in multiple sclerosis. *Magn Reson Med* 1991; **18**:199–206.

74 Hanefeld F, Bauer HJ, Christen HJ, Kruse B, Bruhn H, Frahm J. Multiple sclerosis in childhood: Report of 15 cases. *Brain Develop* 1991; **13**:410–16.

75 Miller DH, Austin SJ, Connelly AL, Youl BD, Gadian DG, McDonald WI. Proton magnetic resonance spectroscopy of an acute and chronic lesion in multiple sclerosis. *Lancet* 1991; **337**:58–9.

76 Bruhn H, Frahm J, Merboldt KD, et al. Multiple sclerosis in children: Cerebral metabolic alterations monitored by localized proton magnetic resonance spectroscopy *in vivo*. *Ann Neurol* 1992; **32**:140–50.

77 Grossman RI, Lenkinski RE, Ramer KN, Gonzalez-Scarano F, Cohen JA. MR proton spectroscopy in multiple sclerosis. *AJNR* 1992; **13**:1535–43.

78 Matthews PM, Francis G, Antel J, Arnold DL. Proton magnetic resonance spectroscopy for metabolic characterization of plaques in multiple sclerosis. *Neurology* 1991; **41**:1251–6 [published erratum appears in *Neurology* 1991; **41**:1828].

79 Davie CA, Hawkins CP, Barker GJ, et al. Serial proton MRS in demyelination. *Proc Soc Magn Reson Med* 1993; **1**:133 [abstract].

80 Posse S, Schuknecht B, Smith ME, van Zijl PC, Herschkowitz N, Moonen CT. Short echo time proton MR spectroscopic imaging. *J Comput Assist Tomogr* 1993; **17**:1–14.

81 Narayana PA, Wolinsky JS, Jackson EF, McCarthy M. Proton MR spectroscopy of gadolinium-enhanced multiple sclerosis plaques. *J Magn Reson Imaging* 1992; **2**:263–70.

82 Wolinsky JS, Narayana PA, Fenstermacher MJ. Proton magnetic resonance spectroscopy in multiple sclerosis. *Neurology* 1990; **40**:1764–9.

83 Husted C, Goodin DS, Maudsley AA, Tsuruda JS, Weiner MW. Biochemical alterations in multiple

sclerosis lesion and normal white matter detected in brain by *in vivo* ^{31}P and ^{1}H magnetic resonance spectroscopic imaging. *Neurology* 1993; **43**:A182 [abstract].

84 Cadoux-Hudson TA, Kermode A, Rajagopalan B, et al. Biochemical changes within a multiple sclerosis plaque in vivo. *J Neurol Neurosurg Psychiatry* 1991; **54**: 1004–6.

85 Newcombe J, Hawkins CP, Henderson CL. Histopathology of multiple sclerosis lesions detected by magnetic resonance imaging in unfixed postmortem central nervous system tissues. *Brain* 1991; **114**: 1013–23.

86 Arnold DL, Riess GT, Matthews PM, et al. Use of proton magnetic resonance spectroscopy for monitoring disease progression in multiple sclerosis. *Ann Neurol* 1994; **36**:76–82.

87 Adams CWA. The general pathology of multiple sclerosis: morphological and chemical aspects of the lesions. In: Hallpike JF, Adams CWA, Tourtellotte WW, eds. *Multiple Sclerosis. Pathology, Diagnosis, and Management.* Williams and Wilkins: Baltimore, 1983:203–40.

88 Arnold DL. Reversible reduction of N-acetylaspartate after acute central nervous system in damage. *Proc Soc Magn Reson Med* 1992; **1**:643 [abstract].

89 De Stefano N, Francis G, Antel JP, Arnold DL. Reversible decreases of N-acetylaspartate in the brain of patients with relapsing remitting multiple sclerosis. *Proc Soc Magn Reson Med* 1993; **3**:1139 [abstract].

90 Arnold DL, Shoubridge EA, Feindel W, Villemure JG. Metabolic changes in cerebral gliomas within hours of treatment with intra-arterial BCNU demonstrated by phosphorus magnetic resonance spectroscopy. *Can J Neurol Sci* 1987; **14**:570–5.

91 Anonymous. Tumor assessment and response to therapy studied by magnetic resonance spectroscopy. Proceedings of the 17th L. H. Gray Conference. Canterbury, 13–16 April 1992. *NMR Biomed* 1992; **5**:215–328.

92 Bruhn H, Frahm J, Gyngell ML, et al. Noninvasive differentiation of tumors with use of localized H-1 MR spectroscopy in vivo: initial experience in patients with cerebral tumors. *Radiology* 1989; **172**:541–8.

93 Gill SS, Thomas DG, Van Bruggen N, et al. Proton MR spectroscopy of intracranial tumours: in vivo and in vitro studies. *J Comput Assist Tomogr* 1990; **14**:497–504.

94 Demaerel P, Johannik K, Van Hecke P, et al. Localized 1H NMR spectroscopy in fifty cases of newly diagnosed intracranial tumors. *J Comput Assist Tomogr* 1991; **15**:67–76.

95 Henriksen O, Wieslander S, Gjerris F, Jensen KM. In vivo 1H-spectroscopy of human intracranial tumors at 1.5 tesla. Preliminary experience at a clinical installation. *Acta Radiol* 1991; **32**:95–9.

96 Kugel H, Heindel W, Ernestus RI, Bunke J, du Mesnil R, Friedmann G. Human brain tumors: spectral patterns detected with localized H-1 MR spectroscopy. *Radiology* 1992; **183**:701–9.

97 Glickson JD. Clinical NMR spectroscopy of tumors. Current status and future directions. *Invest Radiol* 1989; **24**:1011–16.

98 Ott D, Hennig J, Ernst T. Human brain tumors: assessment with in vivo proton MR spectroscopy. *Radiology* 1993; **186**:745–52.

99 Fulham MJ, Bizzi A, Dietz MJ, et at. Mapping of brain tumor metabolites with proton MR spectroscopic imaging: clinical relevance. *Radiology* 1992; **185**: 675–86.

100 Segebarth CM, Baleriaux DF, Luyten PR, den Hollander JA. Detection of metabolic heterogeneity of human intracranial tumors in vivo by 1H NMR spectroscopic imaging. *Magn Reson Med* 1990; **13**: 62–76.

101 Preul M, Collins D, Ethier R, Feindel W, Arnold DL. Classification of major intracranial tumor types using ^{1}H MR spectroscopic imaging and feature space for spectral pattern recognition. *Proc Soc Magn Reson Med* 1993; **1**:64 [abstract].

102 Alger JR, Frank JA, Bizzi A, et al. Metabolism of human gliomas: assessment with H-1 MR spectroscopy and F-18 fluorodeoxyglucose PET [see comments]. *Radiology* 1990; **177**:633–41.

103 Sotak CH, Alger JR. A pitfall associated with lactate detection using stimulated-echo proton spectroscopy. *Magn Reson Med* 1991; **17**:533–8.

104 Herholz K, Heindel W, Luyten PR, et al. In vivo imaging of glucose consumption and lactate concen-tration in human gliomas. *Ann Neurol* 1992; **31**:319–27.

105 den Hollander JA, Luyten PR, Marien AJ, et al. Potentials of quantitative image-localized human 31P nuclear magnetic resonance spectroscopy in the clinical evaluation of intracranial tumors. *Magn Reson Q* 1989; **5**:152–68.

106 Twieg DB, Meyerhoff DJ, Hubesch B, et al. Phosphorus-31 magnetic resonance spectroscopy in humans by spectroscopic imaging: localized

spectroscopy and metabolite imaging. *Magn Reson Med* 1989; **12**:291–305.
107. Hubesch B, Sappey-Marinier D, Roth K, Meyerhoff DJ, Matson GB, Weiner MW. P-31 MR spectroscopy of normal human brain and brain tumors. *Radiology* 1990; **174**:401–9.
108. Sutton LN, Lenkinski RE, Cohen BH, Packer RJ, Zimmerman RA. Localized 31P magnetic resonance spectroscopy of large pediatric brain tumors. *J Neurosurg* 1990; **72**:65–70.
109. Arnold DL, Emrich JF, Shoubridge EA, Villemure JG, Feindel W. Characterization of astrocytomas, meningiomas, and pituitary adenomas by phosphorus magnetic resonance spectroscopy. *J Neurosurg* 1991; **74**:447–53.
110. Heindel W, Bunke J, Glathe S, Steinbrich W, Mollevanger L. Combined 1H-MR imaging and localized 31P-spectroscopy of intracranial tumors in 43 patients. *J Comput Assist Tomogr* 1988; **12**:907–16.
111. Cadoux-Hudson TA, Blackledge MJ, Rajagopalan B, Taylor DJ, Radda GK. Human primary brain tumour metabolism in vivo: a phosphorus magnetic resonance spectroscopy study. *Br J Cancer* 1989; **60**:430–6.
112. Heiss WD, Heindel W, Herholz K, et al. Positron emission tomography of fluorine-18-deoxyglucose and image-guided phosphorus-31 magnetic resonance spectroscopy in brain tumors. *J Nucl Med* 1990; **31**:302–10.
113. Younkin DP, Delivoria-Papadopoulos M, Maris J, Donlon E, Clancy R, Chance B. Cerebral metabolic effects of neonatal seizures measured with in vivo ^{31}P NMR spectroscopy. *Ann Neurol* 1986; **20**:513–19.
114. Matthews PM, Andermann F, Arnold DL. A proton magnetic resonance spectroscopy study of focal epilepsy in humans. *Neurology* 1990; **40**:985–9.
115. Weiner MW, Hetherington H, Hubesch B, et al. Clinical magnetic resonance spectroscopy of brain, heart, liver, kidney, and cancer. A quantitative approach. *NMR Biomed* 1989; **2**:290–7.
116. Connelly A, Gadian DG, Jackson GD, et al. ^{1}H spectroscopy in the investigation of intractable temporal lobe epilepsy. *Proc Soc Magn Reson Med* 1992; **1**:234 [abstract].
117. Cendes F, Andermann F, Preul MC, Arnold DL. Lateralization of temporal lobe epilepsy based on regional metabolic abnormalities in proton magnetic resonance spectroscopic images. *Ann Neurol* 1994; **35**:211–16.
118. Hugg JW, Laxer KD, Matson GB, Maudsley AA, Weiner MW. Neuron loss localizes human focal epilepsy by *in vivo* proton MR spectroscopic imaging. *Ann Neurol* 1993; (in press)
119. Ng TC, Comair Y, Xue M, et al. Proton chemical shift imaging for the presurgical localization of temporal lobe epilepsy. *Proc Soc Magn Reson Med* 1993; **1**:428 [abstract].
120. Gadian DG, Connelly A, Cross JH, Jackson GD, Isaacs EB, Vargha-Khadem F. Relationship of cognitive dysfunction to ^{1}H MRS assessment of temporal lobe epilepsy. *Proc Soc Magn Reson Med* 1993; **1**:430 [abstract].
121. Hugg JW, Laxer KD, Matson GB, Maudsley AA, Husted CA, Weinder MW. Lateralization of human focal epilepsy by ^{31}P magnetic resonance spectroscopic imaging. *Neurology* 1992; **42**:2011–18.
122. Eleff SM, Barker PB, Blackband SJ, et al. Phosphorus magnetic resonance spectroscopy of patients with mitochondrial cytopathies demonstrates decreased levels of brain phosphocreatine. *Ann Neurol* 1990; **27**:626–30.
123. Cortelli P, Montagna P, Avoni P, et al. Leber's hereditary optic neuropathy: Genetic, biochemical, and phosphorus magnetic resonance spectroscopy study in an Italian family. *Neurology* 1991; **41**:1211–15.
124. Matthews PM, Berkovic SF, Shoubridge EA, et al. In vivo magnetic resonance spectroscopy of brain and muscle in type of mitochondrial encephalomyopathy (MERRF). *Ann Neurol* 1991; **29**:435–8.
125. Matthews PM, Andermann F, Silver K, Karpati G, Arnold DL. Proton MR spectroscopic characterization of differences in regional brain metabolic abnormalities in mitochondrial encephalomyopathies. *Neurology* 1993; **43**:2484–90.
126. Detre JA, Wang ZY, Bogdan AR, et al. Regional variation in brain lactate in Leigh syndrome by localized 1H magnetic resonance spectroscopy. *Ann Neurol* 1991; **29**:218–21.
127. Sylvain M, Mitchell GA, Shevell MI, et al. Muscle and brain magnetic resonance spectroscopy and imaging in children with Leigh's syndrome associated with cytochrome *c* oxidase deficiency: Dependence of findings on clinical status. *Ann Neurol* 1993; **34**:464 [abstract].
128. Kreis R, Ross BD. Cerebral metabolic disturbances in patients with subacute and chronic diabetes mellitus: detection with proton MR spectroscopy. *Radiology* 1992; **184**:123–30.
129. Heindel W, Kugel H, Roth B. Noninvasive detection of increased glycine content by proton MR spectroscopy in the brains of two infants with nonketotic

hyperglycinemia. *AJNR* 1993; **14**:629–35.

130 Shevell MI, Matthews PM, Scriver CR, et al. Cerebral dysgenesis and lactic acidemia: An MRI/MRS phenotype associated with pyruvate dehydrogenase deficiency. *Pediatrics* 1994; **10**:228–32.

131 Austin SJ, Connelly A, Gadian DG, Benton JS, Brett EM. Localized 1H NMR spectroscopy in Canavan's disease: a report of two cases. *Magn Reson Med* 1991; **19**:439–45.

132 Grodd W, Krageloh-Mann I, Petersen D, Trefz FK, Harzer K. In vivo assessment of N-acetylaspartate in brain in spongy degeneration (Canavan's disease) by proton spectroscopy. *Lancet* 1990; **336**:437–8 [letter].

133 Marks HG, Caro PA, Wang ZY, et al. Use of computed tomography, magnetic resonance imaging, and localized 1H magnetic resonance spectroscopy in Canavan's disease: a case report. *Ann Neurol* 1991; **30**:106–10.

134 Kreis R, Ross BD, Farrow NA, Ackerman Z. Metabolic disorders of the brain in chronic hepatic encephalopathy detected with H-1 MR spectroscopy. *Radiology* 1992; **182**:19–27.

135 Bruhn H, Kruse B, Korenke GC, et al. Proton NMR spectroscopy of cerebral metabolic alterations in infantile peroxisomal disorders. *J Comput Assist Tomogr* 1992; **16**:335–44.

Practical aspects of localized *in vivo* ^{1}H NMR spectroscopy and spectroscopy imaging of the human brain

Jan A den Hollander, Jan Willem C van der Veen, Peter R Luyten

INTRODUCTION

The first high resolution ^{1}H NMR spectroscopy studies of the animal brain were performed in 1983.[1] Those studies demonstrated that ^{1}H NMR spectroscopy allows *in vivo* observation of a number of brain compounds, including different aminoacids, (phospho)creatine, trimethylamine resonances, *myo*inositol, and lactate. Those initial studies have given rise to a large number of studies in which ^{1}H NMR spectroscopy has been applied to examine different aspects of brain metabolism, under various perturbations. It appears that a wealth of information can be obtained about brain metabolism and physiology by ^{1}H NMR spectroscopy.

Before ^{1}H NMR spectroscopy can be applied successfully to the animal brain, a number of problems have to be faced. Even though the intrinsic sensitivity of ^{1}H NMR spectroscopy is higher than that of ^{31}P NMR spectroscopy for the same number of spins, there are complicating factors that must be addressed in order to obtain spectra of adequate quality. These complicating factors have to do with the presence in the *in vivo* ^{1}H NMR spectrum of two very intense resonances from tissue water and lipids. Thus, any technique used in ^{1}H NMR spectroscopy of the brain has somehow to reduce the intensity of these intense signals in order to uncover signals from the low concentration compounds that are usually of interest. In addition to this problem, the chemical shift range in ^{1}H NMR spectroscopy is small, which leads to severe overlap between resonances originating from different compounds.

In animal studies, these problems are alleviated to some extent by the use of very high magnetic field strengths (up to 9.5T) and sometimes by using invasive techniques to remove overlying tissue so as to reduce interference of intense lipid signals. However, to apply ^{1}H NMR spectroscopy to the human brain, one has to rely entirely on non-invasive techniques.

Notwithstanding these difficulties, the first single voxel ^{1}H NMR spectra of the human brain were obtained only a few years after the first animal experiments.[2–4] Although those early studies did demonstrate the feasibility of *in vivo* ^{1}H NMR spectroscopy of the human brain, they did not lead to a large number of patient studies. The reason was that those techniques were difficult to optimize, and often led to spectra with unacceptable contamination by lipid and water signals. More recently, these problems have been overcome by the use of localization sequences based upon spin-echo techniques, and as a result many ^{1}H NMR studies of the human brain have been conducted over the last few years.

The first successful application of ^{1}H NMR spectroscopic imaging to the animal brain[5] provided the impetus to explore the possibilities for ^{1}H NMR spectroscopic imaging as an extension of the single voxel techniques. ^{1}H NMR spectroscopic imaging of the human brain has a clear advantage over single volume measurements, in that it provides an overview of one or more slices, allows direct comparison between spectra from the affected part of the brain and the contralateral side, and allows presentation of the data as chemical shift (or metabolite) images. However, to obtain high-quality ^{1}H NMR spectroscopic images of the human brain, it is necessary to achieve adequate shimming, as well as water suppression over an entire slice rather than a single volume. Adequate shimming, localization, and water suppression techniques were essential for ^{1}H NMR spectroscopic imaging to be sufficiently developed for patient studies, and even now many fewer spectroscopic imaging studies are being published than studies using single volume techniques. However, with the advent of new NMR scanners, with improved gradient and rf coils, with the incorporation of fast multislice ^{1}H NMR spectroscopic imaging techniques and fully automated preparation steps, it is to be expected that we will witness a shift towards spectroscopic imaging methods in the near future.

Localization techniques

There are a number of localization techniques for obtaining ^{1}H NMR spectra of the human brain. Among the first approaches explored were methods based upon outer volume suppression.[6,7] However, for use with whole head coils, localization sequences that are based upon spin-echo techniques have proven to be more satisfactory. These sequences work by combining three slice-selective excitation and refocusing pulses, combined in such a way that a spatially selected NMR signal is obtained from the volume defined by the intersection of the three selected slices. By avoiding unwanted excitation of outer volume spins as much as possible, the suppression of those signals becomes much less critical than in sequences that are based on outer volume suppression, or on addition–subtraction schemes such as image selective *in vivo* spectroscopy (ISIS). For a certain coil geometry, these spin-echo sequences introduce a fixed ratio between excited spins from the outer volume, which have to be suppressed, and spins in the region of interest.

The stimulated echo sequence[8] for localized NMR spectroscopy[9–12] is composed of three orthogonal slice-selective 90° pulses. A first pulse excites the spins in a slice. During a time, STE/2, the spins dephase in the transverse plane by a magnetic field gradient. At time STE/2, a second slice-selective 90° pulse is applied, distributing the spins in a plane along the longitudinal direction and perpendicular to the axis along which the second 90° pulse was applied. Half of the signal is 'stored' in the longitudinal direction, while the other half will dephase in the presence of a dephasing gradient applied during a time, Tm. After this time Tm, a third slice-selective 90° pulse will be applied, followed by a rephasing gradient, resulting in a stimulated echo signal at a time STE/2 after this final 90° pulse. This signal contains half the original magnetization at the beginning of the sequence. The volume selection for all three orthogonal directions is performed by the same pulses, resulting in a perfectly symmetrical volume of interest. Moreover, all slice-selection pulses are 90° pulses, which results in relative insensitivity for B_1 inhomogeneity and well-defined slice profiles. However, the loss of half the signal intensity is a drawback associated with this approach.

The double spin-echo or point-resolved spectroscopy (PRESS) method[13,14] uses a 90° excitation pulse followed by two refocusing 180° pulses (Fig. 9.1). The first excitation pulse will only affect spins in a particular slice. A 180° pulse applied in the presence of a magnetic field gradient that is orthogonal to the first selection gradient will refocus spins only in the second slice, and will dephase spins outside that slide. Finally, a third pulse combined with a magnetic field gradient orthogonal to the first two gradient directions will result in a spin-echo signal from the intersection of all three slices. This signal originates from a rectangular box, the dimensions of which are determined by the width of the three selected slices.

In principle, the double spin-echo sequence preserves all initial magnetization. For uncoupled spins, the signal strength is attenuated by T_2 and saturation effects only. However, for coupled spins, J-modulation effects tend to dephase resonances within a multiplet, with concomitant loss of signal intensity at longer echo times. In order to optimize the signal obtained from the volume of interest, the sequence requires optimized 180° refocusing pulses to guarantee a well-defined profile of the selected box. Numerically optimized refocusing pulses have been developed[15] to optimize the slice profile in spin-echo imaging. For quantitative purposes, the exact volume-size definition requires an adequate description of the shape and the width of the selected volume. Often the volume-size is defined by the full width at half maximum of the slice profile. In three-dimensional localized spectroscopy, this may lead to deviations from the desired volume-size. Hence, comparing signal-to-noise or quantitative data obtained by different techniques of volume selection becomes meaningless unless a careful calibration of the exact volume-sizes has been made.

Magnetic field homogeneity

Optimization of the magnetic field homogeneity is an absolute requirement for obtaining high-quality ^{1}H NMR spectra of the human brain. The magnetic field of whole body magnets are optimized at installation time over a large spherical volume. However, magnetic susceptibility will affect the magnetic field distribution when the subject is

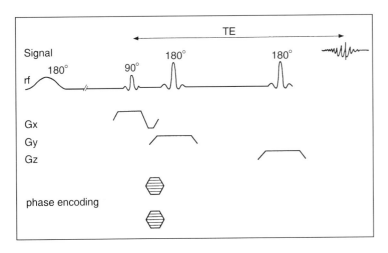

Fig. 9.1. The double echo (PRESS) sequence. By combining one slice-selective 90° pulse with two slice-selective 180° refocusing pulses, a spin echo is generated from a volume defined by the intersection of the three slices. Two-dimensional phase encoding is included, to obtain a spectroscopic image over the selected volume.

placed in the magnet. These effects need to be corrected as much as possible before starting the spectroscopy examination, but with the limits on total examination time when obtaining spectra from human subjects, one needs a fast and reliable method. It is common practice to monitor the ^{1}H NMR signal of the water resonance from the region of interest, and then optimize the magnetic field on the basis of that signal. The technique that has worked out satisfactorily in practice uses the imaging head coil for sending and receiving of the ^{1}H NMR signal. By using the PRESS sequence,[13,14] it has been possible to obtain reproducible localized ^{1}H NMR water signals from the brain in single acquisitions. The PRESS sequence has the advantage of being relatively insensitive to motion, especially for short echo times (≤ 36 ms).

Thus the method used involves selecting a large volume of interest (VOI) centered over the area of interest and monitoring the ^{1}H NMR water signal from that volume using single shot localization.[16] The time-domain signal obtained over that volume is optimized by adjusting the linear and the second-order field corrections. In our approach we have used an automatic shimming algorithm that maximizes the modulus of the time-domain signal,[17] but once a reproducible ^{1}H NMR signal of the myocardium can be obtained, it should be possible to use other shimming strategies as well. At the magnetic field strength of clinical MR instruments (1.5T) the linewidth of the water signal can be optimized to about 0.05 ppm over a slice through the brain, using the procedure outlined here. Recently a method based upon a field mapping method has been applied successfully for shimming of the human brain.[18,19] An NMR imaging approach is used to measure the magnetic field variations across the head and, from that field map, the required shim corrections can be calculated directly.

Water suppression

Many different techniques have been proposed in high resolution NMR spectroscopy to suppress the very intense water resonance in ^{1}H NMR spectra.[20] In principle, all these techniques may be applicable for localized *in vivo* studies. The one important complicating factor is that some water suppression techniques may compromise the requirements for a proper localization with respect to outer volume suppression, or alternatively the localization techniques may compromise the water suppression. A number of water suppression techniques are based upon minimizing excitation at the water resonance while maximizing excitation of resonances around 1.0–3.0 ppm, which includes lipid resonances from subcutaneous fat and bone marrow. These sequences may introduce an undesired interference between localization and water suppression. For example, in ^{1}H NMR brain studies, this would result in excitation of all the lipid signals within the sensitive region of the NMR coil, resulting in an unfavorable ratio between the amount of signal that should be suppressed and the signal from the volume of interest. Excitation pulses that are both spatially selective and chemical-shift selective avoid this problem, in that these pulses do not compromise outer volume suppression as a result of water suppression.[21]

Another important requirement for water suppression is that there should be a uniform excitation of the spectral region of interest. Techniques that meet this requirement and do not interfere with the localization requirements include selective excitations and selective inversions of the water signal.[22-24] Selective excitations consist of a series of 90° pulses at the water frequency followed by dephasing gradients to dephase the water resonance. Selective inversions of the water signal can be used to minimize the longitudinal magnetization by a selective inversion pulse followed by a suitable inversion delay.[25] Excitation of the volume of interest starts at the zero crossing point of the water resonance, resulting in a suppressed water signal in the localized NMR spectrum.

Phase encoding and magnetic resonance spectroscopic imaging

The most obvious limitation of the single volume localization techniques based upon slice-selection pulses is that the spectroscopic information originates from only one (or sometimes two) selected volumes. For some applications this approach is entirely appropriate; in particular it is appropriate for applications involving global metabolic effects in the brain. However, in many instances one is interested in the examination of focal disorders of the brain, such as multiple sclerosis (MS), brain tumors, or stroke. When examining small focal lesions one often wants to obtain spectra from small volumes of interest, for which the single volume approach may not be the most suitable. More importantly, there may be differences between, for example the rim of the lesion and its center. In these circumstances, techniques are needed that generate spectra from multiple volumes, with the highest spatial resolution possible.

Usually, localization of the volume of interest is done on the basis of a NMR scout image. Particularly for ^{1}H NMR, both spectroscopy and imaging can be performed using exactly the same equipment, including the rf coil. Therefore, the patient can be imaged and spectroscopically examined without any repositioning. Since the NMR imaging measurement makes use of the same gradient coils as volume selection for spectroscopy, the rectangular volume of interest can be positioned very accurately. The correspondence between imaging and spectroscopy becomes even more persuasive when using phase-encoding gradients for localization. Spectroscopic imaging sequences include phase-encoding gradients, similar to standard NMR imaging sequences.[26,27] Thus, these measurements lead to the acquisition of spectra from a large number of voxels in a single experimental session.

^{1}H NMR spectroscopic imaging shares all the problems of single volume techniques, in that it is necessary to suppress the intense water and lipid signals. The magnetic field homogeneity over the whole head is such that it is not possible to achieve proper water suppression by using selective excitation or refocusing pulses. For this reason, ^{1}H NMR spectroscopic imaging is only possible over either a selected volume or a selected slice through the head. One approach that is often used for ^{1}H NMR spectroscopic imaging of the human brain selects a slice-shaped box of interest by using a volume-selection technique as described above, to exclude lipid signals from the skull as needed.[28,29] The magnetic field homogeneity is then optimized over that selected volume of interest. Subsequently the water suppression is adjusted, and ^{1}H MRS imaging is measured over that selected volume.

Recently, there has been an effort to perform multiple-slice ^{1}H MRSI in order to obtain spectroscopic data over a large region of the human brain in a single protocol.[30,31] This requires a slice-selection approach, since volume-selection methods will affect regions outside the selected volume. Therefore, suppression of lipids from the skull needs to be performed by outer volume suppression rather than volume selection. Even though lipid suppression is less than ideal, this approach has been shown to provide satisfactory lipid suppression.

Single volume ^{1}H NMR spectroscopy of the human brain

A series of single volume spectra of a large subcortical region of the human brain, above the ventricles, is shown in Fig. 9.2. A 1.5T clinical MRI scanner was used, with a regular quadrature send–receive head coil. The spectra were obtained by using the double echo sequence, using different total echo times, TE = 34 ms, 68 ms, 136 ms, and 272 ms. An excitation pulse was used for water suppression, while the residual water signal was eliminated by removing the signal components in a 1 ppm window around the water resonance by means of a linear fitting technique in the time domain, Hankel singular value decomposition (HSVD).[32] The spectra show a number of resonances typically found in the ^{1}H NMR spectrum of the normal human brain, in particular the acetyl group of N-acetyl aspartate (NAA) at 2.02 ppm, the methylene resonances of the aspartyl residue of NAA at 2.56 ppm, and the methyl groups of creatine and phosphocreatine (Cr) at 3.0 ppm. At 3.2 ppm, the methyl group of choline residues is observed (Cho); this is the sum of different choline-containing compounds that remain unresolved at 1.5T. Furthermore, the CH$_2$ resonances of glutamate and glutamine show up in the spectrum obtained at TE = 34 ms, as well as a resonance which has been assigned to *myo*inositol. The long TE spectra also show a small doublet from lactic acid; the doublet is positive in the TE = 272 spectrum, and negative in the TE = 136 spectrum, as expected for its J-modulation behavior. The lactic acid signals are easily observed in the spectra obtained at long TE, but are not so easily recognized in the shorter TE spectra. This demonstrates the need for obtaining spectra at different TE, since different signals may be observed at the different echo times, because of J-modulation behavior and differences in T$_2$.

The spectra shown in Fig. 9.2 demonstrate the excellent spectral resolution that can be attained on a clinical scanner. However, many spin systems are strongly coupled at this relatively low magnetic field, leading to complicated coupling patterns, such as for the glutamate–glutamine resonances in the range of 2.0–2.4 ppm. As a result, the ^{1}H NMR spectrum of the human brain is very crowded, with many overlapping resonances. Indeed, in these high signal-to-noise spectra, a multitude of small overlapping signals are observed throughout the spectrum between 2 and 4 ppm. This is true for both the short TE and the long TE spectra. Careful inspection shows that the peak positions of these small resonances differ for the different echo times; this demonstrates that these signals actually consist of overlapping signals with different J-modulation patterns. Therefore, the precise assignment of all these small resonances remains difficult. In particular, this series of spectra demonstrates that the signal that has tentatively been assigned to *myo*inositol occurs in a very crowded part of the spectrum, and that several components with different J-modulation behavior may contribute to that signal.

At higher magnetic fields, these spin systems reduce to simpler coupling patterns, and may therefore be better resolved than at 1.5T. In addition, spectral editing techniques and two-dimensional spectroscopy provides an alternative way to resolve and assign these overlapping signals. In particular, a multiple quantum filter has been applied to separate the signal of lactate from overlapping lipid signals.[33,34] This approach has been applied to obtain lactate images, without the need for additional lipid suppression. The importance of the multiple quantum filter approach is not only that it allows one to obtain lactate signals

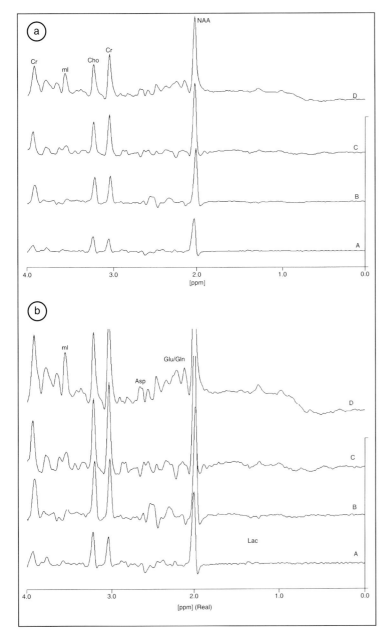

Fig. 9.2. (A) Localized ^{1}H NMR spectra from a large subcortical region of the human brain (75 × 60 × 25 mm), obtained by the double echo sequence from a normal volunteer at 1.5T. The same volume was measured for four different echo times: 272 ms (A), 136 ms (B), 68 ms (C), and 34 ms (D). The spectra show intense signals from the acyl group of N-acetyl aspartate (NAA), (phospho)creatine (Cr), choline compounds (Cho), and myoinositol (mI).

(B) The spectra of Fig. 9.2A were plotted on a higher scale, to show the smaller resonances. Resonances are observed throughout the region between 2 and 4 ppm. The positions of glutamate–glutamine (Glu–Gln), the aspartyl residue of NAA (Asp), and myoinositol are indicated. The long echo time spectra A and B also show the doublet of lactate (Lac), which is positive in the TE = 272 ms spectrum (A), and negative at TE = 136 ms (B).

even in the presence of intense lipid signals, but also that it provides an additional 'signature' for lactate, which may help in its assignment. This may help avoid misassignment of lipid signals to lactic acid, which may easily happen.

Another compound that is of great interest is γ-aminobutyric acid (GABA), a putative neurotransmitter. Although its concentration is sufficiently high to allow its observation in the ^{1}H NMR spectrum, its signals are completely obscured by the NAA and glutamate–glutamine resonances. Recently, it has been possible to resolve the signals from GABA in ^{1}H NMR spectra of the human brain by using a spectral editing technique.[35] In this editing technique, the γ-protons are either irradiated or not in successive scans; as a result the directly bonded protons show up selectively in the difference spectrum. The approach has been used to monitor changes in GABA level in the brains of patients who were receiving vigabatrin medication.

For the interpretation of the ^{1}H NMR spectrum of the human brain, it is necessary to understand the biochemical role of the different compounds. It is unfortunate that the function of some of the main compounds observed in the ^{1}H NMR spectrum of the human brain is not very well understood. This is particularly the case for the N-acetyl aspartate (NAA) resonance, which happens to be the largest signal in the spectrum. It has been suggested that NAA is concentrated in the neurons, but its particular role is not yet very clear.[36] The localization of NAA in neurons and axons is supported by high-resolution studies of brain extracts that showed that the NAA:Cho ratio is higher in grey matter than it is in white matter. The best evidence for the localization of NAA come from studies on cultured cells. Irrespective of the precise biochemical role of NAA, a relative decrease of this compound is commonly considered to be an indication for neuronal loss. Indeed, a relative decrease of NAA has been found in various disorders of the brain, including brain tumors, cerebral ischemia, multiple sclerosis, Alzheimer's disease.

The choline compounds contributing to the choline signal in the ^{1}H NMR spectrum of the human brain are intermediates of phospholipid metabolism. It is thought that the Cho signal in the ^{1}H NMR spectrum arises mainly from small molecules, and that the ^{1}H NMR spectra of choline residues of phosphatidyl cholines incorporated in membranes are broadened such that they do not contribute to the *in vivo* ^{1}H NMR spectrum of the human brain. The Cho signal in the ^{1}H NMR spectrum of the brain is expected to be related to phospholipid metabolism, and changes in this signal may indicate changes in phospholipid metabolism.

The role of creatine and phosphocreatine is much better understood. These compounds play a role in high-energy metabolism. Creatine is phosphorylated to phosphocreatine through the creatine kinase reaction. However, because we are unable to resolve the creatine and phosphocreatine signals in the spectrum of the human brain it is not possible to determine the degree of phosphorylation of the total creatine pool by ^{1}H NMR spectroscopy alone. Thus, a decrease in phosphocreatine due to a perturbation of high-energy phosphate metabolism is not expected to have an immediate effect on the total creatine signal in the ^{1}H NMR spectrum. Only when the total creatine pool changes as a result of a prolonged perturbation is this expected to show up as a reduction of the ^{1}H NMR signal of total creatine. Both in stroke and in tumors, the Cr signal has been found to diminish compared to unaffected tissue.

A much more sensitive indicator of perturbations of energy metabolism is provided by the ^{1}H NMR signal of lactate. In the healthy brain the basal level of lactate can be barely observed by *in vivo* ^{1}H NMR spectroscopy, as shown in Fig. 9.2B. However, lactate is expected to increase in a number of conditions, either because of increased glycolysis or because of a diminished availability of oxygen, or both. Regional increases of lactate have been observed, both in intracranial tumors and in stroke patients. However, in healthy subjects a small increase in cerebral lactate can also be observed by ^{1}H NMR spectroscopy under suitable perturbations. Lactate has been observed to increase during voluntary hyperventilation,[37] presumably as a response to a change in cerebral pH. Also, lactate has been observed to increase in normal subjects, owing to visual[38] and auditory[39] stimulation.

The *myo*inositol resonance, which can be observed in the short TE ^{1}H NMR spectrum, has been observed to increase in Alzheimer's disease and in hepatic encephalopathy. The reason for this increase remains to be investigated.

Localized ^{1}H NMR spectroscopic imaging of the human brain

The result of an ^{1}H NMR spectroscopic imaging examination of a normal volunteer (a 30-year-old female volunteer), performed on a 1.5 tesla MRI scanner, is shown in Fig. 9.3. A large volume of interest was selected, to exclude only part of skull but to include the whole brain. Volume size was 165 mm anterior–posterior (ap), 125 mm left–right (lr), and 15 mm caudal–cranial (cc). 32 × 32 phase-encoding steps were collected, one measurement for each phase-encoding step field of view (FOV) was 230 mm, leading to an in-plane resolution of 7 mm. With a slice width of 1.5 mm, this leads to a nominal voxel size of 0.75 ml. A reference data set without water suppression was measured, for retrospective correction of residual magnetic field inhomogeneity.[40,41] Echo time was 272 ms, and whole echoes were collected by starting the acquisition immediately following the second refocusing pulse of the PRESS sequence. Total measuring time was 34 minutes. Residual water was removed by a numerical approach based upon a linear time-domain fitting technique.[32] The data set was processed by appropriate filtering in the time domain, and Fourier transformation. The modulus spectrum was calculated to obtain the absorption mode spectrum. Chemical shift images were obtained by integration of individual peaks in the frequency domain spectra.

Figure 9.3A shows a turbo spin-echo MRI scan of the slice over which the ^{1}H NMR spectroscopic imaging data set was obtained (TE = 120 ms, TR = 2800 ms). Figure 9.3B shows the NAA image, the Cho image, the Cr image, and the ratio image of Cr:Cho. A suitable threshold was used in obtaining the Cr:Cho ratio image, in order to exclude contributions from outside the brain. The NAA image clearly depicts the ventricles, but outside the ventricles little variation in NAA intensity is observed; there is a slight contribution from lipid from the skull to the NAA image, but that does not interfere at all with its interpretation. However, the Cho and Cr images show more detail. In the Cho image we observe that periventricular white matter has a higher intensity than grey matter. The Cr image, on the other hand, shows a relative higher intensity for grey matter. This variation in Cr and Cho intensity is particularly clear in the Cr:Cho ratio image. In this ratio image, occipital grey matter in particular shows a higher intensity, as does cortical grey matter to a smaller extent. These variations in Cho:Cr ratio are further demonstrated in the selected spectra of Figure 9.3C. Spectra 1–4 were selected from occipital grey matter, and these spectra show that the Cr peak is higher than the Cho peak. Spectra 5–8 were selected from cortical grey matter, and show a Cho slightly higher than the Cr peak. Spectra 9–14 were selected from periventricular white matter, showing a Cho signal considerably higher than the Cr peak. These variations reflect different contributions of grey and white matter to the spectra selected from the different regions, and demonstrate the variation in signal intensity which can be observed in the normal human brain.

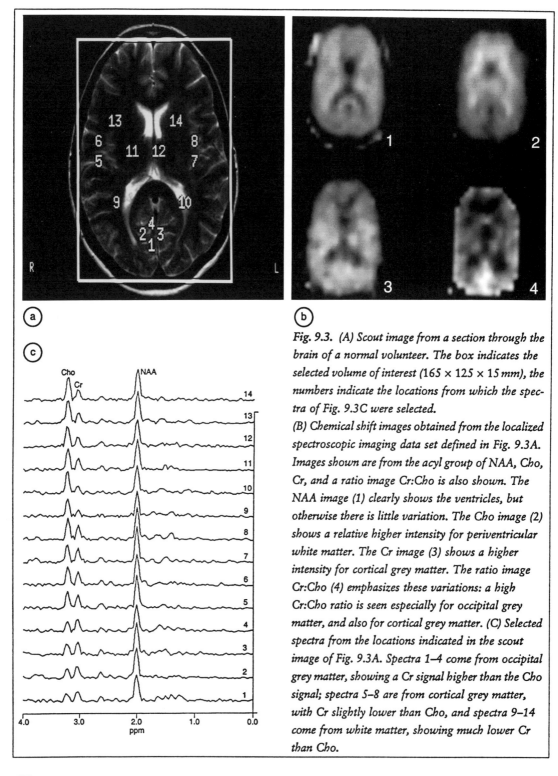

Fig. 9.3. (A) Scout image from a section through the brain of a normal volunteer. The box indicates the selected volume of interest (165 × 125 × 15 mm), the numbers indicate the locations from which the spectra of Fig. 9.3C were selected.

(B) Chemical shift images obtained from the localized spectroscopic imaging data set defined in Fig. 9.3A. Images shown are from the acyl group of NAA, Cho, Cr, and a ratio image Cr:Cho is also shown. The NAA image (1) clearly shows the ventricles, but otherwise there is little variation. The Cho image (2) shows a relative higher intensity for periventricular white matter. The Cr image (3) shows a higher intensity for cortical grey matter. The ratio image Cr:Cho (4) emphasizes these variations: a high Cr:Cho ratio is seen especially for occipital grey matter, and also for cortical grey matter. (C) Selected spectra from the locations indicated in the scout image of Fig. 9.3A. Spectra 1–4 come from occipital grey matter, showing a Cr signal higher than the Cho signal; spectra 5–8 are from cortical grey matter, with Cr slightly lower than Cho, and spectra 9–14 come from white matter, showing much lower Cr than Cho.

In different disease states, large variations have been found in the chemical shift images obtained by ^{1}H NMR spectroscopic imaging. In many focal disorders (brain tumors, stroke, multiple sclerosis) a void is observed in the NAA image at the site of the lesion, which is usually interpreted as being indicative of neuronal loss. A focal increase of lactate has been observed in stroke and in tumors. Cho is often seen to increase in tumors. However, these changes do not occur in exactly the same location; there can be considerable metabolic heterogeneity across the lesion.[28,29] This observation emphasizes the need for obtaining spectroscopic images of focal abnormalities of the brain. Indeed, such metabolic heterogeneity is not easily explored by single volume techniques, because of time limitations.

Conclusion

Since the first attempts to localize in vivo ^{1}H NMR spectroscopy of the human brain,[2–4] there has been impressive progress. About ten years ago, I (JAdH) heard statements that ^{1}H NMR spectroscopy of the human brain would never work in practice, because of the problems with intense water and lipid signals. Now, the ^{1}H NMR spectroscopy techniques have evolved to the point where obtaining localized spectra and spectroscopic images of the human brain is virtually routine and can in fact easily be performed by a trained technician. User-friendly interfaces have been developed, with graphics tools to enable quick and accurate selection of the volume of interest. Preparation steps such as rf optimization, frequency offset determination, localized magnetic field shimming, and adjustment of water suppression is now fully automated. Reconstruction of metabolite maps from spectroscopic imaging data sets has become much easier, owing to the use of water referencing;[41] because of this, the integration boundaries for the different metabolites remain the same always and can be incorporated in an automatic reconstruction algorithm.

Yet, further improvements are under way. For single volume ^{1}H NMR spectroscopy, the main limitation is the lack of spectral specificity – many interesting signals are obscured by larger resonances. The advent of spectral editing,[35] multiple quantum filters,[33,34] and two-dimension spectroscopy is expected to solve some of those limitations. Also, the use of high magnetic fields (4T or higher) is expected to help in this regard.

Part of the problem of spectroscopic imaging in its present implementation is that a single slice is examined, in a total examination time of close to 1 hour, despite all the automated optimization steps. This may not be a serious limitation in a research setting, but is hardly acceptable in a routine clinical environment. Recent results have explored multislice and multiecho approaches to ^{1}H NMR spectroscopic imaging.[30,31] The reduction in total examination time, and the increase of the amount of data obtained in a single examination constitute important steps towards broader acceptance of ^{1}H NMR spectroscopic imaging in the clinical setting.

REFERENCES

1 Behar KL, den Hollander JA, Stromski ME, et al. High-resolution ^{1}H nuclear magnetic resonance study of cerebral hypoxia. *Proc Natl Acad Sci U S A* 1983; 80: 4945–8.

2 Bottomley PA, Edelstein WA, Foster TH, Adams WA. In vivo solvent-suppressed localized hydrogen nuclear magnetic resonance spectroscopy: a window to metabolism? *Proc Natl Acad Sci U S A* 1985; 82:2148–52.

3 Luyten PR, den Hollander JA. Observation of

metabolites in the human brain by MR spectroscopy. *Radiology* 1986; **161**:795–8.

4 den Hollander JA, Luyten PR. Image-guided localized ^{1}H and ^{31}P NMR spectroscopy of humans. *Ann N Y Acad Sci* 1987; **508**:386–98.

5 de Graaf AA, Bovee WMMJ, Deutz NEP, Chamuleau RAFM. In vivo ^{1}H NMR procedure to determine several rat cerebral metabolite levels simultaneously, undisturbed by water and lipid signals. *Magn Reson Imaging* 1988; **6**:255–61.

6 Luyten PR, Mariën AJH, Sijtsma B, den Hollander JA. Solvent suppressed spatially resolved spectroscopy: an approach to high resolution NMR on a whole body MR system. *J Magn Reson* 1986; **67**:148–55.

7 Hanstock CC, Rothman DL, Jue T, Shulman RG. Volume-selected proton spectroscopy in the human brain. *J Magn Reson* 1988; **77**:583–8.

8 Hahn EL. Spin echoes. *Phys Rev* 1950; **80**:580–94.

9 Granot J. Selected volume excitation using stimulated echoes (VEST). Applications to spatially localized spectroscopy and Imaging. *J Magn Reson* 1986; **70**:488–92.

10 Frahm J, Merboldt KD, Hänicke W. Localized proton spectroscopy using stimulated echoes. *J Magn Reson* 1987; **72**:502–8.

11 van Zijl PCM, Moonen CTW, Alger JR, Cohen JS, Chesnick SA. High field localized proton spectroscopy in small volumes: greatly improved localization and shimming using shielded strong gradients. *Magn Reson Med* 1989; **10**:256–65.

12 Frahm J, Bruhn H, Gyngell ML, Merboldt KD, Hänicke W, Sauter R. Localized proton NMR spectroscopy in different regions of the human brain in vivo. Relaxation times and concentrations of cerebral metabolites. *Magn Reson Med* 1989; **11**:47–63.

13 Ordidge RJ, Bendall MR, Gordon RE, Connelly A. Volume selection for in-vivo spectroscopy. In: Govil G, Khetrapal CL, Saran A, (eds). *Magnetic Resonance in Biology and Medicine*. New Delhi: Tata-McGraw-Hill, 1985; 387–97.

14 Bottomley PA. Spatial localization in NMR spectroscopy *in vivo*. *Ann N Y Acad Sci* 1987; **508**:333–48.

15 Groen JP, Rongen P, in den Kleef J. Improved rf pulses for selective inversion and echo refocussing. *Proc Fifth Annual Meeting, Soc Magn Reson Med*; 1986; 1442–3.

16 Luyten PR, Mariën AJH, den Hollander JA. Acquisition and quantitation in proton spectroscopy. *NMR Biomed* 1991; **4**:64–9.

17 Vermeulen JWAH, den Hollander JA, Mariën AJH. Automatic shimming for localized human NMR spectroscopy. *Proc Eighth Annual Meeting, Soc Magn Reson Med* 1989; 626.

18 Webb PG, Macovski A. Rapid, fully automatic arbitrary-volume *in vivo* shimming. *Magn Reson Med* 1991; **20**:113–22.

19 Webb PG, Sailasuta N, Kohler SJ, Raidy T, Moats RA, Hurd RE. Automated single-voxel proton MRS: technical development and multisite verification. *Magn Reson Med* 1994; **31**:365–73.

20 Hore PJ. Solvent suppression in Fourier transform nuclear magnetic resonance. *J Magn Reson* 1983; **55**:283–300.

21 Meyer CH, Pauly JM, Macovski A, Nishimura DG. Simultaneous spatial and spectral selective excitation. *Magn Reson Med* 1990; **15**:287–304.

22 Haase A, Frahm J, Hänicke W, Matthaei D. ^{1}H NMR chemical shift selective (CHESS) imaging. *Phys Med Biol* 1985; **30**:341–4.

23 Moonen CTW, van Zijl PCM. Highly effective water suppression for *in vivo* proton NMR spectroscopy (DRYSTEAM). *J Magn Reson* 1990; **88**:28–41.

24 Griffey RH, Flamig DP. VAPOR for solvent-suppressed, short-echo, volume-localized proton spectroscopy. *J Magn Reson* 1990; **88**:161–6.

25 Patt SL, Sykes BD. T_1 water eliminated Fourier transform NMR spectroscopy. *J Chem Phys* 1972; **56**:3182–4.

26 Brown TR, Kincaid BM, Uğurbil K. NMR chemical shift imaging in three dimensions. *Proc Natl Acad Sci U S A* 1982; **79**:3523–6.

27 Maudsley AA. Sensitivity in Fourier imaging. *J Magn Reson* 1986; **68**:363–6.

28 Segebarth CM, Balériaux DF, Luyten PR, den Hollander JA. Detection of metabolic heterogeneity of human intracranial tumors in vivo by ^{1}H NMR spectroscopic imaging. *Magn Reson Med* 1990; **13**:62–76.

29 Luyten PR, Mariën AJH, Heindel W, et al. Metabolic imaging of patients with intracranial tumors: H-1 MR spectroscopic imaging and PET. *Radiology* 1990; **176**:791–9.

30 Duyn JH, Gillen J, Sobering G, van Zijl PCM, Moonen CTW. Multisection proton MR spectroscopic imaging of the brain. *Radiology* 1993; **188**:277–82.

31 Duyn JH, Moonen CTW. Fast proton spectroscopic imaging of human brain using multiple spin-echoes. *Magn Reson Med* 1993; **30**:409–14.

32 de Beer R, Michels F, van Ormondt D, van Tongeren BPO, Luyten PR, van Vroonhoven H. Reduced lipid

contamination in in vivo ^{1}H MRSI using time-domain fitting and neural network classification. *Magn Reson Imaging* 1993; **11**:1019–26.
33 Hurd RE, Freeman DM. Metabolite specific proton magnetic resonance imaging. *Proc Natl Acad Sci U S A* 1989; **86**:4402–6.
34 de Graaf AA, Luyten PR, den Hollander JA, Heindel W, Bovée WMMJ. Lactate imaging of the human brain at 1.5T using a double quantum filter. *Magn Reson Med* 1993; **30**:231–5.
35 Rothman DL, Petroff OAC, Behar KL, Mattson RH. Localized ^{1}H NMR measurements of γ-aminobutyric acid in human brain *in vivo*. *Proc Natl Acad Sci U S A* 1993; **90**:5662–6.
36 Birken DL, Oldendorf WH. N-acetyl-L-aspartic acid: a literature review of a compound prominent in ^{1}H-NMR spectroscopic studies of brain. *Neurosci Biobehav Rev* 1989; **13**:23–31.
37 van Rijen PC, Luyten PR, Berkelbach van der Sprenkel JW, Kraaier VJ, van Huffelen AC, Tulleken CAF, den Hollander JA. ^{1}H and ^{31}P measurement of cerebral lactate, high energy phosphate levels and pH in humans during hyperventilation: associated EEG, capnographic and Doppler findings. *Magn Reson Med* 1989; **10**:182–93.
38 Prichard JW, Rothman DL, Novotny EJ, et al. Lactate rise detected by ^{1}H NMR in human visual cortex during physiological stimulation. *Proc Natl Acad Sci. U S A* 1991; **88**:5829–31.
39 Singh M, Brechner RR, Terk MR, Colletti PM, Kim H, Huang H, et al. Increased lactate in the stimulated human auditory cortex. *Proc Tenth Annual Meeting, Soc Magn Reson Med* 1991; 1008.
40 Ordidge RJ, Cresshull ID. The correction of transient B_0 field shifts following the application of pulsed gradients by phase correction in the time domain. *J Magn Reson* 1986; **69**:151–5.
41 den Hollander JA, Oosterwaal LJMJ, van Vroonhoven H, Luyten PR. Elimination of magnetic field distortions in ^{1}H NMR spectroscopic imaging. Proc Tenth Annual Meeting, Soc Magn Reson Med 1991; 472.

10 MRS in transplantation

Simon D Taylor-Robinson, Maria L Barnard, Claude D Marcus

INTRODUCTION

Organ transplantation is an expanding surgical field. In the UK and Ireland, the total number of recipients has increased from 438 in 1972 to 5342 in 1993 (Table 10.1). Increased patient survival rates have been achieved because perioperative patient management, anaesthetic and surgical techniques, antirejection chemotherapy, and methods for harvesting and preserving donor organs have all improved.

The demand for transplantation has increased markedly as the clinical outcome has become more reliable, but a limiting factor on the number of operations performed is the supply of donor organs. In January 1994, 6054 patients were awaiting transplantation in the British Isles[1] and these numbers were increasing by 10% per year.[2]

The aim of any transplant program must be to utilize all available organs. This necessitates an efficient matching and retrieval system. The

Table 10.1 Organ transplants in the British Isles in 1993 and waiting list at year end. Data supplied by UKTSSA.

Organ	Patients transplanted	Waiting list
Kidney[a]	1795	4792
Kidney + Pancreas	17	38
Pancreas	0	3
Heart[b]	309	293
Heart–Lung	35	190
Lung[c]	95	151
Liver	538	119
Cornea	2553	468
TOTAL	5342	6054

a. Figures include 125 kidneys from live donors.
b. Figures include 13 hearts from live donors, available when a recipient in need of a lung transplant receives a heart–lung transplant. The recipient's original heart is then available for another patient.
c. Figures include one liver from a live donor (partial hepatectomy).

United Kingdom Transplant Support Services Authority (UKTSSA) coordinates organ transplantation in the British Isles and a similar function is provided by Eurotransplant, France Transplant, and The United Network for Organ Sharing (UNOS) in the USA.

The success of any transplant procedure is dependent on the quality of the donor graft and this is a reflection of organ storage methods. Renal transplantation was revolutionized in the late 1960s with the finding that donor kidneys could be safely preserved for up to 30 hours by simple cold storage in a preservation fluid.[3] This allowed operations to be undertaken on an elective or semi-elective basis, rather than as an emergency procedure. Liver, heart, heart–lung, pancreas and small bowel are more complicated to transplant than kidneys and survival rates remained poor until the advent of cyclosporin as an immunosuppressive agent in the 1980s.[4] The use of more physiological preservation fluids such as University of Wisconsin (UW) solution[5] has led to increased storage times for these organs, particularly the liver. This allows transport between widely separated hospitals and maximizes the availability to potential recipients.[6]

Despite these advances, tissue damage caused by harvesting techniques, the cold storage process, and reperfusion at the recipient operation is an important factor in patient morbidity and mortality.[7] A reliable assessment of stored organ viability is important to prevent wastage of organs and to prevent the usage of organs that fail to function postoperatively (primary non-function, or PNF). The general shortage of organs means that donors older than 50 years are often considered,[8] and in renal transplantation, non-heart beating donors may be used.[9]

Magnetic resonance spectroscopy (MRS) may provide a non-invasive assessment of the viability of the isolated donor organ prior to transplantation. However, the window of opportunity is greater in renal and hepatic transplantation than cardiac procedures (Table 10.2).

Tissue damage in organ preservation

Tissue injury may occur during the initial harvesting of the donor organ, cold preservation, prolonged warm ischaemia at implantation and during reperfusion.[7,10] All these factors lead to hypoxic–ischaemic damage to the graft and may contribute to PNF or impaired initial function.

Current preservation techniques involve hypothermia, which slows cellular metabolism. By reducing the temperature from 37°C to 4°C, enzyme kinetics are reduced 12-fold,[5] but ATP depletion in isolated donor organs is not prevented

Table 10.2 Time limits of organ preservation by simple cold storage. (Information from Belzer[b].)

	Kidney	Liver	Pancreas	Heart	Lung	Intestine
Routine (Hours)	24	12	16	3	3	6
Maximum (Hours)	50	36	30	8	8	12

by this process.[10] Cold preservation may induce cell swelling.[11,12] Extracellular Na$^+$ enters each cell because reduced or absent ATP leads to a failure of the Na$^+$–K$^+$ ATPase-dependent Na$^+$ pump. Cellular swelling occurs because of water accumulation consequent to Na$^+$ influx.

Lactic acidosis results from anaerobic glycolysis and may cause lysosomal instability and activate lysosomal enzymes.[5,13] Further tissue damage may occur as ATP degenerates through ADP, AMP and adenosine to hypoxanthine (Fig. 10.1). The enzyme xanthine oxidase, a catalyst for the further degradation of hypoxanthine to xanthine, is a potent generator of free radicals that cause cell injury in cold ischaemia[13,14] and subsequent reperfusion.[7]

Preservation fluid design

Most donor organs are preserved by simple cold storage with a fluid such as UW solution, although some centres advocate the use of intermittent or continuous machine perfusion during cold ischaemia.[5]

The aim of any preservation solution is to minimize cell swelling induced by hypothermia, using osmotically active substances such as raffinose or lactobionate. Cellular acidosis can be buffered with phosphate or citrate. Specific cell injury can be minimized with the addition of the xanthine oxidase inhibitor, allopurinol, and oxygen free-radical scavengers, such as glutathione. Adenosine may be added to act as a

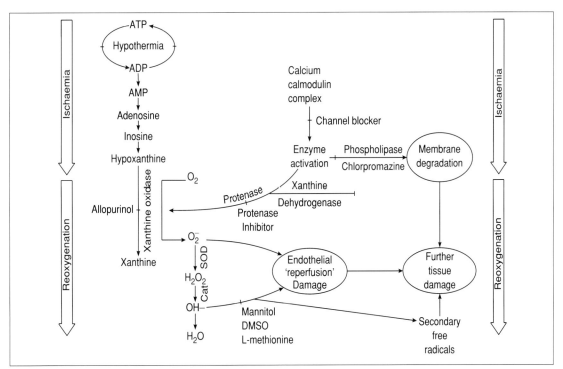

Fig. 10.1. Pathways of ischaemic and reperfusion tissue injury. SOD – superoxide dismutase; CAT – catalase; AMP – adenosine monophosphate; ADP – adenosine diphosphate; ATP – adenosine triphosphate. (Reproduced from Bretan,[124] by permission of Williams and Wilkins, Baltimore, MD, USA.)

substrate for ATP regeneration on reperfusion.[6] There is increasing support for the use of calcium-entry blockers (CEB), such as verapamil, which block cell-membrane calcium channels.[15] These prevent calcium-dependent enzyme activation of phospholipases (thereby preventing subsequent membrane degradation) and of xanthine dehydrogenase (thereby preventing activation of xanthine oxidase). In addition, calcium accumulation in the mitochondria uncouples oxidative phosphorylation, which would abolish rephosphorylation of ADP to ATP.

Further protection from reperfusion injury can be provided by rinsing the organ after preservation in Carolina Rinse, a solution that contains agents to support metabolism, control calcium entry and prevent oxidation.[16] However, a similar improvement in kidney function after late reperfusion with an unmodified Collins solution suggests that this process simply flushes out harmful metabolic end products, such as hypoxanthine.[17]

Measurement of organ viability

Organ donor selection is often based on subjective criteria. These include age, history, length of stay in the intensive therapy unit (ITU), and macroscopic appearances. Standard biochemical parameters may be unreliable.[18] PNF remains a frequent complication despite improvements in organ preservation. A proportion of organs discarded on biological or macroscopic grounds may be suitable for transplantation.

The ultimate measure of donor organ viability is postoperative graft function. However, with organs at a premium, donors are sometimes considered where viability may be questioned.[2,8,9,19–21] In such cases, the donor may be older than 50 years, organ harvesting may have been complicated, preservation times prolonged, or in the case of renal transplantation the graft may have come from a non-heart beating donor.[9,21] There is therefore a need for a reliable, rapidly performed test to prove tissue viability in order to avoid implanting organs that are never going to work.

ATP plays a pivotal role in cellular bioenergetics. The hydrolysis of ATP releases energy from high-energy phosphate (HEP) bonds for all activities involved in maintaining intracellular homeostasis. ATP levels in donor organs have been correlated with patient survival.[22,23] Kamiike et al[10] used high pressure liquid chromatography (HPLC) to measure total adenine nucleotides (TAN) in liver biopsies taken from 30 donor livers during cold ischaemia. Poor graft function was correlated with low TAN.

^{31}P MRS has two advantages over HPLC and enzymatic assays of ATP. It is non-invasive and can be performed without undue prolongation of cold storage time in renal, pancreatic, and liver transplantation.[24] Preservation times for heart and heart–lung are still short (see Table 10.2), so preoperative MRS may not be possible.[6]

^{31}P MRS and organ viability

A typical *in vivo* ^{31}P NMR spectrum contains seven resonances (Fig. 10.2). Phospholipid cell membrane precursors, AMP, and glycolytic intermediates (sugar phosphates) contribute to the phosphomonoester (PME) peak. Phospholipid cell membrane degradation products and endoplasmic reticulum contribute to the phosphodiester (PDE) peak.[25] Information on tissue bioenergetics can be obtained from inorganic phosphate (Pi), phosphocreatine (PCr), and the three NTP

MRS in transplantation

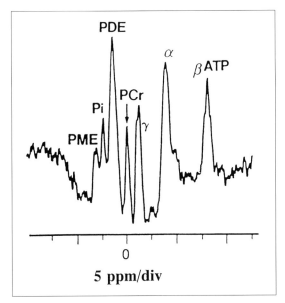

Fig. 10.2. A typical ^{31}P MR spectrum from the liver of a healthy volunteer; 2-D CSI, TR 5s.
PME – phosphomonoester; Pi – inorganic phosphate; PDE – phosphodiester; PCr – phosphocreatine; γ-, α-, β-ATP – NTP resonances.
The PCr resonances arise from overlying muscle and not from the liver itself.

resonances. The α- and γ-NTP resonances contain contributions from ATP and ADP, but the majority of ADP is not detectable using MR methods. This is because of tissue binding.[26] The β-NTP peak contains information almost exclusively from ATP and therefore has been used for purposes of relative quantitation. A measurement of intracellular pH (pHi) can be calculated from the chemical shift of the Pi peak.[27] Peak area ratios have been used for rapid relative quantitation of the various resonances.

Absolute or relative quantitation of ATP levels may be made using ^{31}P MRS, but ATP begins to degenerate to ADP and Pi immediately after each organ has been harvested (see Fig. 10.1). With time, ADP further degenerates to AMP, which con-

tributes to the PME peak.[25] After cold storage, viable organs should be able to rephosphorylate AMP to ATP on reperfusion. The PME:Pi ratio has therefore been suggested as an MR index of organ viability.[23,24]

The PCr:Pi and PCr:ATP ratios have also been used as MR indices of tissue energy reserve.[28] ATP is present in all viable tissue *in vivo*, but PCr (which acts as an energy reservoir for ATP resynthesis) is confined to skeletal, cardiac, and smooth muscle, brain, and a very small pool in the kidneys. Normal hepatocytes contain no PCr. The renal PCr resonance is too small to use quantitatively.

Organ viability post-transplantation

Tissue biopsy remains the gold standard for the diagnosis of graft rejection. At present, no reliable non-invasive diagnostic method of assessment exists. MRS has been studied in cardiac, renal, and hepatic transplants in an attempt to identify specific markers of rejection.

MRS AND RENAL TRANSPLANTATION

The past three decades have seen a dramatic increase in kidney transplantation. The problems faced by transplant teams in the UK reflect the same difficulties seen in the USA and Europe. The numbers of cadaveric kidney transplants performed in the UK have risen from 470 in 1973 to 1687 in 1993 (Table 10.3). The shortage of donor organs is becoming a major concern and is partly responsible for the fall in cadaveric renal transplants that has occurred over the last four years (from 1873 in 1990 to 1687 in 1993). This is

Table 10.3 Cadaveric kidney transplants in the British Isles 1972–1993, and number of patients waiting for kidney transplant at year end 1978–1993. Results include kidneys transplanted as part of kidney–pancreas and multiple organ transplants. (Data supplied by UKTSSA.)

Year	Kidney transplants	Kidney waiting list
1972	438	—
1973	470	—
1974	571	—
1975	618	—
1976	608	—
1977	729	—
1978	765	1274
1979	839	1592
1980	932	1923
1981	854	2273
1982	1085	2481
1983	1182	2693
1984	1506	2780
1985	1388	3443
1986	1585	3468
1987	1558	3565
1988	1650	3684
1989	1837	3705
1990	1873	3834
1991	1766	4905
1992	1768	4464
1993	1687	4830

despite the growing numbers of patients awaiting renal transplantation (3834 in 1990 rising to 4830 in 1993). Both for these patients and for the transplant teams waiting to help them, the mood has been described as one 'approaching desperation'.[2]

Most organ donations are from patients on ventilatory support in ITU in whom brainstem death has been confirmed: the cadaveric 'heart-beating' donors. With the shortfall in donor kidneys, transplant teams have started to re-examine the use of asystolic 'non-heart beating' donors.[9,21] A group from Leicester, UK recently reported their experience using donors referred within 30 minutes of unsuccessful resuscitation.[21] They used rapid *in situ* cooling perfusion of the donor kidney while permission for organ removal was being obtained. PNF was the rule in these renal transplants. Around 88% of grafts functioned after a mean delay of 21 days, the delay being secondary to reversible acute tubular necrosis. These organs contributed 38% of all transplanted kidneys in this series.

The group from King's College Hospital, London[9] used grafts from non-heart beating donors, obtained within 45 minutes of sudden accidental or medical death or from hospice patients dying of a primary cerebral tumour. These organs contributed 28% of total cadaveric renal grafts in this series. On examining their overall transplant results, this group found that 26% of kidneys from non-heart beating donors never functioned compared to 1.4% of those from heart beating donors. Of the kidneys that eventually functioned, PNF occurred in 47% from non-heart beating donors and 7% from heart beating donors. The long-term survival of the grafts from non-heart beating donors was also worse: 55% at two years compared to 71% for grafts from heart beating donors. Non-heart beating donors can make a significant contribution to transplant programmes. However, the implantation of non-viable kidneys not only results in graft nephrectomy, but also risks sensitizing patients to future grafts. The criteria for donor selection are continually being pushed back and the need for a rapid test of tissue viability is becoming paramount.

MRS assessment of kidney viability

The kidney was one of the first organs studied by MRS. The first ^{31}P MRS studies of the kidney were reported by Sehr et al.[29,30] They showed depletion of HEP in isolated rat kidneys during prolonged ischaemia at 4°C, with recovery of HEP on subsequent reperfusion. These authors were the first to suggest that ^{31}P MRS could be used to monitor kidney viability during cold storage for transplantation. It is perhaps because of this early recognition of the potential of MRS that the technique has been developed and assessed in kidney transplantation in the clinical setting.

The feasibility of using ^{31}P MRS to measure organ viability has been studied by Bretan and colleagues,[31] who measured definitive metabolite ratios in *ex vivo* isolated rat kidneys. These were correlated with viability as assessed by electron microscopy (EM). Subsequent animal survival and kidney function in rats subjected to similar periods of renal ischaemic injury were also linked to the metabolite ratios (Fig. 10.3). The isolated kidneys gave no signals from ATP or ADP, at 5.6 T, so that the PME:Pi and NAD:Pi ratios were analysed. These showed a progressive decay with increasing periods of both *in situ* warm ischaemia (0, 20, 60 or 120 minutes) and cold storage (0 to 72 hours), in a time-dependent manner. A PME:Pi ratio of 0.45 and a NAD:Pi ratio of 0.17 corresponded to moderate ischaemic cell damage, with irreversible change on EM, and non-fatal azotaemia after ischaemic injury. A PME:Pi ratio of 0.3 and a NAD:Pi ratio of 0.13 were associated with histological gross cell death and a non-functioning kidney with fatal uraemia. These results suggested the value of ^{31}P MRS as a quantitative measure of viability.

This work was extended[32] to examine the effects of warm and cold ischaemia and subsequent reperfusion on renal HEP levels using a canine model, examined at 5.6 T. Normal kidneys showed a high PME:Pi ratio (0.87 ± 0.12). After 45 minutes of *in situ* warm ischaemia, the renal PME:Pi ratio was reduced (0.50 ± 0.12), but this increased after reperfusion (1.0 ± 0.07). In a third group of kidneys excised and preserved by simple cold storage for 24 hours, the PME:Pi ratio fell to 0.54 ± 0.08. These organs were then transplanted, and after reperfusion the PME:Pi ratio increased to 0.77 ± 0.15. Unsuccessful reperfusion and trans-

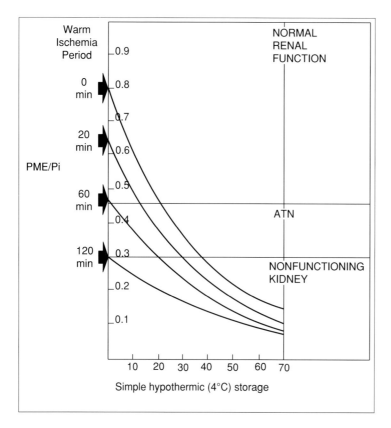

Fig. 10.3. Graph of PME:Pi vs length of cold storage. A graded decay proportional to renal harvest warm ischaemia as well as a time dependent decay over 70 hours of hypothermic storage is noted in the PME:Pi, enabling assessment of viability in a transplantation setting. PME:Pi of 0.45 is associated with acute tubular necrosis (ATN); PME:Pi of 0.30 is associated with gross renal cell death. (Reproduced from Bretan et al,[23] by permission of Williams and Wilkins, Baltimore, MD, USA.)

plantation produced a further fall in the PME:Pi ratio to non-viable levels of 0.28 ± 0.12 within four hours. These findings confirmed that ^{31}P MRS is a marker of organ viability and is predictive of post-transplant function in larger mammals. The regeneration of PME during reperfusion also supported the hypothesis that the PME peak in the kidney represents a precursor energy pool for ADP and ATP, validating the PME:Pi ratio as a viability parameter.

Further studies in dogs[33] showed the regeneration of ATP during reperfusion of kidneys subjected to warm and cold ischaemia. ATP regeneration was related to the PME:Pi ratio. This correlation established the PME:Pi ratio as a strong renal transplant viability parameter.

Similar studies have been performed in other animal models. Kunikata et al[34] examined isolated rabbit kidneys during simple cold storage and subsequent reperfusion, using ^{31}P MRS at 9.4 T. They again showed a fall in ATP:Pi, PME:Pi, and pH during cold storage. The PME:Pi ratio recovered fully after reperfusion of organs preserved for 24 hours; kidneys stored for 48 hours had a lower PME:Pi ratio, which did not recover fully during reperfusion. These results again confirm the PME:Pi ratio as a parameter of organ viability.

Lietzenmayer et al[35] have particularly addressed the problem of designing an animal model. They used slaughterhouse-harvested swine kidneys to develop a simple, low cost *in vitro* perfusion model for metabolic studies with ^{31}P MRS. This may decrease the need for animal sacrifice and allow investigation of a variety of preservation protocols.

Kidney preservation

It is of the utmost importance that all available donor kidneys are used for transplantation. ^{31}P MRS appears to be of value in assessing organ viability, and it may be of particular merit for evaluating kidneys from non-heart beating donors or organs transported over long distances. However, in contrast to liver and heart storage, clinical kidney preservation methods are generally established and undisputed. ^{31}P MRS has therefore not been used extensively in studies evaluating different preservation parameters.

The preservation techniques using simple cold storage or continuous hypothermic perfusion were developed in the late 1960s and have been used almost unchanged for the past 25 years. Most transplant centres in Europe and the United States use simple cold storage, and the kidney has the longest time limits for organ preservation compared to other grafts (see Table 10.2). Kidney storage appears to meet most of the objectives of organ preservation: the methods are simple, reliable, and inexpensive, and they provide an adequate preservation time for organ distribution, enabling the scheduling of 'non-emergency' transplant procedures. The other major advantage of kidney transplantation is that in the event of graft failure, dialysis can replace organ function, a facility not available for long-term treatment of other failed organ transplants.

Given the shortfall in donors, it is essential that all aspects of storage are optimized. There are still some areas of uncertainty. Several storage solutions have been advocated, containing different impermeants to suppress cell swelling, and the final choice is often dependent on the preference of the transplant team. UW and EuroCollins (EC) solutions are the most commonly used preservation fluids, but there have been few prospective studies to assess performance. Information on the adequacy of kidney preservation is difficult both to obtain and to interpret. It could be judged by the percentage of immediate function versus the need for dialysis and by the one-month function rate. Unfortunately, there is no central registry for this data in the UK or USA, and UKTSSA and UNOS do not record this information. One multicentre study suggested that rate of postoperative dialysis in the United States for kidneys preserved by simple cold storage is 27%.[6] However, the causes of delayed graft function and the need for postoperative dialysis are multifactorial and are not solely determined by preservation methods. Since the rate of postoperative dialysis with simple cold storage varies from 20–50% between different centres, it seems unlikely these differences are related solely to preservation factors.[6] The method and experience of the harvesting team, age of donor and recipient, individual policies regarding when to dialyse a patient, and immunological considerations could all influence delayed graft function.

There is also still a question whether initial function is optimized by simple cold storage or by hypothermic perfusion of the kidney. On evaluating the results from Wisconsin, USA, a centre that uses perfusion with UW gluconate solution, Belzer[6] found a postoperative dialysis incidence of 9%. The relationship between dialysis and previous storage conditions was again not clear. Graft survival was inferior in the delayed function group and the lack of immediate function may relate to immunological factors.

Within these limitations, a recent prospective, multicentre trial showed that the need for dialysis after transplantation was 10% lower if organs were preserved by simple storage in UW compared with those preserved by EC solution, using multivariate analysis.[36] ^{31}P MRS has been applied in studies evaluating preservation fluids, and authors have started to examine the importance to renal transplantation of various modifications being investigated for the storage of other organs.

^{31}P MRS studies of preservation fluids

Bretan et al[7] compared a modified intracellular flush solution (PB-2) with the commonly used Collins solutions in a canine renal transplant model, with ^{31}P MRS peformed at 4.7 T. PB-2 was designed to limit adenine nucleotide catabolism and post-transplantation reperfusion injury. Adenine and magnesium were added to enhance the regeneration of ATP. Magnesium also exerts a vasodilator effect. Mannitol was substituted for dextrose; mannitol having multiple actions, including scavenging for and detoxifying free radicals. PME:Pi and NADP:Pi ratios were better preserved in kidneys stored in PB-2 than in Collins solution. HPLC confirmed these findings, showing that TAN (ATP, ADP, and AMP) were better preserved during cold storage in PB-2 than they were in Collins solution. After PB-2 storage, kidneys showed regeneration of TAN within 45 minutes of reperfusion at transplantation. Cold storage in PB-2 also gave less reperfusion injury as judged by electron microscopy and resulted in a significant improvement in renal recovery and viability.

Ciancabilla et al,[37] however, reported that care needs to be taken in using the PME:Pi ratio to compare different preservation fluids. They found a higher PME:Pi ratio in kidneys stored in phosphate-free Ringer solution (Rg) compared to UW or EC solution. This is contrary to findings that have shown poor organ preservation by Rg compared to UW and EC solutions. It is thought that the buffer Pi may passively enter the renal cells. The PME:Pi ratio is a valid index of viability when organs are stored under the same conditions. However, when comparing different preservation fluids by ^{31}P MRS, with different Pi content, the possibility that this buffer Pi is affecting the PME:Pi ratio must be considered. It may be necessary to re-establish normal and pathological values of this ratio for each storage solution under investigation.

^{31}P MRS clinical studies

The major series assessing the application of ^{31}P MRS to analyse donor kidney viability in a clinical renal transplantation setting was reported by Bretan et al.[23] They examined 40 cadaveric kidneys during simple hypothermic storage in Collins solution, without removal from or disruption of the sterile organ storage container. MRS examination was performed at 28 ± 10 hours and transplantation at 35 ± 10 hours. ^{31}P MR spectra were acquired at 1.4 T and identified the renal intracellular phosphate groups. In a second spectrum, taken without the kidney centred directly on the coil, a separate resonance was also detected from the buffer Pi (Fig. 10.4). The pH of the buffer was subsequently measured with a pH meter and this allowed the chemical shift of the renal Pi to be used to calculate intracellular pH. The ^{31}P MRS results were correlated with subsequent clinical parameters of renal function in graft recipients, measured one week post-transplant.

Only 11 kidneys (27.5%) had detectable α-ATP and NADP, with mean values of 0.34 ± 0.07 for α-ATP:Pi and 0.08 ± 0.11 for NADP:Pi. These ratios proved to be insensitive but rather specific

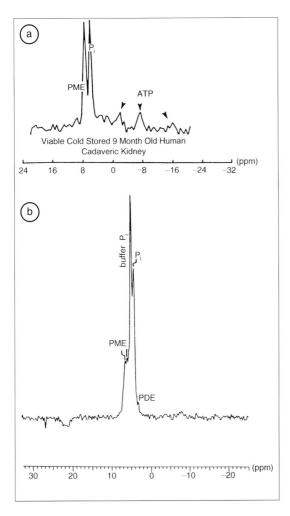

Fig. 10.4. Ex situ ^{31}P *MR spectrum of a viable paediatric (a) and adult (b) kidney. Paediatric kidneys have high levels of PME and care should be taken in using the PME:Pi ratio as an index of organ viability in this group.*
PME – phosphomonoester; Pi – inorganic phosphate, from the preservation fluid or intracellular; PDE – phosphodiester, ATP – adenosine triphosphate. (Reproduced from Bretan et al,[23] by permission of Williams and Wilkins, Baltimore, MD, USA.)

viability markers, as they were associated with the best subsequent renal function. For example, 36% of the patients receiving kidneys with detectable ATP or NADP required dialysis, compared to over 71% of patients receiving kidneys without these MRS detectable metabolites.

The PME:Pi ratio (0.61 ± 0.45) was therefore analysed and showed a significant correlation with subsequent renal function ($p<0.001$), whereas cold storage times or intrarenal pH (7.10 ± 0.31) did not. The lower PME:Pi ratios (0.36–0.10) were significantly associated with prolonged non-function post-transplantation, secondary to acute tubular necrosis. Using the PME:Pi value of <0.5 to predict the need for dialysis and >0.5 to predict no need for dialysis in the recipients, the pretransplant MRS sensitivity was 0.75 and the specificity was 0.87. The PME:Pi ratio appears to be sensitive and specific enough to act as a quantitative marker of renal viability. These results strongly suggest that MRS is a valuable non-invasive, non-destructive method for assessing renal viability and the subsequent effects on clinical outcome.

The extension of this *in vitro* work is the challenge of examining organ viability and determining the effects of transplantation on *in vivo* renal metabolism in man. Boska et al[38] reported the first ^{31}P MRS study of healthy normal and transplanted kidneys in human subjects. There are particular problems in performing kidney MRS. The normal (non-transplanted) kidney lies deep in the body, covered by a thick layer of muscle and fat. The kidney moves with respiration, making it difficult to acquire spectra that are localized solely to this organ during long MRS studies. The kidney is also made up of different internal structures (cortex, medulla, hilum), which complicates the interpretation of spectra acquired across the whole kidney.

Boska et al[38] examined healthy subjects and patients with well functioning allografts, at 2.0 T.

They attempted to overcome the problems of spectral acquisition first by limiting motion by careful positioning of the healthy subjects. Further improvement came from the ability to perform localized shimming with stimulated echo techniques. After taking an initial locator MR image, they used image-selective *in vivo* spectroscopy (ISIS) localization to examine a selected volume of interest in the kidney. Spectra showed little or no PCr. This indicated that the spectra were not significantly contaminated by signal either from overlying skeletal muscle or from intestinal smooth muscle. The study also used external standards to give correct phasing of the spectrum. Phasing of the spectrum is an important variable parameter that can affect signal integration, but is otherwise open to observer bias.

Boska et al[38] were then able to measure several renal metabolite ratios, including PME:Pi (0.9 ± 0.6) and Pi:ATP (1.2 ± 0.6), and the chemical shift of Pi as an index of intracellular pH. They identified no significant difference between the normal healthy kidneys and the well-functioning renal grafts. However, the low sensitivity remained a problem and limited the number of successful examinations in normal subjects to seven out of 14, using a minimum acquisition of 30 minutes.

These studies demonstrated the feasibility of performing localized ^{31}P MRS on the human kidney and measuring metabolite ratios. As techniques improve and clinical systems operating at higher field strengths become available, it may become possible to acquire localized signals from specific zones in the kidney. The currently available techniques have yet to be applied to studies of HEP metabolism in the immediate post-transplant period in humans. However, studies on the effects of graft rejection on kidney metabolites, using ^{31}P MRS, have commenced.

Graft rejection and renal dysfunction

The accurate diagnosis of diminished graft function after transplantation is essential for patient management. Renal graft failure may be caused by a number of pathologies, including:
- rejection;
- cyclosporin nephrotoxicity;
- acute tubular necrosis;
- obstruction;
- acute ischaemic injury; and
- renal artery stenosis.

Differentiation between these remains a problem and there is no non-invasive diagnostic test that will definitively determine post-transplant graft failure when obstruction has been excluded. Renal biopsy is the only reliable method for differentiating between rejection, acute tubular necrosis, and cyclosporin toxicity. MRS may detect metabolic change and, on the basis that different renal pathology may give different metabolic patterns, the technique has been investigated as a non-invasive diagnostic test for evaluating disease in this setting.

Animal studies

Shapiro et al[39] investigated the possibility of differentiating pathology post-transplantation in a rat model. *In vivo* ^{31}P MR spectra were acquired at 1.89 T with a probe placed directly on the exposed transplanted kidney. They found clear distinction between the causes of induced graft dysfunction (Table 10.4).

Kidneys subjected to ischaemia showed an increase in Pi and decrease in β-ATP associated with a decrease in pH. Rejected kidneys also showed a marked increase in Pi and decrease in β-ATP, but the intracellular pH was not different from controls. These similar findings to ischaemia may be expected, as vascular involvement is

Table 10.4 Metabolite ratios detected by ^{31}P MRS of rat kidney transplant allografts subjected to different insults. (Information from Shapiro et al[39]).

^{31}P MRS ratios	Control	Ischaemia	Rejection	Obstruction	Cyclosporin toxicity
Pi:β-ATP	0.37 ± 0.07	2.86 ± 0.19**	2.30 ± 0.32**	0.35 ± 0.09	0.32 ± 0.05
PD + UP:β-ATP	0.76 ± 0.04	1.70 ± 0.11**	2.01 ± 0.28**	3.80 ± 0.44**	0.69 ± 0.06
pH	7.33 ± 0.07	7.00 ± 0.05*	7.33 ± 0.07	7.46 ± 0.06	7.21 ± 0.10

Pi – inorganic phosphate; PD + UP – phosphodiesters and urine phosphate; β-ATP – β-phosphate of ATP.
Ratios were calculated from the percentage of total peak area.
Results expressed as mean + SEM; * p<0.05, ** p<0.01 from control.

prominent in rejection across major histocompatibility mismatches in the rat. The absence of acidosis may be due to patchy areas of ischaemia, giving a low pH, being averaged with areas of cellular infiltration and proliferation occurring in cellular rejection, which have a high pH.[40,41]

Obstructed kidneys were notable for their dramatic increases in the PDE peak. This was mainly due to the increase in signal from urinary phosphate, the position or chemical shift of the peak being consistent with the urine pH.

After toxic doses of cyclosporin, no spectroscopic differences were found compared to normal controls. The reason for this was unclear, but this had been noted with other nephrotoxic drugs, such as gentamicin. In the early transplant period, the major differential is between rejection, urinary obstruction, or PNF due to ischaemia, whereas cyclosporin toxicity is more likely later. The authors therefore suggested that ^{31}P MRS may distinguish between the causes of renal graft dysfunction.

Bretan and colleagues[42] used a canine model and found that ^{31}P MRS was a sensitive but not specific diagnostic technique post-transplantation. It did not differentiate between severe ischaemia and severe acute rejection. MRS of transplanted kidney was performed using a surface coil at 2 T and autografts showed excellent viability, with high PME:Pi and ATP:Pi ratios. Rejecting allografts or autografts subjected to vascular ischaemia and thrombosis both showed reduced viability, with reduced PME:Pi and ATP:Pi ratios. There was also a similar chemical shift change for the Pi resonance in rejected and ischaemic renal transplants. This conflicts with the findings in the rat, which showed no change in pH in rejecting kidneys.[39] ^{31}P MRS may possibly distinguish a more minor degree of rejection that is not associated with global renal ischaemia. However, when severe vascular rejection occurs, the resulting ischaemia becomes difficult to separate from thrombotic events.

Human clinical studies

It is hoped that these animal findings can be extended into non-invasive patient diagnosis by ^{31}P MRS. As yet, there are few studies. Wenzel et al[43] examined a small series of subjects after kidney transplantation, comparing patients with well-functioning grafts (controls) to patients with different pathological renal states. Spectra were acquired at 1.5 T. Large spectral differences from controls were not detected in most pathologies. However, all patients were on maximum therapy at the time of MRS and indeed, because of treatment, extreme tissue damage will usually be prevented in patients. The diseased kidneys had a tendency towards higher Pi:α-ATP values, particularly in chronic rejection, and spectral differences were sometimes detected despite normal clinical parameters, such as serum creatinine.

Grist et al[44] examined prospectively a larger series of renal transplant patients, using ^{31}P MRS performed at 1.5 T. They compared recipients with normal graft function (12 controls) to patients biopsied for allograft dysfunction. The subjects undergoing biopsy were divided into those showing rejection (21 patients) and those showing no rejection (7 patients). The PDE:PME ratio differed significantly in patients with rejection (1.43 ± 0.21) compared to those showing no rejection (0.79 ± 0.42) or control patients (0.47 ± 0.11). The Pi:ATP ratio was also significantly increased in patients with rejection (1.10 ± 0.21) compared to those with no rejection (0.41 ± 0.07). Further analysis calculated that a PDE:PME ratio exceeding 0.8 had a sensitivity of 100% and a specificity of 86% for predicting rejection. A Pi:ATP ratio greater than 0.6 had a sensitivity of 72% and a specificity of 86% for predicting rejection. If dual requirements of a PDE:PME ratio over 0.8 and a Pi:ATP ratio greater than 0.6 are met, the sensitivity and specificity are both 100%.

The elevation in the Pi:ATP ratio observed in the patients with rejection is thought to be the result of tissue ischaemia. An elevation in the PDE:PME ratio has also been found in animal models of rejection,[39] but the cause remains unclear. This elevation results mainly from an increase in PDE. This resonance has contributions from several compounds, including glycerophosphorylcholine (GPC), glycerophosphoethanolamine (GPE), the mobile fractions of large membrane phospholipid molecules, and Pi found in acidotic urine (principally in the obstructed kidney).

It is possible that the increased PDE:PME ratio may be related to an increase in GPC and GPE. The metabolites within the PME and PDE resonances are found on the pathways of phospholipid synthesis and catabolism. Changes in these metabolites may be related to alterations in cell growth, and stimulus for cell growth is associated with renal dysfunction. Alternatively, the elevation in PDE may be secondary to changes in the contribution from membrane phospholipids. This is related to an alteration in membrane metabolism associated with the inflammatory changes occurring in rejection. The PDE:PME ratio showed an increasing trend with increasing severity of rejection, as graded by histology, and this may represent the sequelae of membrane breakdown or a stimulus for cellular hypertrophy.[44]

^{31}P MRS may therefore be of diagnostic value in graft dysfunction post-transplantation. However, future prospective studies must encompass larger patient numbers and results must be compared to other techniques, such as Doppler ultrasound examination and Indium-111 platelet studies.

High-resolution urine studies

Cyclosporin A (CsA) has been a major factor contributing to prolonged graft survival in many fields of organ transplantation. With renal transplants, there is a particular problem. The use of CsA has made the diagnosis of acute rejection difficult, not only because it masks the clinical features of acute rejection, but also because it is itself nephrotoxic.[45] With acute renal graft dysfunction, it is therefore difficult to distinguish clinically between

MRS in transplantation

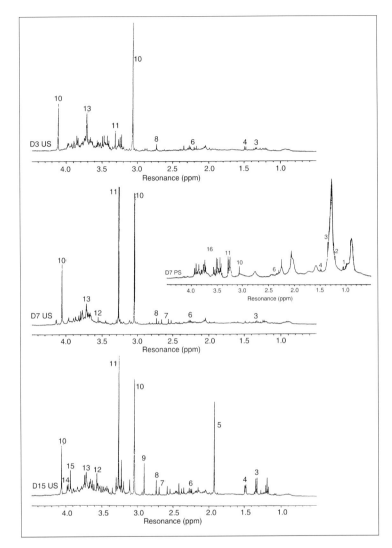

Fig. 10.5. Aliphatic region of urine ^1H MR spectra from a renal transplanted patient at day 3 (D3), day 7 (D7), and day 15 (D15), with the plasma spectrum (PS) at day 7. The patient presented with acute rejection at day 6 and was dialysed till day 11. Urine spectra show a strong signal at 3.7 ppm (13) at day 3 and increased TMAO (11) at day 7. In plasma, TMAO (11) was present at day 7.
Assignment of resonances:
1 – Valine; 2 – hydroxybutyrate; 3 – lactate; 4 – alanine; 5 – acetate; 6 – glutamine and glutamate; 7 – citrate; 8 – dimethylamine (DMA); 9 – dimethylglycine; 10 – creatinine; 11 – trimethylamine-N-oxide (TMAO); 12 – glycine; 13 – unidentified peak P3.7; 14 – hippurate; 15 – betaine; 16 – glucose. (Reproduced from Le Moyec et al,[48] by permission of S Karger, Basel, Switzerland.)

rejection and CsA toxicity, diagnoses that require opposite treatments.

Reliable markers of rejection or CsA nephrotoxicity are as yet undetermined. Furthermore, blood levels of CsA are usually monitored by measurement before and four hours after the oral CsA dose. The timing of the peak concentration may vary and show great inter- and intra-patient variability. It is therefore often difficult to correlate the blood CsA levels with biological effects, such as nephrotoxicity or overdosage.

High resolution proton (^{1}H) MRS has been applied for rapid, concurrent analysis of multiple low molecular weight metabolites in plasma and urine in chronic renal failure.[46] It was further noted that, after dosing rats with CsA, the urinary ^{1}H MR spectra showed an as yet unidentified peak at 3.7 ppm.[47] Le Moyec et al[48] therefore examined whether this technique could distinguish between different causes of renal dysfunction after transplantation in the human clinical setting. They found that the overall pattern of change in three parameters could distinguish between rejection (Fig. 10.5), CsA nephrotoxicity or CsA overdose (Table 10.5):

- P3.7:Ct – the ratio of the unidentified resonance at 3.7 ppm to the creatinine peak in urine;
- TMAO:Ct – the ratio of the trimethylamine-N-oxide resonance (TMAO) to the creatinine peak in urine; and
- TMAO in plasma.

The unidentified resonance at 3.7 ppm remains poorly characterized, but it may be due to the presence in urine of polyethylene glycol molecules released from Labafril, the solvent of CsA. The plasma CsA concentrations may be lower in CsA nephrotoxicity than in CsA overdosage. The raised urinary P3.7:Ct in CsA nephrotoxicity may therefore represent an alternative marker to manage CsA dosage in this condition.

TMAO is thought to be an osmolyte molecule synthesized in the medullary cells of the kidney, to counteract perturbations from high urea concentrations. Plasma TMAO is not usually detected, but has been found in patients on chronic dialysis.[46] It is similarly not found in normal human urine. Its excretion may be related to enhanced choline degradation or leakage from renal medulla cells. In transplant patients, the increase in TMAO:Ct in urine and the presence of TMAO in plasma may be related to medullary ischaemia, possibly associated with cold storage times. Notably, medullary ischaemia may be involved in CsA toxicity, and urinary TMAO:Ct was increased in these patients.

This MRS method could be routinely applied in the clinical renal transplant setting to give rapid screening of urine and plasma with minimal sample preparation. Future studies should confirm the ability of high resolution ^{1}H MRS to act as a diagnostic test to distinguish between rejection and CsA toxicity or overdosage.

MRS ASSESSMENT OF HEPATIC VIABILITY

Hepatic transplantation has become the treatment of choice in end-stage liver disease. The introduc-

Table 10.5 Discrimination between causes of renal impairment in patients after transplantation, by analysis of metabolite patterns in urine and plasma using high resolution ^{1}H MRS. (Information from Le Moyec et al.[48])

^{1}H MRS parameters	Acute rejection	Cyclosporin nephrotoxicity	Cyclosporin overdosage
P3.7:Ct[a]	Low	High	Moderate
(mmolH+/mol)	(387 ± 42)	(974 ± 186)	(644 ± 123)
TMAO:Ct[b]	Moderate	High	Low
(mmol/mol)	(175 ± 48)	(339 ± 78)	(130 ± 25)
TMAOp[c]	100%	87%	46%
(frequency)			

a. P3.7:Ct is the ratio of an unidentified resonance at 3.7 ppm, which is possibly related to the pharmaceutical cyclosporin solvent, to the creatinine peak in urine.
b. TMAO:Ct is the ratio of the trimethylamine-N-oxide resonance to the creatinine peak in urine.
c. TMAOp is the presence of TMAO in the plasma, expressed as the percentage of subjects showing this resonance.

tion of cyclosporin and UW solution.[49] in the 1980s have radically changed the clinical outcome.

UW solution has extended the workable preservation time for cold stored livers to 24 hours, allowing operations to be undertaken on a semi-elective basis. Donor livers can be matched with recipients over greater geographical areas, maximizing the use of the donor organ pool. The shortage of transplantable livers (Table 10.6) sometimes necessitates the consideration of non-ideal donors, where age, history and preservation time may not be optimal.[19,20,50] Cold ischaemic damage may be greater with extended preservation times. Artificial liver support systems are not yet available, and therefore immediate graft function is needed. PNF still accounts for 6–13% of initial graft failure,[51–53] and therefore a reliable means of assessing organ viability is required. During 1992, eight liver recipients in the UK received a retransplant for PNF in the first 7 days following liver transplantation. This figure had risen to 14 recipients in 1993.[1]

At present, the selection of donors for liver transplantation remains largely subjective. Organs are accepted or rejected on the basis of donor history, the length of ITU stay, standard liver function tests, and macroscopic appearance at laparotomy.[54] Steatosis has been correlated with PNF.[55] It is therefore standard practice to discard livers that are macroscopically fatty or that show steatosis on donor liver histology.

Oellerich et al[56] reported on a rapid method of assessment of *in situ* donor hepatic function. This method relies on the conversion of an administered lignocaine bolus to monoethylglycinexylidide (MEGX). They correlated graft survival with plasma MEGX levels of above 90 μg/l. Levels of less than 90 μg/l were associated with poorer outcome. The formation of MEGX is dependent on hepatic blood flow, hepatocyte function, and the integrity of the cytochrome P450 system.[57] Rosenlof et al[53] could not distinguish histological differences in livers with high or low MEGX values. They suggested that the MEGX test should

Table 10.6 Cadaveric liver transplants in the British Isles 1983–1993, and number of patients waiting for liver transplants at year end 1990–1993. (Data supplied by UKTSSA.)

Year	Liver transplant	Waiting list
1983	20	—
1984	51	—
1985	88	—
1986	127	—
1987	175	—
1988	244	—
1989	297	—
1990	359	57
1991	420	83
1992	506	83
1993	538	119

not be used as the sole discriminatory factor when deciding whether to accept a donor for transplantation. The MEGX test may provide information on initial hepatic function of the donor liver *in situ*, but cannot provide information on any tissue damage sustained during cold storage.

Adequate postoperative graft function results in the resumption of urea and bile production and protein synthesis. These processes require ATP, but this is rapidly depleted during harvesting, cold storage and the subsequent transplant procedure. The ability of the liver to maintain or regenerate ATP has been found to determine both liver viability and subsequent survival. This has been shown in experimental transplantation models[58] and in HPLC studies of biopsies taken from human donor liver during cold ischaemia.[10,22] ^{31}P MRS has been used non-invasively in animal and human studies to assess hepatic viability following warm ischaemia and during cold storage.

MRS studies on small animals

The bulk of work using MRS to assess viability in isolated livers has been performed on small animals using small bore magnets, usually operating at a magnetic field strength of 4.7 T or above. Work on pigs and dogs has been performed as a prelude to human examinations, which have employed wide bore clinical systems at 1.5 T in the majority of ongoing patient studies.

Preservation techniques
The vast majority of human livers are preserved by simple cold storage on a bed of ice. Rossaro et al[59] assessed simple cold storage and continuous hypothermic perfusion of rat livers. ATP levels and pHi were measured before, during and after cold preservation in UW or EC solutions at 4.7 T. In simple cold storage, ATP declined to undetectable levels. After reperfusion, the livers stored in UW solution showed a better recovery of ATP than those stored in EC solution. In continuous hypothermic perfusion, ATP levels were still detectable after 24 hours preservation, but the perfusion process caused damage to the vascular endothelium with resultant portal vein thrombosis. Only livers stored in UW solution with simple cold storage had significant postoperative survival, which was correlated with ATP levels measured by ^{31}P MRS.

Lanir et al[60] compared mouse livers preserved by simple cold storage, flushed with preservation fluid, with livers subjected to hypothermic pulsatile perfusion. ^{31}P MRS was used to follow ATP levels. The ATP decay was shown to be much slower with lower temperatures in simple storage. With pulsatile perfusion, ATP levels were stable for at least six hours. Lanir et al suggested that continuous or intermittent perfusion methods should be developed for clinical studies, but acknowledged that technical difficulties have limited the development of such systems. The subsequent introduction of more physiological preservation fluids, such as UW solution, may have obviated this need, because preservation times have been extended well beyond those achieved with the available fluids at the time of the study.

In a subsequent publication, the same group[61] used ^{31}P MRS in the evaluation of intermittent perfusion during cold storage of mouse livers. ATP levels were measured during modifications of the perfusion technique and the highest levels were correlated with the longest flush times. They concluded that, although intermittent perfusion is more effective at maintaining ATP levels in the isolated liver than simple cold storage, this

Warm and cold ischaemia

Bowers et al[62] used ^{31}P MRS to measure NTP (ATP) levels in rat livers subjected to different lengths of warm ischaemia times. Subsequent survival was correlated with ATP recovery post-transplantation. The longer the warm ischaemic time, the lower the MR measurable ATP levels and the smaller the ATP recovery postoperatively. This was associated with poor outcome. The authors concluded that the ability of the graft to restore ATP levels was an overall indicator of graft damage sustained during storage.

Orii et al[63] used both ^{31}P and ^{23}Na MRS to study rat livers subjected to 24–48 hours of simple cold storage with UW solution. Postoperative fructose loading was used as an assessment of graft integrity. The hepatic conversion of fructose to fructose-1-phosphate can be detected indirectly, as this sugar phosphate resonates in the PME region of the ^{31}P MR spectrum. The subsequent rise in the PME peak was less in livers preserved for 48 hours than in those preserved for 24 hours. Such findings are not applicable to the human situation, in which a rapid diagnostic assessment of viability during cold storage is required. ^{23}Na MRS measured the intracellular and extracellular sodium compartments. Membrane function was found to be significantly impaired with prolonged cold ischaemic time. Subsequent study by the same group[64] showed similar results, but it would appear that ^{31}P MRS measurements are probably more relevant to viability assessment in the human transplant situation.

Comparison of preservation fluids

Gulian et al[65] used ^{31}P MRS, HPLC and light microscopy to compare hepatic preservation in cold storage with EC and UW solutions in rats. After 25 hours cold storage, no measurable NTP or NDP was seen in the ^{31}P MR spectra with either storage solution. This correlated with HPLC and enzymatic assays. The formation of ATP breakdown products was inhibited by the presence of allopurinol in the UW solution, preventing the generation of free radicals by xanthine oxidase. The recovery of ATP following reperfusion, as seen in the ^{31}P MR spectra, was much better for the livers stored in UW solution than for those stored in EC solution. There was no histological damage in the UW-preserved livers, but EC-preserved livers showed cell swelling. This study used ^{31}P MRS to confirm the superior preservation of UW over EC solution for liver transplantation.

Nedelec et al[66] correlated ^{31}P MRS and biochemical assays in mouse livers reperfused after 24–48 hours cold storage in UW and Collins solutions. ATP measured in the ^{31}P MR spectra was observed to recover better postoperatively in livers subjected to shorter preservation times. UW solution was superior to Collins solution when ATP recovery rates were assessed, the MRS findings were confirmed by biochemical assays.

Reflushing after cold storage

Reperfusion injury has been shown to be reduced in both animal[16] and humans[67] by the use of Carolina Rinse to flush the liver at the end of cold storage. Busza et al[68] used ^{31}P MRS to investigate the consequences of reflushing rat livers with fresh perfusate prior to implantation. They found that ATP levels fell with cold storage, with a concomitant rise in Pi. Subsequent reflushing led to a recovery in ATP. A paper from the same group[69] confirmed that reflushing after prolonged cold ischaemia permitted resuscitation of ATP to *in vivo* levels prior to hepatectomy.

Other MR indices of viability

The temporal changes in the ^{31}P MR spectrum during ischaemia were measured in mouse livers by Fuchinoue et al,[70] who found that ATP measurement was not a perfect MR parameter of organ viability in itself, as viable organs can rephosphorylate AMP to ATP on reperfusion. The PME:Pi ratio was found to be a good index of subsequent graft outcome.

^{1}H MRS studies

Holzmüller et al[71] used ^{1}H MRS to obtain T_1 values on rat livers stored at 4°C for varying periods in five different preservation fluids. These were correlated with other parameters of viability, including bile flow following reperfusion and HPLC determination of HEP content. Livers were preserved in five solutions: UW, EC, Bretschneider's histidine-tryptophan-ketoglutarate (HTK), St Thomas's Hospital (ST), and Krebs-Henseleit (KHB) solutions. KHB has no protective ability, but was used as a control solution. It was found that T_1 values were affected by the length of cold storage and the composition of the preservation fluid. With KHB solution, simple cold storage resulted in a rapid rise in T_1 at 1–7 hours after harvesting. These such changes were attributed to tissue oedema. Livers stored in EC solution showed a rise in T_1 largely between 14 and 24 hours of cold storage. These changes were attributed to cell swelling, tissue oedema, and disturbance of cell integrity. They were correlated with direct measurements of liver water content before and after desiccation. Such measurements were in agreement with the T_1 values, which suggested tissue oedema. The best preservation appeared to be obtained with UW solution, where T_1 values actually decreased over the storage period. This was attributed to the osmotic pressure exerted by this preservation fluid. There was good correlation with direct measures of water content.

A similar study by the same group[72] looked at T_2 values of rat livers stored in the same five preservation fluids. A link between T_2 values and tissue water content was obtained. This was correlated with the same parameters of viability as in the previous study.[71] Livers stored in KHB and EC showed a decrease in T_2 values associated with cell swelling, but livers stored in UW showed a decrease in T_2 values, which were again attributed to the osmotic effects of this preservation fluid. It was concluded that ^{1}H MR relaxometry may be a promising viability parameter and is a rapid method for water content assessment.

^{13}C MRS and markers of hepatic viability

Nedelec et al[66] were unable to draw any conclusions on ^{13}C MRS of storage solutions after 24–48 hours cold storage with mouse livers. Vine et al[73] reported on a small number of observations using ^{13}C MRS to investigate hypothermic storage of porcine livers. Signals from glycogen were seen for up to 24 hours during preservation. ^{31}P remains the most suitable MR nucleus for the assessment of transplant viability in cold-stored organs. From the animal work, assessment of ATP levels or measurement of PME:Pi ratio seem to be the best indicators of viability for human liver studies.

Human and large animal studies

The ^{31}P MR spectra obtained at 1.5 T from isolated human livers in cold storage and those from large animals preserved under the same conditions (Fig. 10.6) contain resonances attributable to PME and PDE, which diminish with time. Variable amounts of ATP may be present depending on the length of cold storage. A large, broad Pi resonance, which arises mainly from the UW solution, dominates the spectrum, and at 1.5 T it cannot be easily separated from hepatic Pi. As the PME and PDE peaks diminish, the Pi peak increases (Fig. 10.7). Henze et al[24] used modified transport containers to look at two human livers, one perfused with UW solution and one stored without any preservation solution. Both livers were stored on ice and were studied on a 40 cm horizontal bore system, operating at 4.7 T. ATP was observable in the UW-preserved liver for longer than the unperfused liver. The PME:Pi ratio in the UW liver was greater than in the unperfused organ. It was concluded

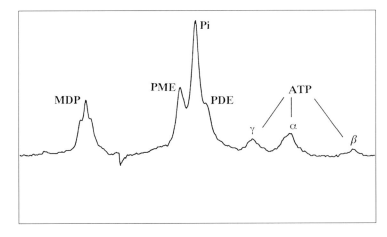

Fig. 10.6. *[31]P MR spectrum from an isolated pig liver after 2 hours in simple cold storage.*
MDP – methylene diphosphonate (external reference standard); PME – phosphomonoester; Pi – broad inorganic phosphate peak arising both from University of Wisconsin (UW) preservation fluid and from the liver itself; PDE – phosphodiester; ATP – adenosine triphosphate.

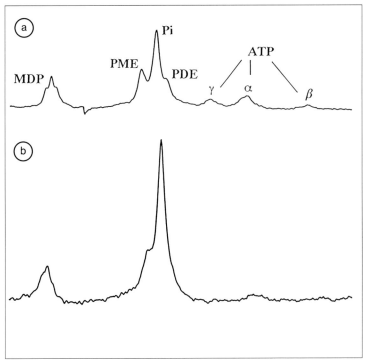

Fig. 10.7. *[31]P MR spectrum from an isolated pig liver, preserved in University of Wisconsin (UW) preservation fluid (a) after two hours simple cold storage and (b) after 48 hours simple cold storage.*
MDP – methylene diphosphonate (external reference standard); PME – phosphomonoester; Pi – broad inorganic phosphate peak arising both from University of Wisconsin (UW) preservation fluid and from the liver itself; PDE – phosphodiester; ATP – adenosine triphosphate. After 48 hours, the ATP is no longer evident, the PDE peak is barely discernible, and the PME peak is left as a shoulder on the much enlarged Pi resonance.

that, because ATP could still be detected by ^{31}P MRS six hours after harvesting, absolute levels of ATP or the PME:Pi ratio were both useful MR indices of hepatic viability.

Wolf et al[74] used ^{31}P MRS on a wide bore clinical system, operating at 1.5 T, to look at a series of 25 human livers preserved by simple cold storage in UW solution. Each liver was examined using a coil system that did not disturb the cold storage arrangements (Fig. 10.8). ATP, when present, could only be demonstrated in the spectra in very small amounts. It did not necessarily correlate with graft function. NADH, which resonates in the α-ATP region of the spectrum, was present in the absence of definable ATP. PME was clearly demonstrated in each stored liver. This peak was observed to decrease with time. The authors suggested that, if the liver was viable, it could rephosphorylate AMP to ATP on reperfusion.[75] The changes in the PME peak reflected changes in AMP levels and were therefore proposed as an MR indicator of viability. Poor function was observed in three of the 25 transplanted organs. Two of these livers showed an elevated PDE level. Since phospholipid breakdown products contribute to the PDE peak, the study concluded that the elevated PDE resonance may reflect cell membrane damage sustained during cold ischaemia.

Our own preliminary studies both in human livers considered to be unsuitable for transplantation on macroscopic grounds and in a pig model, harvested and stored under the same conditions as the clinical transplant programme (Taylor-Robinson SD; unpublished work), have confirmed the time-dependent reductions in the PME, PDE, and ATP peaks (see Fig. 10.7) found by Wolf et al.[74] Each liver was perfused with UW solution and stored on ice. Spectra were obtained on a wide-bore Picker prototype spectroscopy system, operating at 1.5 T. We found that ATP disappears within the first 6 hours after harvesting, but NADH remains in the β-ATP region for up to 48 hours. The PME initially increases over the first few hours concomitant with the ATP reduction. This is presumably due to an increased contribution from AMP caused by ATP hydrolysis. There is then a slow reduction in PME, but this resonance is still evident after 24–36 hours. These findings were similar for both human and pig livers stored under the same conditions. The results from these studies are only preliminary, but the larger body of small animal work suggests that ^{31}P MRS may be useful in the assessment of cold-storage damage.

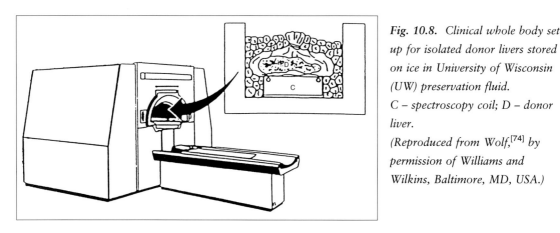

Fig. 10.8. Clinical whole body set up for isolated donor livers stored on ice in University of Wisconsin (UW) preservation fluid.
C – spectroscopy coil; D – donor liver.
(Reproduced from Wolf,[74] by permission of Williams and Wilkins, Baltimore, MD, USA.)

^{31}P MRS and graft rejection

The only method of diagnosing hepatic rejection is histology. The changes of rejection may not be uniform throughout the liver. Percutaneous liver biopsy may therefore be subject to sampling errors and it itself is not without morbidity or even mortality. Ongoing studies[76] are using ^{31}P MRS to look for markers of rejection. No marker of acute rejection has yet been found. Chronic rejection is characterized by bile duct loss.[77] Results to date show an elevated PDE in patients with this condition (Fig. 10.9). This may be related to biliary stasis and may not be a specific indicator of rejection in the liver.

MRS IN CARDIAC TRANSPLANTATION

As with all other organs, the inadequate supply of donors remains a limitation in cardiac transplantation (Table 10.7). The particular challenge in this field is to improve cardiac preservation methods. Current storage techniques limit viability to a maximum of eight hours (see Table 10.2). This restricts the distribution of donor hearts for utilization in distant centres. However, the usual upper clinical limit for cold storage is 4–5 hours in cardiac transplantation. This has been extended up to 9.6 hours in paediatric surgery because of a critical lack in donor organs of suitable size.[78] The other major area of research is the development of a non-invasive technique for diagnosis of graft rejection.

Evaluation of myocardial preservation

The majority of studies have been performed in animals or in extracted human tissue, using small bore magnets, usually operating at 4.7 T.

Hypothermia

Hypothermia allows preservation of cellular metabolites but may itself lead to cell damage. Prior to harvesting, the heart is flushed with a potassium-rich cardioplegic solution. The majority of hearts are preserved by simple cold storage at 4°C in saline or cardioplegic solution.[79] Several studies

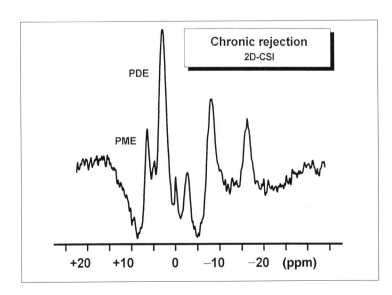

Fig. 10.9. ^{31}P MR spectrum from the liver of a patient with chronic rejection.
PME – phosphomonoester;
PDE – phosphodiester.
The PDE resonance is elevated.

Table 10.7 Thoracic transplants in the British Isles 1983–1993, and number of patients waiting for thoracic transplants at year end 1990–1993. (Data supplied by UKTSSA.)

	Number of transplants			Waiting list		
Year	Heart	Heart–lung	Lung	Heart	Heart–lung	Lung
1983	53	1	—	—	—	—
1984	116	10	—	—	—	—
1985	137	37	—	—	—	—
1986	176	51	3	—	—	—
1987	244	72	4	—	—	—
1988	280	101	16	—	—	—
1989	303	94	39	—	—	—
1990	348	94	52	239	219	69
1991	299	79	72	284	219	111
1992	340	53	89	325	236	145
1993	309	35	95	293	190	151

have tried to define the optimal temperature for cold storage. MRS analysis and functional studies, as described below, suggest a range between 4°C and 12°C.

In isolated rat hearts, storage at 4°C appears to preserve ATP better than at higher temperatures and enhances post-ischaemic haemodynamic recovery, as assessed by left ventricular pressure measurement.[80] However, cold preservation is known to cause cell swelling.[11] Askenasy et al[81] confirmed a greater accumulation of intracellular Na^+ in isolated rat hearts at 4°C using ^{23}Na MRS. In contrast, ^{31}P MRS of this model showed that ATP was better preserved and left ventricular functional recovery was higher after storage at 4°C compared to 15°C or 20°C.

The acute shortage of donor hearts means that studies of organ viability have generally been limited to animal hearts. An alternative source of human tissue for experimental research is atrial appendages obtained from donors at cardiac surgery. There are conflicting results from these human atrial trabecular studies. Deslauriers et al[82] reported optimal ATP preservation at 4°C. They also measured lactate by ^1H MRS and found the highest concentrations at 20°C, which correlated to the lowest pHi on concurrent ^{31}P MRS. This was due to the increased anaerobic glycolysis activity during ischaemia. There is no consensus of opinion, because other authors have found higher ATP levels[83] or better recovery of contractile function after storage at 12°C or 20°C than at 4°C.[84]

PCr can normally be detected in healthy cardiac muscle *in vivo*. Isolated hearts and human atrial trabeculae exhibit an early and marked reduction in PCr for all temperatures between 0°C and 20°C. PCr therefore does not appear to be a practical marker of viability during cold storage in cardiac donor organs. However, it is useful to study PCr recovery after reperfusion as an indicator of graft function.

Improvement in cardioplegic solutions

There is no specific cardioplegic solution for myocardial preservation,[15] and there are still controversies surrounding the ideal cardioplegic and storage solutions. MRS has been used in attempts to clarify this debate.

As with other donor organs, the ideal preservation solution should limit the effects of ischaemia, protecting the heart against HEP depletion[85] and acidosis.[86] The preservative should also guard against free radical and calcium overload.[87,88] The latter is particularly important in cardiac transplantation, since calcium overload has been related to arrhythmias and stunning during reperfusion.[89]

Several common[90] or modified[91] cardioplegic and storage solutions have been tested using ^{31}P MRS during cold ischaemia, to study their effects on myocardial ATP reduction and degree of acidosis. Their efficacy in restoring ATP and PCr after reperfusion has also been assessed.[92] Different constituents of the preservation solutions and parameters of the storage protocol have been altered with varying success. All these studies have been performed with ^{31}P MRS in animal models.

Aussedat et al[93] added a pharmacological anti-ischaemic agent, trimetazidine, to the cardioplegic solution and found a smaller reduction in the ATP:Pi ratio. Malhotra et al[87] added precursors for adenine nucleotide synthesis and glutathione, a free radical scavenger. They found the ATP:Pi ratios and PCr:Pi were better preserved, but the benefit in cardiac functional recovery remained limited.

An alternative approach was used by Wikman-Coffelt et al,[86] who studied the effect of adding alcohol and pyruvate to the preservation solution. Pyruvate is thought to protect the heart against ischaemia, acidosis, and calcium overload. Alcohol may affect transmembrane ion fluxes, decrease intracellular Na$^+$ and prevent oedema. This solution resulted in higher myocardial PCr but no differences in ATP and Pi after *in vitro* reperfusion, as compared to control solutions.

Some of the modifications have not shown any beneficial effect. Transient hypocalcaemic reperfusion[89] or storage in medium containing verapamil, a CEB,[87] did not improve myocardial HEP content or postischaemic functional recovery *in vitro*. A high concentration of pH buffer was added to the cardioplegic solution by Tian et al[94] aiming to reduce acidosis. However, this did not improve HEP preservation or functional recovery.

Low flow perfusion has also been compared to storage by simple immersion.[93] It was thought that this would wash out lactate, thereby reducing intracellular acidosis, but functional recovery was poorer in the perfused grafts than in the immersed hearts.

MRS has been shown to be useful in improving HEP preservation in cardiac storage. Metabolic function analysed by ^{31}P MRS does not fully correlate with mechanical recovery. Therefore this method cannot be used alone to assess organ viability.

Graft rejection

Despite recent advances in immunosuppressive therapy, heart rejection still remains a major cause of death after transplantation.[95] Acute rejection of the graft requires early diagnosis and treatment before the onset of heart failure. Endomyocardial biopsy is the reference method for detecting graft rejection, and it entails a harsh regime of transjugular biopsies, which are often performed once a week for the first two postoperative months.[96] Several non-invasive techniques have been explored to supplement diagnosis, including immunological markers.[97] Echocardiography[98] and radionuclide scintigraphy[97] provide information on left ventricular function, but findings are not specific for rejection. Recently, antimyosin antibody labelled with Indium-111 has distinguished between mild, moderate, or severe rejection.[99]

Animal studies have extensively investigated the non-invasive role of MRS to detect graft rejection. Haug et al[100] found that PCr:ATP, PCr:Pi, and ATP:Pi ratios were markedly decreased in a rat model cardiac rejection. The PCr:Pi ratio was well correlated with histological grading of rejection and residual myocardial function.[101]. These changes preceded histological diagnosis of rejection.[102] A rise in PDE was also found by Fraser et al.[103] This metabolite partly reflects products of cell membrane degradation that are related to immunological or reperfusion injury.

MRS may also be used to monitor the effectiveness of antirejection regimes. Spectra from allografted rats treated with immunosuppressive therapy were identical to the isografted controls.[100] Furthermore, abnormalities in metabolite ratios owing to rejection were reversible under treatment in the dog.[104]

Few human studies have been undertaken. Bottomley et al[105] found that the PCr:ATP ratio was decreased in cardiac rejection, using a 1.5 T system. This change did not correlate with the biopsy score. They were therefore unable to differentiate patients having moderate or mild rejection from those with severe rejection.

In an interesting but preliminary experiment, Mouly-Bandini et al[106] have studied blood plasma from patients with heart transplants using ^{1}H MRS. They found a significant increase in the glycosylated residues of proteins in graft rejecting patients. This may allow rapid screening of rejection using blood samples.

At the moment, more complex transplantation, such as heart–lung grafting, has not been studied by MRS. A new area of research may be graft atherosclerosis. This represents the third most common (between 22% and 50%) cause of death after the second post-operative year.[95] It is possible that myocardial ischaemia associated with the resultant concentric coronary stenosis may be detectable by ^{31}P MRS, which would then represent a valuable, non-invasive diagnostic test.

MRS IN OTHER TRANSPLANTED ORGANS

MRS has been used less extensively in other fields of transplantation. Initial studies again indicate the potential value of this technique for assessing graft viability and rejection.

Pancreatic transplants

Pancreatic transplantation is currently being developed for treatment of diabetes mellitus. The particular advantage over insulin delivery regimes is the excellent continuous glucose homoeostasis that can be achieved.[107] Normoglycaemia may be of benefit in reducing the incidence and severity of diabetic complications.[108]

Henze et al[24] were the first to demonstrate that ^{31}P MRS could be used to give an index of viability of pancreatic donor organs. They adapted the usual storage conditions and transport container to obtain spectra from the human pancreas at 4.7 T in a 40 cm bore spectrometer. Well-defined resonances of γ-, α- and β-ATP were identified. The β-ATP peak declined first, suggesting the β-ATP:γ-ATP ratio as a semiquantitative measure of organ viability. This ratio declined slowly (0.81 at 1 hour, 0.53 at 2 hours, and zero at 20 hours) and this may enable prospective studies of organ viability to be performed. The ideal ratio or parameter to be assessed by MRS is still open to debate, and it is yet to be determined whether the complete disappearance of ATP indicates that the pancreas should no longer be implanted.

Patient and graft survival in pancreatic transplantation have markedly improved over the last 5 years. Belzer[6] found a low incidence of PNF (0.5%) among his series of combined pancreatic and kidney tranplants in Wisconsin, USA. Pancreatic organs were stored in UW solution for an average of 18 hours and it was suggested that the objectives for effective clinical pancreas preservation have been met. MRS may therefore be worthwhile for evaluating pancreatic graft viability when storage conditions are suboptimal.

Pancreatic graft rejection

A particular problem following pancreatic transplantation is the early diagnosis of acute rejection at a potentially reversible stage. There is currently no completely satisfactory diagnostic test. Recurrent hyperglycaemia is a late feature.[109] A decrease in urinary amylase has been reported in rejection of grafts drained through pancreaticocystostomy, but wide fluctuations occur.[110] The problem is partially overcome by combining pancreatic with renal transplantation. Renal rejection is more easily diagnosed and treatment will deal with pancreatic rejection at the same time. However, this approach tends to limit pancreatic transplantation to diabetics with end-stage renal failure.

The early stages of acute graft rejection may be due to ischaemic damage from microthrombi and platelet aggregates.[111] The possibility of using ^{31}P MRS as a diagnostic test of rejection has therefore been examined.[112] A rat model was used to compare functioning pancreatic isografts with allografts performed across a major histocompatibility barrier, to produce acute rejection. At 5 days post-transplant, invasive in vivo ^{31}P MR pancreatic spectra were acquired at 8.5 T, with the surface coil placed directly on the graft. Compared to the isografts, the rejecting allografts showed a decrease in the ATP:Pi ratio (0.73 ± 0.50 vs 1.43 ± 0.62) and a decrease in pH (7.01 ± 0.12 vs 7.22 ± 0.09). These changes are characteristic of partial ischaemia and were supported by in vitro measurement. The spectral changes occurred prior to any detectable alterations in blood sugar. There is therefore clear potential in developing ^{31}P MRS for the diagnosis of acute rejection in humans. However, this may prove difficult because the small bulk and anatomical position of the human pancreas lead to problems in localization.

Pancreatic islet cell transplantation

An alternative approach to whole organ grafting is pancreatic islet cell transplantation, which aims to implant sufficient insulin-producing cells. The insulin secretion test is the accepted method of evaluating islet cell viability, but it is time and labour consuming. Danis et al[113] applied in vitro ^{31}P MRS to assess viability of porcine pancreatic islets using the PDE:PME ratio. Islets were examined in Hanks' solution in a glass cuvette, within an ice containing vessel placed on a surface coil, at 4.7 T. They characterized three types of tissue preparations:

- vital (PDE:PME ratios 0.5 to 0.9);
- damaged (PDE:PME ratio <0.2); and
- necrotic (no PDE, no PME).

These findings showed excellent correlation with function as assessed by insulin secretion after glucose challenge, and with histological estimation of necrosis. The authors suggested that ^{31}P MRS offers a rapid, practical, and suitable method for classifying isolated pancreatic islet viability.

Corneal transplants

Donor corneas are currently examined by microscopic techniques that evaluate tissue viability at the histological level. Assessment is by biomicroscopic examination of clarity and anatomic abnormalities. Endothelial specular microscopy also determines endothelial cell counts, size, and morphologic condition. These methods do not measure metabolic viability of the cornea at the biochemical level. MRS has been suggested as an analytical technique for verifying viability of eye bank tissue before keratoplastic procedures.

Greiner et al[114] were the first to show the feasibility of ^{31}P MRS for monitoring the metabolic status of a single intact cornea preserved in McCarey–Kaufman (MK) solution. They obtained a ^{31}P MR spectrum from a porcine cornea at 4.7 T. The spectra were acquired over 3.72 hours, in order to improve signal to noise. A numerical index of corneal viability (spectral energy modulus) was determined by calculating the ratio of HEP signals (the three NTP resonances) to low energy phosphate signals (Pi and sugar phosphates). This technique has been successfully applied to human corneas,[115,116] and intracorneal pH, ATP:Pi and PME:Pi ratios have also been measured. These studies provided baseline data for the eventual determination of optimal parameters of eye bank corneal viability.

^{13}C MRS studies of corneal viability
Ninety per cent of high energy phosphates are found in the corneal epithelium. Glucose metabolism, via the pentose shunt and glycolytic pathways, provides the energy and metabolites to maintain corneal transparency and integrity. Gottsch et al[117] therefore suggested an alternative MRS-based method of evaluating corneal viability. They used *in vitro* ^{13}C MRS to measure glycolytic activity after adding [1-^{13}C]-glucose to the incubation buffer. Glucose utilization and lactate formation were found to be reduced in the de-epithelialized human corneas (Fig. 10.10). This MRS analysis may be a useful means of assessing the viability of human donor corneas.

Bone marrow transplants

There are few studies applying MRS in bone marrow transplantation. It has been suggested that ^{1}H MRS of red bone marrow may have important future applications.[118] Ballon et al[119] have used volume-selective ^{1}H MRS of the posterior iliac crest to estimate bone marrow cellularity using the lipid -(CH$_2$)$_n$- and water signals. This technique has not been applied in bone marrow transplantation, but it could potentially be developed to follow the success of transplantation and increasing cellularity.

Graft-versus-host disease
Initial studies have also been performed to evaluate MRS as a non-invasive diagnostic test of graft-versus-host disease (GVHD) following bone marrow transplant. The liver is often involved, but it is biopsied infrequently because of an unacceptable

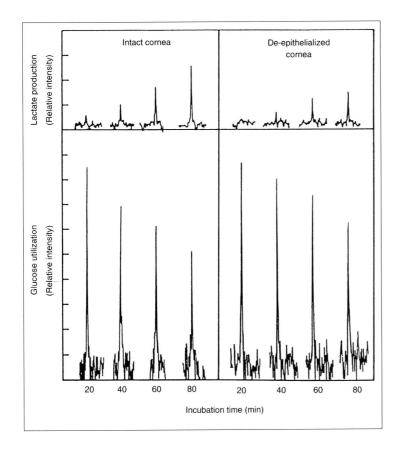

Fig. 10.10. Time course of glucose utilization and lactate production in intact and de-epithelialized human cornea. Intact and de-epithelialized human cornea were incubated in 0.275mM [1–^{13}C]-glucose and nine spectra in array were taken with 10 minutes per spectrum. Resonances for β-glucose at 20-minute intervals are shown that progressively decrease with time. Resonances for lactate progressively increase with time. De-epithelialized corneas show a reduced rate of glycolysis and lactate formation. (Reproduction from Gottsch et al,[117] by permission of Arch Opthalmol, Chicago, IL, USA.)

haemorrhagic risk in thrombocytopenic patients. Acute hepatic GVHD is often inferred from abnormal serum liver enzymes and biopsy-proven GVHD elsewhere in the body (e.g. the skin). Blatter et al[120] therefore assessed the sensitivity of ^{1}H MRS to detect acute hepatic GVHD in an established mouse model. They induced graded intensities of GVHD in these animals and found that T_2 relaxation times correlated directly with the degree of hepatic involvement on histology. Many other conditions may alter tissue water content and thereby change T_2 values. Further work is therefore needed to define diagnostic hepatic MR indices.

Small intestinal transplantation

Extensive small bowel disease, such as Crohn's disease, may destroy the absorptive capacity of the small intestine, leading to the requirement for long-term total parenteral nutrition. Although this affects only a small proportion of patients, the cost in social and economic terms is great. Transplant centres are actively pursuing the possibility of performing reliable and successful small bowel transplantation.

There are few studies using MRS to assess tissue viability for gut transplantation. Kasperk et al[121] have studied rat intestines with ^{31}P MRS at 8.5 T to compare the effects of three preservation fluids (EC, HTK, and Ringer's lactate) on HEP preservation, both at room temperature and at 4°C. EC solution delayed the HEP breakdown and concomitant rise in Pi. It also slowed the development of acidosis. These MRS findings were confirmed by Burgmann et al.[122] They used HPLC to measure HEP and showed that EC solution provided the best small bowel cold preservation of the currently available fluids.

CONCLUSIONS

Organ transplantation has proven to be a safe and effective means of treatment for end-stage organ failure. Improvements in preservation fluids, harvesting and storage techniques have been made in recent years, but despite these advances, organs are sometimes transplanted that function poorly or fail to work altogether. There is a need for an efficient means of assessing organ viability and the ischaemic damage sustained during cold storage. MRS is a non-invasive technique that can be performed in the time-frame of the average cold storage period for most organs. It remains largely a research tool at present. The studies outlined show that ^{31}P MRS may be applied to clinical transplant programmes, and MR indices of viability, such as the PME:Pi ratio, have been correlated with patient outcome.

There are drawbacks to MRS, including expense and the need for specialist staff.[123] However, most transplant centres have clinical MR systems which may be adapted for spectroscopy if these facilities do not already exist. The insensitivity of the technique is a potential problem, which has been addressed successfully in the preliminary studies outlined in this chapter.

The application of MRS to clinical and experimental transplant programmes may require adaptation of existing equipment to avoid either apparent or concealed ferromagnetic metal objects. These may include surgical clips, components of the transport container, and – if used – the perfusion pump.

The body of work on renal grafts has established defined MR parameters of viability. However, the storage and subsequent transplantation of kidneys is generally successful. MRS may therefore have a particular role in assessing viability of the stored liver, where PNF is a potentially fatal complication of transplantation. The lack of alternative non-invasive measures of tissue viability is profound. The expense of MRS becomes insignificant when compared to the cost of retransplantation for PNF, both to society and to the individual.

ACKNOWLEDGEMENTS

We thank Mrs. Julia Warren, Publicity Officer at the United Kingdom Transplant Support Services Authority, Fox Den Road, Bristol BS12 6RR for her assistance in the provision of the UK and the Republic of Ireland transplant statistics for the tables in this chapter.

REFERENCES

1 UKTSSA. *Yearly Transplant Statistics*. (United Kingdom Transplant Support Services Authority: Bristol, UK, 1994).
2 Wing AJ, Chang RWS. Non-heart beating donors as a source of kidneys. *BMJ* 1994; **308**:549–50.
3 Collins GM, Bravo-Shugarman M, Terasaki PI. Kidney preservation for transportation: initial perfusion and 30 hours ice storage. *Lancet* 1969; **2**:1219–23.

4 Southard JH. Advances in organ preservation. *Transplant Proc* 1989; **21**:1195–6.
5 Belzer FO, Southard JH. Principles of solid organ preservation by cold storage. *Transplantation* 1988; **45**:673–6.
6 Belzer FO. Evaluation of preservation of the intra-abdominal organs. *Transplant Proc* 1993; **25**:2527–30.
7 Bretan PN, Baldwin N, Martinez A, Stowe N, Scarpa A, Easley K, et al. Improved renal transplant preservation using a modified intracellular flush solution (PB-2). Characterisation of mechanisms by renal clearance, high performance liquid chromatography, phosphorus-31 magnetic resonance spectroscopy, and electron microscopy studies. *Urol Res* 1991; **19**:73–80.
8 Mor E, Klintmalm GB, Gonwa TA, Solomon H, Holman MJ, Gibbs JF, et al. The use of marginal donors of liver transplantation. *Transplantation* 1992; **53**:383–6.
9 Phillips AO, Snowden SA, Hills AN, Bewick M. Renal grafts from non-heart beating donors. *BMJ* 1994; **308**:575–6.
10 Kamiike W, Burdelski M, Steinhoff G. Adenine nucleotide metabolism and its relation to organ viability in human liver transplantation. *Transplantation* 1988; **45**:138–43.
11 MacKnight ADC, Leaf A. Regulation of cellular volume. *Physiol Rev* 1997; **57**:510–73.
12 Momii S, Koga A. Time related morphological changes in cold-stored rat livers. A comparison of Euro-Collins solution with U.W. solution. *Transplantation* 1990; **50**:745–50.
13 Freeman BA, Crapo JD. Biology of disease. Free radicals and tissue injury. *Lab Invest* 1982; **47**:412–26.
14 Parks DA, Bulkley GB, Granger DN. Role of oxygen free radicals in shock, ischemia and organ preservation. *Surgery* 1983; **94**:428–32.
15 Collins GM, Wicomb WN. New organ preservation solutions. *Kidney Int* 1992; **S38**:S197–S202.
16 Currin RT, Toole JG, Thurman RG, Lemasters JJ. Evidence that Carolina Rinse solution protects sinusoidal endothelial cells against reperfusion injury after cold ischemic storage of rat liver. *Transplantation* 1990; **50**:1076–9.
17 Parrott NR, Forsythe JLR, Matthews JNS, Lennard TW, Rigg KM, Proud G, et al. Late perfusion. A simple remedy for renal allograft primary nonfunction. *Transplantation* 1990; **49**:913–15.
18 Makowka L, Gordon RD, Todo S, Ohkochi N, Marsh JW, Tzakis AG, et al. Analysis of donor criteria for the prediction of outcome in clinical liver transplantation. *Transplant Proc* 1987; **19**:2378–82.
19 Adam R, Astarcioglu I, Azoulay D, Chiche L, Bao YM, Castaing D, et al. Liver transplantation from elderly donors. *Transplant Proc* 1993; **25**:1556–7.
20 Wight C, Rogers CA, Friend PJ. Utilization of available paediatric donor livers in the United Kingdom. *Transplant Proc* 1993; **25**:1550–1.
21 Varty K, Veitch PS, Morgan JDT, Kehinde EO, Donnelly PK, Bell PRF. Response to organ shortage: kidney retrieval programme using non-heart beating donors. *BMJ* 1994; **308**:575.
22 Lanir A, Jenkins RL, Caldwell C, Lee RG, Khettry U, Clouse ME. Hepatic transplantation survival: Correlation with adenine nucleotide level in donor liver. *Hepatology* 1988; **8**:471–5.
23 Bretan PN, Baldwin N, Novick AC, Majors A, Easley K, Ng T, et al. Petransplantation assessment of renal viability by phosphorus-31 magnetic resonance spectroscopy. *Transplantation* 1989; **48**:48–53.
24 Henze E, Lietzenmayer R, Kunz R, Schnur G, Clausen M, Schoser K, et al. Prerequisites and initial experience for the non-invasive routine evaluation of viability of experimental and human organ transplants by magnetic resonance spectroscopy. *Transplant Proc* 1990; **22**:2404–8.
25 Bell JD, Cox IJ, Sargentoni J, Peden CJ, Menon DK, Foster CS, et al. A ^{31}P and ^{1}H NMR investigation of normal and abnormal human liver. *Biochim Biophys Acta* 1993; **1225**:71–7.
26 Iles RA, Griffiths JR, Stevens AN. Perturbations in hepatic energy metabolism. *Biochem Soc Trans* 1985; **13**:843–5.
27 Dhasmana JP, Digerness SB, Geckle JM, Ng TC, Glickson JD, Blackstone EH. The effect of adenosine deaminase inhibitors on the heart's functional and biochemical recovery from ischemia. *J Cardiovasc Pharmacol* 1983; **5**:1040–7.
28 Neubauer S, Krahe T, Schindler R, Horn M, Hillenbrand H, Entzeroth C, et al. ^{31}P Magnetic resonance spectroscopy in dilated cardiomyopathy and coronary artery disease. Altered cardiac high energy phosphate metabolism in heart failure. *Circulation* 1992; **86**:1810–18.
29 Sehr PA, Radda GK. A model kidney transplant studied by phosphorus nuclear magnetic resonance. *Biochem Biophys Res Commun* 1977; **77**:195–202.
30 Sehr PA, Bore PJ, Papatheofanis J, Radda GK. Nondestructive measurement of metabolites and tissue

pH in the kidney by ^{31}P nuclear magnetic resonance. *Br J Exp Pathol* 1979; 60:632–41.

31 Bretan PN, Vigneron DB, James TL, Williams RD, Hricak H, Juenemann KP, et al. Assessment of renal viability by phosphorus-31 magnetic resonance spectroscopy. *J Urol* 1986; 135:866–71.

32 Bretan PN, Vigneron DB, Hricak H, Juenemann KP, Williams RD, Tanagho EA, et al. (1986b). Assessment of renal preservation by phosphorus-31 magnetic resonance spectroscopy: *in vivo* normothermic blood perfusion. *J Urol*, 136:1356–59.

33 Bretan PN, Vigneron DB, Hricak H, Collins GM, Price DC, Tanagho EA, et al. Assessment of clinical renal preservation by phosphorus-31 magnetic resonance spectroscopy. *J Urol* 1987; 137:146–50.

34 Kunikata S, Ishii T, Nishioka T, Uemura T, Kanda H, Matsuura T, et al. Measurement of viability in preserved kidneys with ^{31}P NMR. *Transplant Proc* 1989; 21:1269–71.

35 Lietzenmayer R, Henze E, Knorpp R, Schwamborn C, Clausen M, Schnur G, et al. Application of P-31 nuclear magnetic resonance spectroscopy to a new experimental kidney perfusion model using cadaveric porcine kidney from slaughterhouse. *Nephron* 1991; 57:340–8.

36 Ploeg RJ, van Bocket JH, Langendiijk PTH, Groenewegen M, van der Woude FJ, Persijn GG, et al. Effect of preservation solution on results of cadaveric kidney transplantation. The European Multicentre Study Group. *Lancet* 1992; 340:129–37

37 Ciancabilla FG, Pincemail JF, Defraigne JO, Franssen CL, Carlier PG. The effect of technical conditions and storage medium composition on the phosphomonoesters to inorganic phosphate ratio determined by ^{31}P nuclear resonance spectroscopy in rabbit kidney. *Transplantation* 1993; 56:696–9.

38 Boska MD, Meyerhoff DJ, Twieg DB, Karczmar GS, Matson GB, Weiner MW. Image-guided ^{31}P magnetic resonance spectroscopy of normal and transplanted human kidneys. *Kidney Int* 1990; 38:294–300.

39 Shapiro JI, Haug CE, Weil R, Chan L. ^{31}P nuclear magnetic resonance study of acute renal dysfunction. *Magn Reson Med* 1987; 5:346–52.

40 Shapiro JI, Haug CE, Shanley PF, Weil R, Chan L. ^{31}P nuclear magnetic resonance study of renal allograft rejection in the rat. *Transplantation* 1988; 45:17–21.

41 Chan L, Shapiro JI. Contributions of nuclear magnetic resonance to study of acute renal failure. *Ren Fail* 1989; 11:79–89.

42 Bretan PN, Vigneron DB, Hricak H, Price DC, Yen TSB, Luo JA, et al. Assessment of *in situ* renal transplant viability by ^{31}P MRS: experimental study in canines. *Am Surg* 1993; 59:182–7.

43 Wenzel F, Kugel H, Heindel W, Stippel D, Pichlmaier H, Lackner K. ^{31}P-NMR spectroscopy of the normal and pathologic human transplant kidney. *Proceedings of the 12th Annual Meeting of the Society of Magnetic Resonance Medicine* 1993; 1:206.

44 Grist TM, Charles CH, Sostman HD. Renal transplant rejection: diagnosis with ^{31}P MR spectroscopy. *Am J Roentgenol* 1991; 156:105–12.

45 Kasiske BL, Heim-Duthoy K, Venkatenswara Rao K, Awni WM. The relationship between cyclosporin pharamacokinetic parameters and subsequent acute rejection in renal transplant recipients. *Transplantation* 1988; 46:716–22.

46 Bell JD, Lee JA, Lee HA, Sadler PJ, Wilkie DR, Woodham RH. Nuclear magnetic resonance studies of blood plasma and urine from subjects with chronic renal failure: Identification of trimethylamine-N-oxide. *Biochim Biophys Acta* 1991; 1096:101–7.

47 Chan L, Suleymanlar G, Malhotra D, Lien YH, Shapiro, JI. Nuclear magnetic resonance and biochemical studies to determine the synergistic detrimental effects of renal ischemia on cyclosporin nephrotoxicity in the rat. *Transplant Proc* 1991; 23:711–13.

48 Le Moyec L, Pruna A, Eugene M, Bedrossian J, Idatte JM, Huneau JF, et al. Proton nuclear magnetic resonance spectroscopy of urine and plasma in renal transplantation follow-up. *Nephron* 1993; 65:433–9.

49 Jamieson NV, Sundberg R, Lindell S, Claesson K, Moen J, Vreugdenhil PK, et al. Preservation of the canine liver for 24-48 hours using simple cold storage with U.W. Solution. *Transplantation* 1988; 46:517–22.

50 Adam R, Astarcioglu I, Azoulay D, Morino M, Bao YM, Castaing D, et al. Age greater than 50 years is not a contraindication for liver donation. *Transplant Proc* 1991; 23:2602–3.

51 Gubernatis G, Tusch G, Ringe B, Bunzendahl H, Pichlmayer R. Score-aided decision making in patients with severe liver damage after hepatic transplantation. *World J Surg* 1989; 13:259–65.

52 Schroeder TJ, Pesce AJ, Ryckman FC, Tressler TP, Brunson ME, Pedersen SH, et al. Selection criteria for liver transplant donors. *J Clin Lab Anal* 1991; 5:275–7.

53 Rosenlof LK, Sawyer RG, Broccoli AV, Ishitani MB, Stevenson WC, Pruett TL. Histological comparison of

54 Adam R, Azoulay D, Astarcioglu I, Bao YM, Bonhomme L, Fredj G, et al. Reliability of the MEGX test in the selection of liver grafts. *Transplant Proc* 1991; **23**:2470–1.

55 D'Alessandro AM, Kalayoglu M, Sollinger HW, Hoffman RM, Reed A, Knechtle SJ, et al. The predictive value of donor liver biopsies for the development of primary nonfunction after orthotopic liver transplantation. *Transplantation* 1991; **51**:157–63.

56 Oellerich M, Burdelski M, Ringe B, Lamesch P, Gubernatis G, Bunzendahl H, et al. Lignocaine metabolite formation as a measure of pre-transplant liver function. *Lancet* 1989; **1**:640–1.

57 Oellerich M, Burdelski M, Ringe B, Wittekind C, Lamesch P, Lautz HU, et al. Functional state of the donor liver and early outcome of transplantation. *Transplant Proc* 1991; **23**:1575–8.

58 Marubayashi S, Takenaka M, Dohi K, Ezaki H, Kawasaki T. Adenine nucleotide metabolism during hepatic ischaemia and subsequent blood reflow periods and its relation to organ viability. *Transplantation* 1980; **30**:294–6.

59 Rossaro L, Murase N, Caldwell C, Farghali H, Casavilla A, Starzl TE, et al. Phosphorus-31 nuclear magnetic resonance spectroscopy of rat liver during simple storage or continuous hypothermic perfusion. *J Lab Clin Med* 1992. **120**:559–68.

60 Lanir A, Clouse ME, Lee RGL. Liver preservation for transplant: Evaluation of hepatic energy metabolism by ^{31}P NMR. *Transplantation* 1987; **43**:786–90.

61 Clouse ME, Lanir A, Jones DA, Khettry U, Lee RG. 31P nuclear magnetic resonance evaluation of intermittent perfusion as a method of liver perfusion. *Transplantation* 1988; **45**:1137–8.

62 Bowers JL, Teramoto K, Khettry U, Clouse ME. ^{31}P NMR assessment of orthotopic rat liver transplant viability. The effect of warm ischemia. *Transplantation* 1992; **54**:604–9.

63 Orii T, Ohkohchi N, Satomi S, Taguchi Y, Mori S, Miura I. Assessment of liver graft function after cold preservation using ^{31}P and ^{23}Na magnetic resonance spectroscopy. *Transplantation* 1992; **53**:730–4.

64 Orii T, Ohkohchi N, Satomi S, Taguchi Y, Mori S, Miura I. Assessment of the viability of hepatic cell membrane after cold storage by ^{31}P NMR and ^{23}Na NMR spectroscopy. *Transplant Proc* 1993; **25**:1657–8.

65 Gulian JM, Dalmasso C, Desmoidin F, Scheiner C, Cozzone P. Twenty-four hour hypothermic preservation of rat liver with Eurocollins and UW solutions. A comparative evaluation by ^{31}P NMR spectroscopy, biochemical assays and light microscopy. *Transplantation* 1992; **54**:599–603.

66 Nedelec JF, Capron-Laudereau M, Adam R. Mouse liver metabolism after 24 hours and 48 hours cold preservation using UW, hydroxyethyl starch free UW and Eurocollins solutions: a ^{31}P, ^{13}C NMR spectroscopy and biochemical analysis. *Transplant Proc* 1990; **22**:492–5.

67 Sanchez-Urdazpal L, Gores GJ, Lemasters JJ, Thurman RG, Steers JL, Wahlstrom HE, et al. Carolina Rinse solution decreases liver injury during clinical liver transplantation. *Transplant Proc* 1993; **25**:1574–5.

68 Busza AL, Proctor E, Fuller BJ. Biochemical consequences of reflushing hypothermically stored liver with fresh cold perfusate. Studies on rat liver using ^{31}P NMR spectroscopy. *NMR Biomed* 1989; **2**:115–19.

69 Fuller BJ, Busza AL, Proctor E. Possible resuscitation of liver function by hypothermic reperfusion *in vitro* after prolonged (24 hour) cold preservation – a ^{31}P NMR study. *Transplantation* 1990; **50**:511–35.

70 Fuchinoue S, Teraoka S, Tojimbara T, Nakajima I, Honda H, Ota K. Evaluation of intracellular energy status during liver preservation by ^{31}P NMR spectroscopy. *Transplant Proc* 1988; **20**(suppl 1):953–7.

71 Holzmüller P, Moser E, Reckendorfer H, Burgmann H, Sperlich M. Proton spin-lattice relaxation time as liver transplantation graft viability parameter. *Magn Reson Imag* 1993; **11**:229–39.

72 Holzmüller P, Moser E, Reckendorfer H, Burgmann H, Winklmayr E, Sperlich M. Proton spin–spin relaxation times as liver transplantation graft viability parameter. *Magn Reson Imag* 1993; **11**:749–59.

73 Vine W, Gordon E, Alger J, Flye M. Hepatic preservation assessed by magnetic resonance spectroscopy. *Transplant Proc* 1986; **18**:577–81.

74 Wolf RFE, Kamman RL, Mooyart EL, Haagsma EB, Bleichrodt RP, Slooff MJ. ^{31}P magnetic resonance spectroscopy of the isolated human donor liver – feasibility in routine clinical practice and preliminary findings. *Transplantation* 1993; **55**:949–51.

75 Pontegnie-Istace S, Lambotte L. (1977). Liver adenine nucleotide metabolism during hypothermic anoxia and a recovery period in perfusion. *J Surg Res* 1977; **23**:339–47.

76 Taylor-Robinson S, Menon DK, Sargentoni J, Cox IJ, Bell JD, Bryant DJ, et al. The use of hepatic ^{31}P MRS in patients with cirhosis undergoing liver transplantation: Assessment of liver function pre-transplantation and monitoring of rejection post-transplantation. *Proceedings of the 11th Annual Meeting of the Society of Magnetic Resonance Medicine* 1992; 3:3340.

77 Hubscher SG, Buckels JAC, Elias E. Vanishing bile duct syndrome following liver transplantation – is it reversible? *Transplantation* 1991; 51:1004–10.

78 Alonso De Begona J, Gundry SR, Razzouk AJ, Boucek MM, Bailey LL. Prolonged ischemic times in pediatric heart transplant: Early and late results. *Transplant Proc* 1993; 25:1645–8.

79 McGregor CGA. Cardiac transplantation: Surgical considerations and early postoperative management. *Mayo Clin Proceed* 1992; 67:577–85.

80 Karck M, Vivi A, Tassini M, Schwalb H, Askenasy N, Merchav H, et al. Optimal level of hypothermia for prolonged myocardial protection assessed by ^{31}P Nuclear magnetic resonance. *Ann Thorac Surg* 1992; 54:348–51.

81 Askenasy N, Vivi A, Tassini M, Navon G. Sodium ion transport in rat hearts during cold ischaemic storage: ^{23}Na and ^{31}P NMR study. *Magn Reson Med* 1992; 28:249–63.

82 Deslauriers R, Keon WJ, Lareau S, Moir D, Saunders JK, Smith IC, et al. Preservation of high-energy phosphates in human myocardium. *J Thorac Cardiovasc Surg* 1989; 98:402–12.

83 Lareau S, Keon WJ, Wallace JC, Whitehead K, Mainwood GW, Deslauriers R. Cardiac hypothermia: ^{31}P and ^{1}H NMR spectroscopic studies of the effect of buffer on preservation of human heart atrial appendages. *Can J Physiol Pharmacol* 1991; 69:1726–32.

84 Keon WJ, Hendry PJ, Taichman GC, Mainwood GW. Cardiac transplantation: The ideal myocardial temperature for graft transport. *Ann Thorac Surg* 1988; 46:337–40.

85 Tian G, Biro GP, Butler KW, Xiang B, Vu C, Deslauriers R. The effects of Ca^{2+} on the preservation of myocardial energy and function with University of Wisconsin solution. A ^{31}P nuclear magnetic resonance study of isolated blood perfused Langendorff pig hearts. *J Heart Lung Transplant* 1993; 12:81–8.

86 Wikman-Coffelt J, Wagner S, Wu S, Parmley W. Alcohol and pyruvate cardioplegia. *J Thorac Cardiovasc Surg* 1991; 101:509–16.

87 Malhotra D, Zhou HZ, Konh YL, Shapiro JI, Chan L. Improvement in experimental cardiac preservation based on metabolic considerations. *Transplantation* 1991; 52:1004–8.

88 Aussedat J, Ray A, Lortet S, Reutenauer H, Grably S, Rossi A. Phosphorylated compounds and function in isolated hearts: a ^{31}P-NMR study. *Am J Physiol* 1991; 260:H110–H117.

89 Chambers DJ, Harvey DM, Braimbridge MV, Hearse DJ. Transient hypocalcemic reperfusion does not improve post-ischaemic recovery in the rat heart after preservation with St. Thomas' Hospital cardioplegic solution. *J Thorac Cardiovasc Surg* 1992; 104:344–56.

90 English TA, Foreman J, Gadian DG, Pegg DE, Wheeldon D, Williams SR. Three solutions for preservation of the rabbit heart at 0°C. *J Thorac Cardiovasc Surg* 1988; 96:54–61.

91 Karck M, Vivi A, Tassini M, Schwalb H, Askenasy N, Navon G, et al. The effectiveness of University of Wisconsin solution on prolonged myocardial protection assessed by phosphorus 31-nuclear magnetic resonance spectroscopy and functional recovery. *J Thorac Cardiovasc Surg* 1992; 104:1356–64.

92 Whitman GJR, Roth RA, Kieval RS, Harken AH. Evaluation of myocardial preservation using ^{31}P NMR. *J Surg Res* 1985; 38:154–61.

93 Aussedat J, Kay L, Ray A, Verdys M, Harpey C, Rossi A. Improvement of long term preservation of isolated arrested rat heart: Beneficial effect of the antiischaemic agent trimetazidine. *J Cardiovasc Pharmacol* 1993; 21:128–35.

94 Tian G, Mainwood GW, Biro GP, Smith KE, Butler KW, Lawrence D, et al. The effect of high buffer cardioplegia and secondary cardioplegia on cardiac preservation and postischemic functional recovery: a ^{31}P NMR and functional study in Langendorff perfused pig hearts. *Can J Physiol Pharmacol* 1991; 69:1760–8.

95 Andreone PA, Olivari MT, Ring WS. Clinical considerations of cardiac transplantation in organ transplantation: Preoperative and postoperative evaluation. *Radiol Clin North Am* 1987; 25:357–65.

96 Tazelaar HD, Edwards WD. Pathology of cardiac transplantation: recipient hearts (chronic heart failure) and donor hearts (acute and chronic rejection). *Mayo Clin Proc* 1992; 67:685–96.

97 Kemkes BM. Non-invasive assessment methods to detect rejection following cardiac transplantation. In: Wallwork J, ed. *Heart and Heart–Lung Transplantation*. WB Saunders: Philadelphia, 1989:299–312.

98 Störk T, Möckel M, Eichstädt H, Walkowiak T, Siniawski H, Muller R, et al. Noninvasive diagnosis of cardiac allograft rejection by means of pulsed doppler and M-mode ultrasound. *J Ultrasound Med* 1991; 10:569–75.

99 Latre JM, Arizón JM, Jiménez-Heffernan A, Anguita M, Gonzalez FM, Rubio FL, et al. Noninvasive radioisotopic diagnosis of acute heart rejection. *J Heart Lung Transplant* 1992; 11:453–7.

100 Haug CE, Shapiro JI, Chan L, Weil R. P-31 nuclear magnetic resonance spectroscopic evaluation of heterotopic cardiac allograft rejection in the rat. *Transplantation* 1987; 44:175–8.

101 Walpoth BH, Tschopp A, Lazeyras F, et al. Magnetic resonance spectroscopy for assessing myocardial rejection in the transplanted rat heart. *J Heart Lung Transplant* 1993; 12:271–82.

102 Canby RC, Evanochko WT, Barrett LV, Kirklin JK, McGiffin DC, Sakai TT, et al. Monitoring the bioenergetics of cardiac allograft rejection using *in vivo* P-31 nuclear magnetic resonance spectroscopy. *J Am Coll Cardiol* 1987; 9:1067–74.

103 Fraser CD, Chacko VP, Jacobus WE, Soulen RL, Hutchins GM, Reitz BA, et al. Metabolic changes preceding functional and morphologic indices of rejection in heterotopic cardiac allografts. *Transplantation* 1988; 46:346–51.

104 Fraser CD, Chacko VP, Jacobus WE, Hutchins GM, Glickson J, Reitz BA, et al. Evidence from ^{31}P nuclear magnetic resonance studies of cardiac allografts that early rejection is characterized by reversible biochemical changes. *Transplantation* 1989; 48:1068–70.

105 Bottomley PA, Weiss RG, Hardy CJ, Baumgartner WA. Myocardial high-energy phosphate metabolism and allograft rejection in patients with heart transplants. *Radiology* 1991; 181:67–75.

106 Mouly-Bandini A, Vion-Dury J, Sciaky M, Viout P, Confort-Gouny, S, Mesana T, et al. Résidus glycosylés plasmatiques mis en evidence par spectroscopie de RMN du proton: Intérêt dans la détection du rejet de greffe cardiaque. *Presse Med* 1992; 21:2003–4.

107 Brons IGM, Calne RY, Rolles K, Williams PF, Fishwick NG, Evans DB. Glucose control after simultaneous segmental pancreas and kidney transplantation. *Transplant Proc* 1987; 19:2288–9.

108 Hanssen KJ, Dahl-Jorgensen K, Lauritzen T, Feldt-Rasmussen B, Brinkmann-Hansen O, Deckert T. Diabetic control and microvascular complications: The near normoglycaemic experience. *Diabetologia* 1986; 29:677–84.

109 Sutherland DER, Goetz FC, Najarian JS. Intraperitoneal transplantation of immediately vascularized segmental pancreatic grafts without duct ligation: A clinical trial. *Transplantation* 1979; 28:486–91.

110 Tyden G, Reinholt F, Brattstrom C, Lundgren G, Wilczek H, Bolinder J. Diagnosis of rejection in recipients of pancreatic grafts without enteric exocrine diversion by monitoring pancreatic juices, cytology and amylase excretion. *Transplant Proc* 1987; 19:3892–4.

111 Jurewicz WA, Buckels JAC, Dykes JGA, Chandler ST, Gunson BK, Hawker RJ. 111 Indium platelets in monitoring pancreatic allografts in man. *Br J Surg* 1985; 73:228–31.

112 Morris GE, Williams SR, Proctor E, Gadian DG, Browse NL. ^{31}P nuclear magnetic resonance of rat pancreatic grafts. *Transplantation* 1989; 47:779–84.

113 Danis J, Tunggal B, Weyer J, Meyer G, Saad S, Isselhard W, et al. ^{31}P NMR spectroscopy for *in vitro* viability testing of porcine pancreatic islets. *J Surg Res* 1990; 49:534–8.

114 Greiner JV, Kopp SJ, Glonek T. Non-destructive metabolic analysis of a cornea with the use of phosphorus nuclear magnetic resonance. *Arch Opthalmol* 1984; 102:770–1.

115 Greiner JV, Lass, JH, Glonek T. *Ex vivo* metabolic analysis of eye bank corneas using phosphorus nuclear magnetic resonance. *Arch Opthalmol* 1984; 102:1171–3.

116 Greiner JV, Lass JH, Glonek T. Noninvasive metabolic analysis of eye bank corneas: A magnetic resonance spectroscopy study. *Graefe's Arch Clin Exp Opthalmol* 1989; 227:295–9.

117 Gottsch JD, Chen C-H, Aguayo JB, Cousins JB, Strahlman ER, Stark WJ. Glycolytic activity in the human cornea monitored with nuclear magnetic resonance spectroscopy. *Arch Opthalmol* 1986; 104:886–9.

118 Schick F, Bongers H, Jung WI, Skalej M, Lutz O, Claussen CD. Volume-selective proton MRS in vertebral bodies. *Magn Reson Med* 1992; 26:207–17.

119 Ballon D, Jakubowski A, Gabrilove J, Graham MC, Zakowski M, Sheridan C, et al. *In vivo* measurements of bone marrow cellularity using volume – localized proton NMR spectroscopy. *Magn Reson Med* 1991; 19:85–95.

120 Blatter DD, Crawford JM, Ferrara JLM. Nuclear

magnetic resonance of hepatic graft-versus-host disease in mice. *Transplantation* 1990; **50**:1011–18.

121 **Kasperk R, Kasperk C, Werk W, Leibfritz D.** ^{31}P magnetic resonance studies of isolated rat small intestine: influence of preservation in Ringer's lactate, EuroCollins and Bretschneider solution. *Transplant Proc* 1987; **5**:4122.

122 **Burgmann H, Rekendorfer H, Sperlich M, Spieckermann PG.** Viability testing of cold stored small bowel using 'everted sac' technique. *Transplant Proc* 1992; **24**:1085–6.

123 **Fuller BJ, Busza AL.** The application of nuclear magnetic resonance spectroscopy to assess viability in stored tissues and organs. *Cryobiology* 1989; **26**:248–55.

124 **Bretan PN.** Extracorporeal renal preservation. In: Novick AE, ed. *Stewart's Textbook of Operative Urology*. Williams and Wilkins: Baltimore, 1989.

11 MRS in clinical oncology, with particular reference to applications to monitor therapy

Wolfhard Semmler, Peter Bachert, Gerhard van Kaick

INTRODUCTION

The emergence of *in vivo* nuclear magnetic resonance spectroscopy (MRS) has created a window for the detection of metabolites and the observation of metabolic processes in living tissue. The high spectral resolution of nuclear magnetic resonance allows the identification of a compound by the characteristic resonance frequency of nuclear spins located at a specific site in the molecule (chemical shift).

Experimental studies *in vitro* and *in vivo*,[1–9] showed that MRS can detect different stages of tumor growth and can be used to observe tumor biochemistry and to monitor chemotherapy and radiotherapy. Accordingly, MRS has been applied in a large number of clinical studies in oncology with the aim of exploring the potential of the technique for tumor characterization, staging, and therapy monitoring in patients.[10–27]

This chapter discusses the current status of MRS in oncology, with an emphasis on clinical applications. The choice of studies discussed here partly reflects the personal view of the authors.

DIAGNOSTIC PROBLEMS IN ONCOLOGY

The major diagnostic problems in oncology are early tumor detection, tumor grading, and tumor staging. After diagnosis of a malignant tumorous disease and selection of an appropriate therapy, monitoring is important for optimizing treatment and predicting response to therapy. Since MRI was introduced into clinical practice, MRS has been evaluated for its possible contribution in this field. Whereas the value of MRI for detection and staging of tumors was quickly recognized and MRI was then transferred into the clinical practice, the contribution of MRS to the solution of clinical problems in oncology is not yet decided.

Tissue heterogeneity

Tumor tissue is heterogeneous and contains areas of vital and hypoxic tissue as well as anoxic–necrotic cells, depending on vascularization and local supply of oxygen and nutrients.[3,4] Sostman et al[28] observed gross regional variations of metabolic parameters derived from localized ^{31}P MR spectra of tumor tissue. Similar results were obtained for ^{1}H MRS.[29] Owing to the large detection volumes that are required because of the low sensitivity of the method, in vivo MR spectra can only reflect an overall signature of the metabolic state of the malignant lesion and not the microscopic heterogeneity of the tissue.

Sensitivity

As a consequence of the low energy of the quanta involved in the transitions between nuclear spin states, the sensitivity of MR is extremely low. In the radiofrequency range, a coherent superposition of a tremendous number of photons must be generated to obtain a detectable signal. Accordingly, the detection limit of MRS for metabolite concentrations is fairly high, $\geq 10^{-4}$ mol per liter, much larger than the concentrations of approximately 10^{-12} mol per liter that can be detected by radionuclide tracer techniques. Spatial resolution of MRS is therefore strongly limited, i.e. large tissue volumes (> 30 ml) must be excited in order to obtain localized in vivo MR spectra of nuclei other than protons in acquisition times in the order of 10 minutes.

Localization

Spatial localization techniques for multinuclear in vivo MRS that are useful in clinical studies of tumor patients, can be classified as follows:
- surface coils (also used in combination with B_1-field profiling techniques);[30–35]
- slice-selective excitation or presaturation;[36–38]
- single-voxel acquisition techniques;[39–44] and
- chemical shift imaging (also called spectroscopic or metabolic imaging).[45–47]

Surface coils and slice-selective saturation allow the acquisition of good-quality in vivo MR spectra from superficial tumors, whereas single-voxel acquisition and chemical shift imaging (CSI) are appropriate for MRS of tumors located in the human brain, liver, bone marrow, and other deep lying tumors in soft tissues. In particular, ^{1}H stimulated and double spin-echo, ^{1}H CSI, and ^{31}P CSI are useful localization techniques for in vivo spectroscopy of human brain tumors.

NUCLEI FOR IN VIVO MRS IN HUMANS

Studies in tumor patients by means of in vivo MRS are based on the observation of the nuclei ^{1}H, ^{13}C, ^{19}F, and ^{31}P. In vivo MRS with other nuclei,[48] have hitherto found no application in tumor diagnostics and monitoring of tumor therapy. Whereas for ^{1}H, ^{31}P, and ^{13}C (in fatty acids), the abundance in living tissue is sufficient to acquire good-quality spectra within measurement times ≤ 10 minutes, ^{19}F is a trace element in the tissue with concentration far below the detection limit of MRS (larger amounts of fluorine are immobilized and hence not detectable by MR). However, after administration of ^{19}F-containing drugs, the signal of this nucleus can be detected in vivo, allowing pharmacokinetic studies without the presence of interfering background signal from the body.

Multinuclear spectroscopic studies (i.e. the observation of more than one spin species in the same region) enhance information of in vivo MRS because the variety of detectable metabolites is larger and peak assignments can be mutually verified. Multinuclear experiments may also allow the study of almost the complete pathway of phospholipid

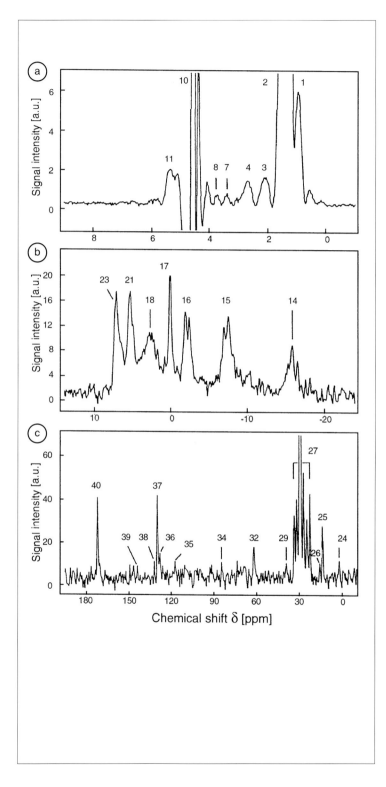

Fig. 11.1. In vivo 1.5 T *nuclear magnetic resonance spectra of a malignant fibrous histiocytoma in the lower leg of a 55-year-old patient. (Modified from Bachert et al.*[49]*)*

(a) *Localized water-suppressed stimulated echo* ^1H *MR spectrum obtained from a 27 ml VOI in the tumor. (1–3 lipids and amino acids; 4 ?; 7 ?; 8 ?; 10 water; 11 oleic acid.)*

(b) ^{31}P *MR spectrum obtained with a double surface coil of 5 cm ⌀ for* ^{31}P *and 10 cm ⌀ for* ^1H *frequency and WALTZ-8* ^1H*-decoupling. (14 β-NTP; 15 α-NTP and α-NDP; 16 γ-NTP and β-NDP; 17 PCr; 18 PDE; 21 P$_i$; 23 PME, mainly PE)*

(c) ^{13}C *MR spectrum obtained with a double surface coil of 5 cm ⌀ for* ^{13}C *and 10 cm ⌀ for* ^1H *frequency and WALTZ-4* ^1H*-decoupling. 24 isoleucine; 25 TAG and PL; 26 isoleucine; 27 TAG and PL; 29 ethanolamine-containing compounds; 32 glycerol backbone of TAG and PL; 34 carbohydrates; 35 aromatic amino acids; 36 polyunsaturated acyl chains of TAG and PL; 37 mono- and polyunsaturated acyl chains of TAG and PL; 38 aromatic amino acids; 39 carbohydrates; 40 TAG and PL* synthesis and degradation in normal and neoplastic tissue.[49] Data obtained in an *in vivo* ^{1}H, ^{31}P-{^{1}H}, and ^{13}C-{^{1}H} MRS study on malignant fibrous histiocytoma are shown in Fig. 11.1.

^{31}Phosphorus

Phosphoryl groups in nucleoside-5′-triphosphate (NTP) and phosphocreatine (PCr) in the living tissue can be detected with good signal-to-noise ratio (SNR) by means of *in vivo* ^{31}P MRS. The major contribution of the NTP and NDP signals originates from adenine ribonucleotides (ATP, ADP). PCr, a temporary storage form of chemical energy, is in equilibrium with ATP. The corresponding turnover rates and enzyme activities (creatine kinase reaction) can be measured non-invasively in the tissue by means of ^{31}P MRS.[50–54]

Broad resonance bands at the low-field and the high-field side of the inorganic phosphate (P_i) signal in ^{31}P MR spectra of brain, liver, and tumor tissue are attributed to phosphomonoesters (PME) and phosphodiesters (PDE) (see Fig. 11.1). Frequency resolution of these complex signals can be improved substantially by means of proton-decoupling, which reveals signals from intermediates of phospholipid metabolism.[49,55] In addition, the application of ^{31}P-{^{1}H} double resonance techniques produces *in vivo* ^{31}P MR signal enhancements of up to 70% (^{31}P-{^{1}H} nuclear Overhauser effect, NOE).[49,56,57] This effect is the consequence of heteronuclear cross-relaxation mediated by intermolecular dipole–dipole interaction between phosphorus nuclei and protons of coordinated water molecules.

In spectra of neoplastic tissue, signal components of PME are assigned to phosphorylcholine (PC) and phosphorylethanolamine (PE), and PDE signals are assigned to glycerophosphorylcholine (GPC) and glycerophosphorylethanolamine (GPE).[2,58,59] Upon ^{1}H-decoupling, the major component of the PDE resonance of leg skeletal muscle tissue collapses into a narrow singlet resonance which is attributed to GPC on the basis of its chemical shift.[49,55] ^{31}P MRS allows the detection of ischemia and hypoxia,[60] and this gives an objective parameter for monitoring tumor therapy. The P_i chemical shift is a measure of the apparent intracellular pH, and it has been found to be elevated in tumors.[10,11,61,62]

Hydrogen

High-resolution *in vivo* ^{1}H MRS has its major field of application in studies of the human brain. In other tissues and organs, inhomogeneous line-broadening, physiological motion, and the presence of large amounts of mobile lipids reduce the information content of localized proton MR spectra. The number of reported studies[63–65] is therefore smaller. Resonances resolved in water-suppressed ^{1}H MR spectra of human brain tissue are assigned to N-acetyl-aspartate and N-acetyl-aspartyl-glutamate (NAA, NAAG), total creatine (tCr – phosphocreatine and creatine), total choline (tCho), glutamate and glutamine (Glu, Gln), aspartate (Asp), inositols (Ins), and lactate (Lac)[42,66] (Fig. 11.2).

Data from MR experiments on cell cultures and tissue extracts indicate that the NAA signal is a marker for intact and active neurons. It increases during brain development and decreases during senescence and when neurons are lost. Strongly increased levels of Lac in ischemic tissue and in tumors can be detected by editing the characteristic doublet signal of β-methyl protons, which resonate at 1.33 ppm. *In vivo* ^{1}H MR spectra obtained at 1.5 T are complex, owing to the superposition of

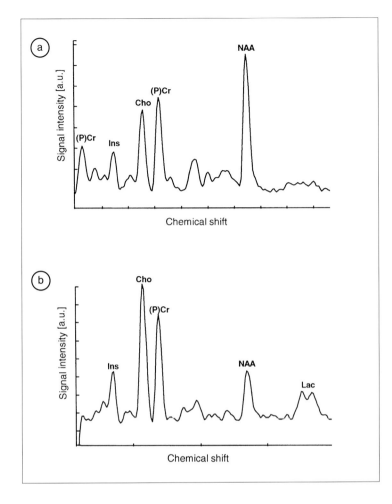

Fig. 11.2. Localized water-suppressed stimulated echo ^{1}H MR spectra obtained at 1.5 T from 8 ml VOIs ($2 \times 2 \times 2$ cm^3).
(a) Normal brain tissue (28-year-old volunteer).
(b) Astrocytoma II (26-year-old patient). The tumor spectrum shows the doublet signal of β-methyl protons of lactic acid (Lac, resonating at 1.3 ppm relative to the residual water resonance at 4.7 ppm), reduced signal intensity of N-acetyl-aspartate (NAA, at 2.0 ppm) and a strong resonance of N-trimethyl protons of choline-containing compounds (Cho, at 3.2 ppm) which exceeds that of N-methyl protons of creatine and phosphocreatine [(P)Cr, at 3.0 ppm].

unresolved multiplets and a multitude of resonances of various metabolites. The interpretation of broad peaks, which could originate from signals of macromolecules,[67] is still under discussion.

^{19}Fluorine

Clinical studies that employ ^{19}F MRS focus on the detection of fluorinated drugs and on the observation of their metabolization.

In particular, the pharmacokinetics of 5-fluorouracil (5-FU), a fluoropyrimidine applied in the chemotherapy of liver metastases of advanced colorectal cancer, has been examined by several authors in patients during their regular treatment course by means of ^{19}F MRS.[68–75] In these studies, the resonances of 5-FU, its major catabolite α-fluoro-β-alanine (FBAL), and, in very rare cases, cytotoxic anabolites (5-FUranuc, 5-FU nucleosides, and nucleotides)[69] and the catabolic intermediate 5-fluoro-5,6-dihydrouracil (DHFU)[75] have been detected in the liver of patients after administration of the drug. A spectrum with 5-FUranuc (resonating at $\delta = \pm 23.6$ ppm), 5-FU ($\delta = \pm 18.6$ ppm), and FBAL ($\delta = 0$ ppm) signal in the liver, obtained after administration of 5-FU, is shown in Fig. 11.3. Although attributed to

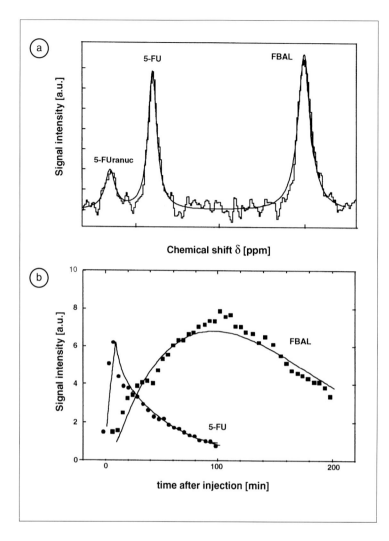

Fig. 11.3. In vivo 60.0 MHz (1.5 T) ^{19}F MR spectrum (a) obtained after 5-fluorouracil (5-FU) administration by means of a 15 cm ⌀ surface coil placed over the liver region of a 48-year-old patient undergoing fluoropyrimidine chemotherapy (modified from Semmler et al.[69]) Resonances of 5-FU, its major catabolite α-fluoro-β-alanine (FBAL) [chemical shift reference (set to 0)], and of 5-fluorouracil nucleotides and nucleosides (5-FUranuc) are found at 18.6 ± 0.3, and 23.6 ± 2.4 ppm, respectively. (b) Integrated peak areas of ^{19}F MR signals of 5-FU and FBAL versus time after start of intra-arterial injection of 5-FU. Exponential functions were fitted to the data points.

FBAL, this broad peak is presumably a superposition of the signals of FBAL and of its direct precursor in the catabolic pathway, α-fluoro-β-ureido propanoic acid (FUPA).

In order to improve the fluorine SNR and fluorine spectral resolution *in vivo*, ^{19}F-[^{1}H] double resonance techniques, i.e. dynamic nuclear polarization (^{19}F-[^{1}H] NOE) and ^{1}H-decoupling, have been investigated.[76,77] Experiments in model solutions showed substantial signal enhancements of the order of 40–50% of the 5-FU and FBAL resonances upon excitation of the proton spin system and a complete removal of FBAL multiplet splittings, which reduces the linewidth of this resonance by a factor of 6.[77,78] ^{1}H-decoupling may allow the resolution of FBAL and FUPA signals *in vivo*.

In vivo fluorine signal localization by means of phase encoding chemical shift imaging (CSI) has been applied successfully in the examination of patients undergoing 5-FU chemotherapy. The localized kinetics of the *in vivo* ^{19}F MR signal intensity of FBAL in 6 × 6 × 4 cm^3 voxels within the liver of the treated patient could be followed in time frames of 12 minutes during a period of 60 minutes after infusion of 5-FU.[78]

^{13}Carbon

Notwithstanding the low MR sensitivity and low natural abundance (1.1%) of the ^{13}C nucleus, good-quality *in vivo* carbon signals of mobile lipids, mainly triacylglycerides, can be acquired in measurement times in the order of 10–20 minutes.[49,79] Proton decoupling is mandatory for suppressing the strong scalar spin–spin couplings of ^{13}C nuclei with directly bound protons (coupling constant $J_{CH} \sim 120$–180 Hz). Accordingly, the experimental equipment for ^{13}C MRS must allow magnetic double resonance, which is also useful for ^{19}F and ^{31}P MRS.[56, 76, 77] Carbon resonances of metabolites other than lipids (e.g. amino acids) can be detected in the tissue when techniques of dynamic nuclear polarization (^{13}C-[^{1}H] NOE) are employed to improve SNR in conjunction with sensitive antenna systems.[79]

A number of studies in volunteers and patients have been performed by means of ^{13}C MRS.[79–86] However, there are only a few clinical studies on ^{13}C MRS in oncology. Halliday et al report on the differentiation of human tumors from non malignant tissue.[87] ^{13}C-[^{1}H] double resonance has been applied in examinations of malignant histiocytoma[49] and breast carcinoma.[88] There is one report on ^{13}C chemical shift imaging of human brain tumors where localized signals from fatty acids in meningioma were able to be mapped.[89]

TUMOR DIAGNOSIS

The main problem of cancer diagnosis is the detection and localization of the tumor in the early phase of the disease. Since the tumor mass is small at this stage, the contribution of MRS is limited because of the large detection volume. In the case of diffuse cancer involvement, however, which can be overlooked by other imaging modalities, MR spectra may provide information on the presence of neoplastic tissue within the suspected region. However, to our knowledge, no case has been reported in which MRS was able to contribute significantly to early cancer detection and was indispensable for the clinical decision.

The same is true for the application of MRS to tumor staging, the other significant diagnostic problem in oncology. A contribution of MRS would be important because the detection of multiple tumor localizations, metastases, and tumor infiltrations into the surrounding tissue drastically changes patient management.

On the other hand, the potential of *in vivo* MRS for tumor characterization is presumably high, though still controversial.[15,90,91] Because metabolism and physiology (e.g. turnover rates of high-energy phosphates and of membrane phospholipids and tissue perfusion) of the neoplastic tissue differ from that of normal tissue of the same region, MR spectra of tumors are expected to contain information on the proportion of vital tumor cells to necrotic cells, on vascularization, oxygen supply, and acidosis.

Brain

Different classification systems for human brain tumors have been developed. The TNM classification specifies the topical relation of the tumor, e.g. supratentorial versus infratentorial localization. In addition, the histomorphology must be taken into account in order to differentiate primary tumors, which result from a transformation of cells in the

brain tissue, and secondary brain tumors (metastases). Zülch[92,93] proposed a classification system that considers brain development:
- neuroepithelial tumors (e.g. medulloblastoma, glioma);
- mesodermal tumors (e.g. meningioma, angioblastoma);
- ectodermal tumors (e.g. craniopharyngioma);
- maldevelopmental tumors (e.g. epidermoid, teratoma);
- tumors from vessel malformations (e.g. angioma, aneurysms); and
- other tumors (e.g. metastases).

Today, brain tumors are commonly graded in 4 groups (I–IV) according to Kernohan's scheme.[94]

Phosphorus MRS

In ^{31}P MR spectra of human brain tumors *in vivo*, no resonances that are not also detected in spectra of normal brain tissue have been found. Differences, however, have been observed in absolute and relative signal intensities of the various phosphorus resonances and in P_i-chemical shift[24,95,96] (Figs 11.4, 11.5). Quantitative evaluation of the spectra, which includes fitting of the resolved resonances and integration of the fit curves, shows elevated signals of phosphomonoester (PME) and inorganic phosphate (P_i) and reduced signal intensities of phosphodiester (PDE) and phosphocreatine (PCr) in the spectrum of meningioma when compared to the spectrum of the contralateral, unaffected hemisphere.[97,98] In contrast, the differences of *in vivo* ^{31}P MR spectra of glioblastoma, neuroepithelial tumor, astrocytoma, and normal brain tissue are less significant.

High-grade gliomas are grossly heterogeneous on macroscopic and microscopic scales as demonstrated by MRI and histopathological studies. Because typical *in vivo* ^{31}P MR detection volumes are > 30 cm^3 in examinations of human organs

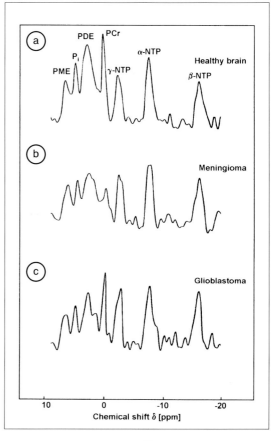

Fig. 11.4. Localized in vivo ^{31}P MR spectra of normal and neoplastic tissue in the human brain obtained by means of the ISIS pulse sequences. (Modified from Heindel et al.[95])
(a) Normal brain tissue.
(b) Meningioma (a mesodermal tumor, which originates from meningial and not from brain tissue).
(c) Glioblastoma.

and tissues, spectra mainly reflect the signal contributions of the different cell fractions (vital, hypoxic, anoxic–necrotic tissue) rather than the microscopic heterogeneity of the tumor tissue. The problem of large voxel sizes of *in vivo* MRS is less severe when the relatively homogeneous meningiomas and pituitary adenomas are examined.

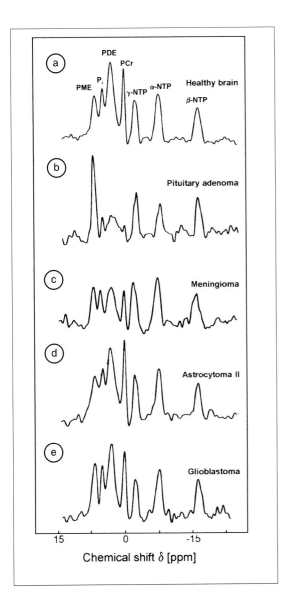

Fig. 11.5. Localized in vivo ^{31}P MR spectra of normal and neoplastic tissue in the human brain obtained by means of the ISIS pulse sequence from voxels of sizes of 41–220 cm^3. (Modified from Arnold et al.[14])
(a) Normal brain tissue.
(b) Pituitary adenoma.
(c) Meningioma.
(d) Astrocytoma grade II.
(e) Glioblastoma.

Negendank[24] reviewed reported data on primary brain tumors with respect to signal variations of in vivo ^{31}P MR resonances and found that 81% of examined high-grade gliomas (Kernohan III–IV) showed low PDE and 47% showed low PCr signal intensities, while 96% of the meningiomas examined showed low PDE and low PCr signal intensities. The apparent tissue pH values of high-grade gliomas and meningiomas were significantly higher than in normal brain tissue. High-grade glial tumors showed decreased PDE:NTP and increased PME:NTP signal intensity ratios.

Hwang et al[99] observed decreased PDE:NTP and PCr:NTP signal intensity ratios and elevated pH in high-grade gliomas and in a menigioma, in agreement with results reported by other groups.[100–102]

The tissue pH value is in the normal range in low-grade glial tumors; it is often elevated in high-grade glial tumors. Heiss et al[96] measured an average pH of 7.17 in these tumors, whereas Arnold et al[14] observed a pH of 7.04, which is in the normal range (for comparison, an average pH value of 7.07 ± 0.02 for muscle tissue was obtained by Lorentzian line fitting of in vivo ^{31}P MR spectra of the calf muscle of 17 volunteers[103] measured with use of WALTZ-8-^{1}H-decoupling and a concentric surface coil system of 5 cm diameter ($\varnothing$) for ^{31}P and 10 cm $\varnothing$ for ^{1}H frequency). The pH value derived from in vivo P_i–PCr chemical shift differences could possibly be used to grade gliomas when those inconsistensies are explained.

Ependymomas also show low PDE:NTP signal intensity ratios, while low PDE and high PME levels are observed in in vivo ^{31}P MR spectra of benign pituitary adenomas. The pH value is elevated in ependymomas and in the normal range in pituitary adenomas when compared to healthy

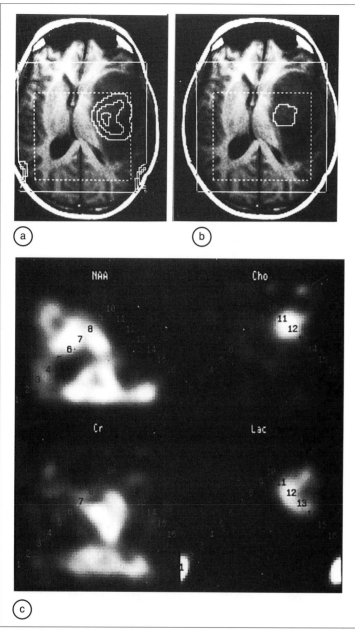

Fig. 11.6. In vivo chemical shift imaging (^{1}H CSI) of a patient with oligodendroglioma of grade II/III. (Modified from den Hollander et al.[190])

(a) Scout view with selected detection volume (dashed box) and contour plot of lactate (Lac) signal derived from metabolic map in (c).

(b) Scout view with selected detection volume (dashed box) and contour plot of choline (Cho) signal derived from metabolic map in (c).

(c) Metabolic maps derived from ^{1}H CSI data in (d) for integrated peak areas of N-acetylaspartate (NAA), choline-containing compounds (Cho), total creatine (Cr) and lactic acid (Lac).

brain tissue. Other tumor entities are studied less extensively and will not be discussed here.

Proton MRS

A number of narrow resonances can be resolved and assigned in localized water-suppressed *in vivo* proton MR spectra of normal and neoplastic tissue of the human brain, in particular those of NAA (methyl protons), tCr (N-methyl), tCho (N-trimethyl), Ins, and Lac (β-methyl).[42,104,105]

In vivo ^{1}H MR spectra of human brain tumors generally show reduced NAA and tCr and

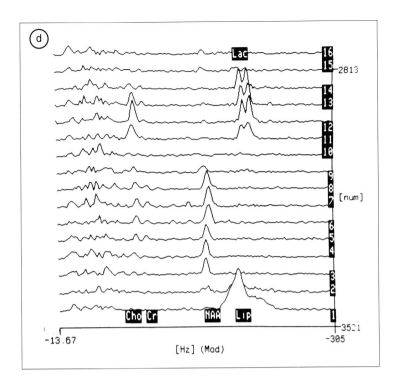

Fig. 11.6 (continued).
(d) Series of ^{1}H MR spectra obtained from the detection volumes shown in (a) and (b).

increased tCho signal intensities compared to spectra of normal brain tissue (Fig. 11.6, see also Fig. 11.2). In spectra of meningiomas, this effect is more pronounced than in spectra of tumors of cerebral origin.[91,106] Low NAA ^{1}H MR signal intensity results from loss of neurons, which is common in brain tumors. However, a low NAA signal is not specific for tumors – ^{1}H MR spectra obtained in patients with stroke[107] and multiple sclerosis[108,109] also show a reduced NAA signal. Enhanced signals of choline-containing compounds are related to cell membrane breakdown,[110] as observed for vital tumor tissue of breast metastases.[111]

The lactate signal (β-methyl proton doublet at 1.33 ppm) was detected in localized ^{1}H MR spectra of neoplastic and ischemic tissue[107,112] (see Fig. 11.2b). The distribution of elevated concentrations of lactate in human brain tumors, cerebrospinal fluid, and cysts could be mapped by means of ^{1}H CSI (see Fig. 11.6).[113] Negendank[24] estimated from data of different clinical studies that lactate is observed in 80% of high-grade gliomas and 65% of low-grade gliomas. The lactate signal was not found in ^{1}H MR spectra of neurinomas and lymphomas. From a clinical study with a small number of patients, Arnold et al and Henriksen et al suggest that high ^{1}H MR signal intensity of lactate could discriminate between grades of astrocytoma,[114–116] while other groups[91,106,117,118] could not find a correlation of lactate levels and tumor grade on the basis of histological data. The presence of lactate is not characteristic for a specific entity. A high concentration of lactate in the tumor tissue appears to correspond to a less favorable outcome of the disease.[106]

Combined positron emission tomography (PET) and ^{1}H MRS studies[119] show that lactate has a tendency to accumulate in necrotic tissue and in cysts, but not in vital tumor tissue with high ^{18}F-deoxyglucose uptake. On the other hand, Alger et al found elevated lactate levels in about 50% of high-grade brain tumors examined by means of PET.[117] They concluded that lactate levels depend not only on lactate production but also on the nature and the regulation of elimination mechanisms, and that they therefore do not correlate with grading. In a study of brain metastases of breast cancer, lactate was only found in large metastases with central necrosis.[111]

Two reports announce the detection of specific ^{1}H MR resonances in human brain tumors *in vivo* which are not present in spectra of normal brain tissue. Naruse et al[120] observed such signals in two of 45 examined patients with brain tumors – a resonance at 3.5 ppm assigned tentatively to glycine and a resonance at 2.4 ppm, which was also detected in the *in vitro* spectrum of the PCA extract and was assigned to pyruvate or succinate. The metabolic characteristic that causes the accumulation of these compounds in the tumor (glioblastoma multiforme) is unknown. Arnold et al[115] observed *in vivo* resonances at 2.1, 1.3, and 0.9 ppm in ^{1}H MR spectra of human brain tumors. These were attributed to Glu or γ-amino butyric acid (GABA), and lipids.

The initial presumption by Bruhn et al[15] of differences of *in vivo* ^{1}H MR spectra of various tumor entities and characteristic spectra of histologically similar tumors is not supported by subsequent data. Problems arise from the heterogeneity of tumor tissue and the complex structure of *in vivo* ^{1}H MR spectra. To enhance the value of *in vivo* ^{1}H MRS, the absolute quantification of metabolite concentrations,[121–123] a reduction of voxel sizes and the application of editing techniques are needed. Localized two-dimensional multiquantum coherence filter could possibly be useful when sensitivity problems are solved.[124–126] Moreover, new MRS criteria have to be established, like the aliphatic signal as marker for necrosis as suggested by Ott et al[91] and Lazeyras et al.[127]

Recently, multicenter studies were performed to investigate the potential of ^{1}H MRS when applied to examinations of human brain tumors.[128,129] These studies show signals in the lipid region in spectra of many, but not all, high-grade astrocytic tumors, suggesting lipids as a prognostic factor.

Breast tumors

The diagnosis of breast malignancy is an important medical problem, especially since an increasing incidence of the disease has been observed in industrialized countries in the last few decades.[130] The differential diagnosis of breast tumors is still a major problem. Imaging techniques such as ultrasound and MRI (with and without application of contrast media) could improve the specificity significantly. However, questions remain and only the examination of the histological specimen, obtained by needle biopsy or surgery, allows definite diagnosis.

Phosphorus MRS

Experimental MRS studies showed increased concentrations of mobile phospholipids in human breast tumor cell lines compared with the levels detected in normal cells.[1] This agrees with findings in patient examinations, where breast tumor spectra often show strongly elevated signals of PME, P_i, and PDE[88,131–133] (Fig. 11.7). In general, SNR of *in vivo* ^{31}P MR spectra of human breast tissue is poor; spectral quality improves when proton decoupling is employed, e.g. WALTZ-8 ^{1}H-irradiation during the acquisition period of the ^{31}P-FID.

Redmond et al.[19] performed a MRS study of 59 women, including 9 patients with breast carcinoma and 4 patients with benign breast tumors. ^{31}P MR data of healthy women were related to menopausal status and age. Spectra from the breast of pre-menopausal women had reduced PCr and increased PME and NTP signals. No significant differences were found in the spectra obtained in the control subjects and in the patients with benign breast tumors. The comparison of spectra from patients with breast carcinoma and of spectra from an age-matched group of volunteers showed reduced PCr and elevated NTP signals and β-NTP:PCr signal intensity ratios in the tumor spectra.

Kalra et al[134] evaluated 31 tumor spectra obtained in a ^{31}P MRS study with 56 patients and found signal intensity ratios of 1.48 for PME and NTP and of 1.65 for PDE and NTP. The measurement of the DNA index (ratio of diploid G0/G1 peak to aneuploid G0/G1 peak) and the S-phase fraction by means of flow cytometry of paraffin-embedded tissue showed that 12 tumors were diploid and 19 aneuploid. There was a significant association between the PME:NTP signal intensity ratio and the S phase fraction ($P = 0.03$), owing to a significant correlation for aneuploid tumors ($P = 0.01$). High-resolution ^{31}P MRS of tissue extracts from 18 tumors (including 7 tumors examined *in vivo*) showed that the PME resonance predominantly consists of signals of phosphoethanolamine (PE) with a smaller signal contribution from phosphocholine (PC). The mean value of the measured PE:PC signal intensity ratios was 3.02.

Proton MRS

Only a few clinical studies of breast cancer have been performed by means of *in vivo* proton MRS. Spectral resolution is lower than in ^{1}H MR spectra of brain and there are broad resonances from methyl and methylene protons of lipids (see Fig. 11.7). In a study with 5 patients with breast carcinoma and 5 healthy women, the water-lipid signal intensity ratio of ^{1}H MR spectra of the breast was 1.2–5.0 for the tumorous and 0.3–0.4 for the unaffected breast tissue.[131] Although large variations of this ratio were observed between individuals, significant changes in the same individual were not seen during therapy.[88]

The signal of choline-containing compounds, a marker for tumor growth and regression in experimental breast tumors, is difficult to resolve in *in vivo* ^{1}H MR spectra of the breast. In experiments with estrogen-sensitive rat mammary tumors, the choline level was found to be higher in regressing tumors than in growing tumors.[9]

Neoplasms of the liver

Maris et al published the first paper on ^{31}P MRS of liver tumors in patients.[135] Increased phosphomonoester signals were observed in spectra obtained in examinations of 2 infants with neuroblastoma.

Elevated PME:β-NTP and in some cases also increased PDE:β-NTP signal intensity ratios in comparison to spectra from healthy liver tissue were found in spectra of liver metastases and of hepatocellular carcinoma[136–141] (Fig. 11.8). High phosphomonoester signals, however, are not unique for liver tumors; they are also found in other pathologies of the liver, e.g. hepatic lymphoma,[142] and in alcoholic liver disease.[143] In the few studies on characterization of liver tumors, benign and malignant neoplastic tissue could not be differentiated by means of *in vivo* ^{31}P MRS.

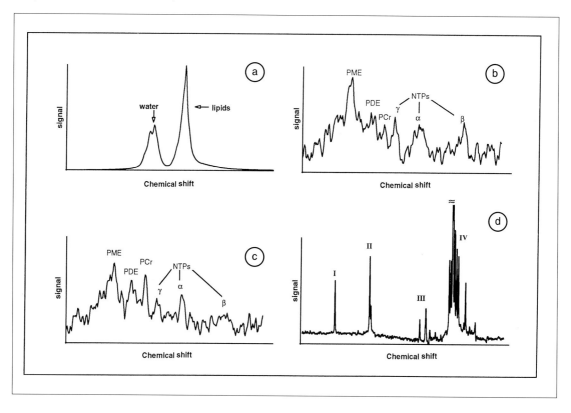

Fig. 11.7. Multinuclear in vivo MRS at 1.5 T of the breast of a 52-year-old female patient with breast carcinoma. Spectra were obtained by means of a concentric surface coil system of 5 cm ⌀ for ^{31}P or ^{13}C and 10 cm ⌀ for ^{1}H frequency. (Modified from Knopp et al.[88])

(a) Non-selective ^{1}H MR spectrum obtained before therapy shows intense lipid peak (at 1.3 ppm) and superposition of water (4.7 ppm) and olefinic proton signals (5.5 ppm).

(b) ^{31}P MR spectrum before therapy with strong resonance of phosphomonoester (PME).

(c) ^{31}P MR spectrum after two cycles of chemotherapy with increased signal of phosphocreatine (PCr).

(d) Broadband ^{1}H-decoupled ^{13}C MR spectrum before therapy with signals from quarternary carbons (I), unsaturated acyl chains (II), glycerol backbone (III), and methylene and methyl carbons (IV) in cellular lipids.

Urogenital tract

Prostate

MRS examinations of the human prostate are based on the observation of the nuclei ^{31}P and ^{1}H. Good-quality in vivo ^{31}P MR spectra of this organ are obtained by means of transrectal coils.[144] Quantitative evaluation yielded signal intensity ratios of PCr:β-NTP of 1.2 ± 0.2, PME:β-NTP of 1.1 ± 0.1, and PME:PCr of 0.9 ± 0.1 in spectra of the normal prostate. Spectra of malignant prostatic tissue demonstrated lower PCr:β-NTP signal intensity ratios (0.7 ± 0.1) than those of normal prostates ($P < 0.02$) or of prostates with benign hyperplasia (BPH) (1.1 ± 0.2, $P < 0.01$). Malignant prostatic lesions

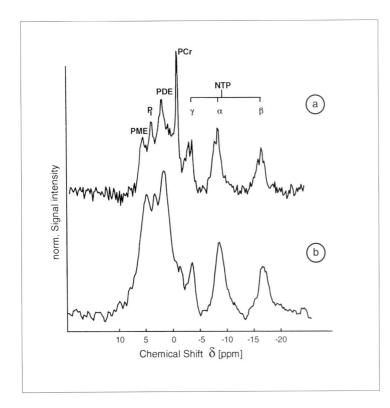

Fig. 11.8. In vivo 1.5 T ^{31}P MR spectra obtained with a 7.5 cm ⌀ surface coil from the liver region of a 68-year-old patient with large liver metastasis.

(a) Non-selective excitation yields strong phosphocreatine (PCr) signal from muscle layer close to the coil.

(b) Selective presaturation of the spin magnetization in a slice of 3 cm thickness adjacent to the surface coil yields localized spectrum with strong resonances of phosphomonoester (PME), inorganic phosphate (P_i), and phosphodiester (PDE) of the metastasis.

demonstrated PME:β-NTP ratios (1.8 ± 0.2) that were higher than those of spectra of normal prostates ($P < 0.02$).

On the basis of the PME:PCr signal intensity ratio, malignant (2.7 ± 0.3) and normal prostate tissues could be differentiated ($P < 0.001$) without overlap of individual ratios. The mean PME:PCr signal intensity ratio (1.5 ± 0.5) of spectra from prostates with benign hyperplasia was between the corresponding values measured for normal and malignant prostatic tissue, while the values of individual PME:PCr signal intensity ratios in spectra of benign hyperplasia tissue overlapped with those of normal and malignant tissue. The results indicate characteristic levels, as measured by ^{31}P MRS, of phosphorylated metabolites in normal, hyperplastic, and malignant prostatic tissue that allow the differentiation of these tissues.[144,145]

It has been shown with biochemical techniques that the normal prostate tissue has a high concentration of citrate. The citrate level is still higher in benign prostate hyperplasia, but strongly reduced in prostatic adenocarcinoma,[146–149] and it can therefore be used as a marker for the differentiation of benign and malignant disease of the human prostate; similar results can be obtained using *in vivo* MRS.[150] Citrate gives well-resolved proton MR signals, so ^{1}H MRS presumably can afford this differentiation and be utilized for early detection of the tumorous disease.[151–153] Proton MR signal intensity ratios of other metabolites were also suggested as spectral markers for malignant tissue of the prostate.[150]

These studies demonstrate that *in vivo* ^{31}P and ^{1}H MRS can detect metabolic differences between normal and malignant prostatic tissue. Malignant lesions are characterized by low levels of PCr and citrate and high levels of PME. These findings agree with the results of biochemical studies on biopsies of prostatic tissues.[154]

Kidney

In vivo ^{31}P and ^{1}H MRS have been applied to monitoring the metabolic status of the kidney after transplantation, renal failure, and obstruction.[155–157] A particular problem is spectral localization of this organ. To our knowledge, there is no clinical study of human kidney tumors by means of *in vivo* MRS.

Testis

Seminoma is the most common malignancy of the testis. Clinical studies of this tumor by means of ^{31}P MRS are very rare. Spectra of seminoma show high P_i and reduced PME:β-NTP signal ratio in comparison to spectra obtained from the normal testis.[158] The authors claim that MRS is most useful in studying infertility.[159]

Tumors of the neck region

Lymph node metastases of squamous tumors of the head and neck region are quite common and are often superficially localized. Therefore, these tumors are easily accessible for surface coil techniques; as a result these tumors were studied early in the beginning of ^{31}P MRS.[10,12,160,161] The latter authors[161] studied 30 patients with squamous cell carcinomas and claim that the ratios of PDE:β-ATP and PME:β-ATP are greater than unity and that the tumors have higher PME than PDE signal intensities. Higher PME in comparison to PDE signal intensities were also observed in a patient with a lymph node metastasis of a squamous cell carcinoma[10] and in 10 patients who were studied during therapy monitoring.[12]

Tumors of the musculoskeletal system

Phosphorus MRS

In vivo ^{31}P MRS by means of surface coils has been applied in clinical studies of muscle tissue and superficial tumors, e.g. soft tissue sarcomas, skin cancers, lymphomas and extremity bone tumors.[10,11,21,22,162–164] Relative signal contributions from the tumor and from the surrounding muscle tissue can be estimated on the basis of the signal intensities of phosphomonoester, phosphodiester, and phosphocreatine. However, when the background tissue has high levels of PME and PDE and reduced or vanishing levels of PCr, which is the case for liver (see Fig. 11.8) and breast tissue, the discrimination of signals from healthy and neoplastic tissue is difficult without appropriate spatial localization technique.

The non-invasive determination of tissue pH value derived from the *in vivo* P_i–PCr ^{31}P chemical shift difference showed elevated pH in soft tissue tumors as well as in other tumors (Fig. 11.9). This unexpected result – according to Warburg's hypothesis, acidic pH values are attributed to neoplastic tissue – has also been observed by other authors.[61,62,165,166] The alkaline pH of tumor tissue can be explained by necrosis.[3]

Proton MRS

Clinical studies of soft tissue and bone tumors by means of ^{1}H MRS are rare. While lipid resonances are absent in localized ^{1}H MR spectra of human brain tumor tissue in most cases, those signals often dominate the metabolite resonances in other tumors.

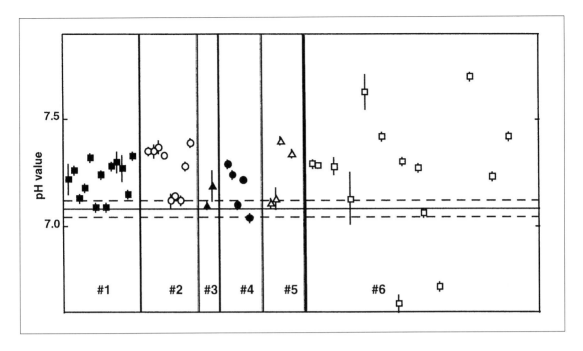

Fig. 11.9. In vivo ^{31}P MRS. Apparent tissue pH values derived from P_i-PCr chemical shift difference for various soft tissue tumors and normal muscle tissue. The reference value determined by means of WALTZ-8 ^1H-decoupled in vivo ^{31}P MRS of the human calf is: mean pH_{muscle} =7.07 ±0.02 (#1 lymph node metastases of squamous cell carcinoma; #2 osteosarcomas; #3 Ewing sarcomas; #4 liposarcomas; #5 histiocytomas; #6 miscellaneous tumors)

In a clinical study, Bongers et al[167] observed methyl and methylene signals from fatty acids in ^1H MR spectra of 16 out of 18 patients examined with tumors of the leg or pelvis. In the spectra of 6 patients lipid and water resonances only were detected. In the spectra of 10 patients, the tCho signal (which was stronger than that of tCr) and additional unassigned resonances at 1.8–2.5 ppm and 3.2–4.0 ppm were found. The authors concluded that the specificity of ^1H MRS appears to be very low, and that in particular there was no correlation between ^1H spectral data and the various tumor entities, and tumor-specific resonances were not observed.

Neoplasms of bone marrow

Localized *in vivo* ^1H MRS has been applied in clinical studies with leukemia patients in order to detect bone marrow infiltrations and to monitor therapy. No resonances other than those of lipids and water were resolved in the spectra.[64,168,169] Spatial localization of bone marrow by means of single-voxel techniques is difficult; moreover, the resonances are broadened by magnetic field inhomogeneities and susceptibility artefacts caused by trabecular bone structures. Proton CSI is an appropriate spectroscopic technique for the non-invasive detection of relative fat and water fractions in neoplastic tissue of the human bone marrow.[170–172]

Conclusions

The review of available data on clinical studies of MRS in oncology shows important results when ^{1}H MRS is applied to examinations of tumors of the human brain and prostate. In particular the NAA, tCho, Lac, and citrate *in vivo* ^{1}H MR signals turned out as markers for neoplastic tissue. A similar role is played by the *in vivo* ^{31}P MR signals of PME and PDE, which are related to membrane phospholipid turnover, and the tissue pH derived from ^{31}P chemical shifts, which typically is elevated in tumors.

Problems that arise when MRS is applied in studies with tumor patients result from the heterogeneity of tumors; the superposition of signals from neoplastic, infiltrated, and unaffected tissue; the localization of the tumor; and the specific limitations of the method (i.e. poor sensitivity); large voxel sizes; and complex structure of *in vivo* MR spectra. Moreover, the comparability of the data is limited as long as the examination protocols are not standardized and different localization techniques and acquisition parameters are employed. The statistical basis of the studies is often poor, because of the small number of patients examined.

On the basis of available MRS data, tumor diagnosis is presently possible only for a few entities, in which the method provides valuable information. The information content of MR spectra can be enhanced when more than one spin species is observed and magnetic double resonance techiques are applied.[49,79] The absolute quantification of metabolite concentrations in the living tissue[75,121–123,173,174] is needed to improve specifity and reliability of MRS data. Multicenter studies are launched in order to establish appropriate examination protocols and to enhance the number of examined cases.

Presently, immunohistological and histomorphological characterization is the gold standard, and a useful scheme of tumor characterization by means of MRS will have to provide at least the same level of confidence.

THERAPY MONITORING

Phosphorus MRS

In vivo ^{31}P MR experiments with tumor-bearing animals showed changes of levels of phosphorus-containing metabolites during tumor growth and therapy. These effects depend on the type of treatment, treatment protocol, and tumor entity. Often high-energy phosphate signal intensities and the pH value decrease, while PME and PDE signal intensities increase.[4,6–8,58,175] In particular, the P_i resonance can increase strongly and dominate the total signal that is observed after the therapeutic intervention[7,8] (Fig. 11.10).

Clinical studies of therapy monitoring in patients by means of ^{31}P MRS followed. Marris et al[135] observed spectral changes in examinations of patients with infiltrative neuroblastoma in the liver and osteosarcoma treated by radiation therapy and chemotherapy. After a transient increase, the PME:β-NTP signal intensity ratio showed normal values after 24 weeks when a complete remission of the tumor was diagnosed. This was the first therapy follow up study in tumor patients by means of *in vivo* MRS.

The majority of subsequent clinical ^{31}P MRS studies on therapy monitoring focused on easily

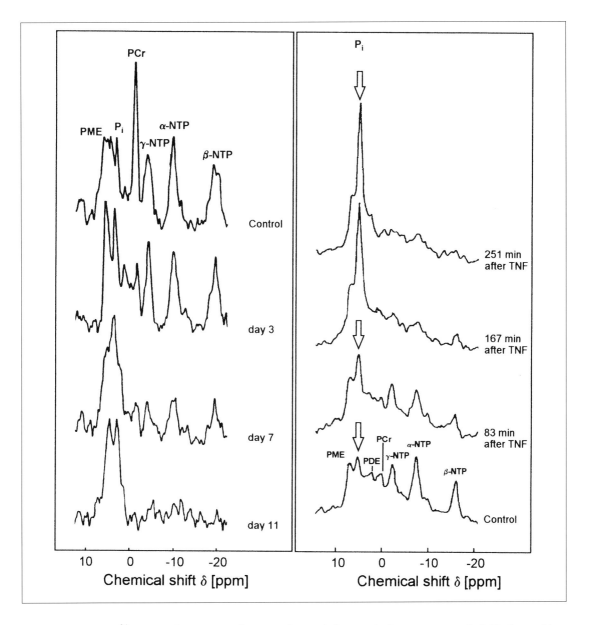

Fig. 11.10. In vivo ^{31}P MRS of experimental tumors. Spectral changes during tumor growth (left) observed in a period of 11 days in a MOPC 104E myeloma implanted subcutaneously: Complete depletion of high-energy phosphate pool (NTP, nucleoside 5′-triphosphate; PCr, phosphocreatine), strong increase of inorganic phosphate (P_i) level, variations of phosphomonoester (PME) signal intensity; and drop of tumor tissue pH value from 7.1 to 6.6. (Modified from Evanochko et al.[58]) Spectral changes (right) of murine methylcholanthrene-induced (Meth-A) sarcoma upon application of recombinant human tumor necrosis factor α (rHuTNF-α): progressive decline of high-energy phosphate signals (NTP, PCr) and strong increase of P_i level. (Modified from Shine et al.[8])

accessible lymph node metastases, bone tumors, lymphomas, and soft tissue tumors.[10,12,17,18, 20–22,62,165,166,176–180] (For recent reviews see Steen[27] and Negendank.[24]) Radiation therapy or chemotherapy or both,[10,95,100,181–183] immunotherapy,[184] or embolization therapy[97,98] have been monitored in patients with brain tumors by means of ^{31}P MRS.

In a patient with non-Hodgkin lymphoma, tumor volume and PDE:NTP signal intensity ratio decreased depending on the applied ^{60}Co γ-radiation dose, while the PME signal did not change.[177] Immediately after the third treatment, the PDE:NTP signal intensity ratio dropped strongly, while the PME:NTP ratio was constant over 200 minutes.

Hyperthermic regional perfusion chemotherapy has been applied in a patient with local recurrence of malignant melanoma in the foot.[11] The therapeutic intervention was monitored by means of *in vivo* ^{31}P MRS with a 4.5 cm diameter surface coil placed upon the metastasis. In this study, strong changes of spectral parameters, i.e. integrated peak areas and P_i-chemical shift, were observed (Fig. 11.11).

After an initial increase of the signal intensities of PME (maximum at the first day after perfusion therapy) and of PDE (maximum at the second

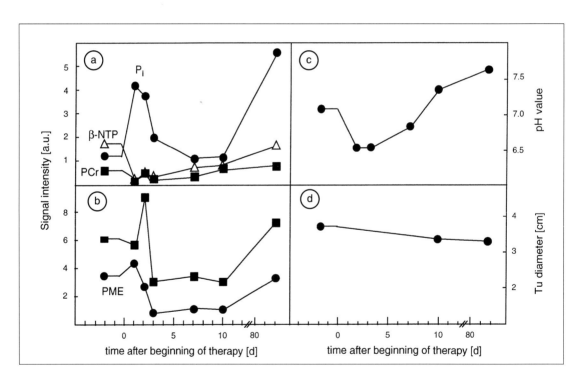

Fig. 11.11. Monitoring hyperthermic locoregional perfusion chemotherapy by means of in vivo ^{31}P *MRS. (Modified from Semmler et al.*[11]*) Examination of a 59-year-old patient with a local recurrence of a malignant melanoma in the soft tissue of the plantar pedis region. Signal intensities of inorganic phosphate (P_i), β-nucleoside 5′-triphosphate (β-NTP), phosphocreatine (PCr), phosphomonoester (PME), and phosphodiester (PDE), tumor tissue pH, and tumor diameter as a function of time after therapy.*

day), a decrease followed, while the P_i signal increased strongly and a complete depletion of the NTP pool occurred.

Tumor tissue pH dropped during the first days and increased later to alkaline values. During this period, the size of the tumor did not change significantly (see Fig. 11.11c, d). Eighty days after therapy and before surgery, another MRS examination was performed; this showed a spectrum similar to the baseline spectrum obtained before therapy, and still alkaline pH. The change of spectral parameters was explained consistently by a simple pathophysiological model.[11] The histology of the tumor mass showed a necrotic tumor, as expected from the alkaline pH value determined *in vivo*.

Lymphoma, breast cancer, sarcoma, and adenocarcinoma were studied by Redmond et al[20] after different regimes of chemotherapy. With the exception of one case, a correlation of response to therapy and decrease of PME signal intensity was found. Low pH after chemotherapy was a good prognostic factor for response to treatment.

Lymph node metastases of head and neck tumors were studied after chemotherapy, radiotherapy, or both.[10,12] The clinical outcome correlates only with an increase of pH values and not with other spectroscopic parameters.[12]

Segebarth et al[181] report on a clinical study of therapy monitoring of brain tumors by means of ^{31}P MRS. They examined a patient with recurrence of prolactinoma before and after bromocriptine chemotherapy, a patient with intracranial lymphoma treated by radiation therapy, and a patient with astrocytoma of grade II after subtotal surgical removal of tumor tissue and subsequent radiation therapy with a dose of 60 Gy. Figure 11.12 shows ^{31}P MR spectra obtained in the patient with prolactinoma before and 1 and 5 weeks after bromocriptine pharmacotherapy.

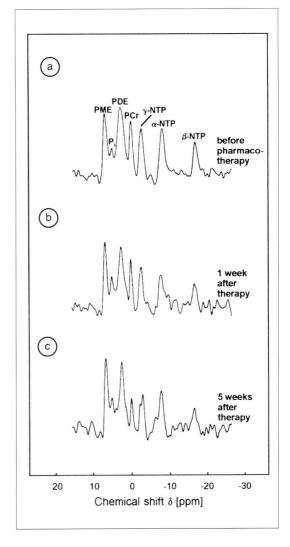

Fig. 11.12. Localized ^{31}P NMR spectra of a prolactinoma obtained in a patient before and 1 and 5 weeks after beginning of bromocriptine pharmacotherapy. (Modified from Segebarth et al.[181]) Changes of relative signal intensities reflect interference of energy metabolism in the course of treatment. Spectra were rescaled to take into account differences of voxel size.

During therapy, the patient improved clinically and the NTP level of the tumor decreased, but the tumor tissue pH value did not change (pH = 7.07, 7.05, and 7.09 before and 1 and 5 weeks after therapy; pH of about 6.92–7.03 in unaffected brain tissue).

More pronounced ^{31}P spectral effects were observed after 2 weeks of radiation therapy in the patient with intracranial lymphoma (Fig. 11.13). The enhanced PME and the reduced PCr signals turned to intensities observed in normal brain tissue after a radiation dose of 24.2 Gy; the pH of tumor tissue did not change. The patient improved clinically during therapy. Similar results were obtained when monitoring treatment of the patient with grade II astrocytoma.

In their ^{31}P MR follow-up study of tumor radiation therapy, Heindel et al[95] examined patients with meningioma, glioblastoma, astrocytoma, and other cerebral tumors. Detected MR signal intensity changes were similar to those of the spectra in Fig. 11.13; in particular, a PME signal reduction was observed for grade II astrocytoma upon radiation therapy (Fig. 11.14).

Arnold et al[100,182] examined 4 patients with glioblastoma after intra-arterial (Fig. 11.15)

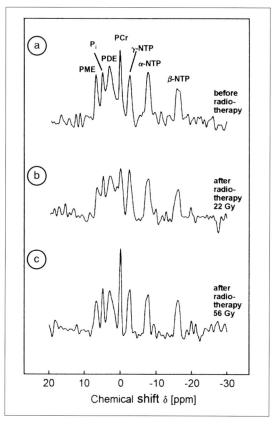

Fig. 11.14. *Monitoring radiation therapy by means of localized in vivo ^{31}P MRS. (Modified from Heindel et al.[95]) Examination of a patient with astrocytoma of grade II: MR spectra were obtained before (a), after 22 Gy (b), and after completion of the radiation therapy (56 Gy) (c). Intensity changes of phosphomonoester (PME) and phosphocreatine (PCr) resonances are observed.*

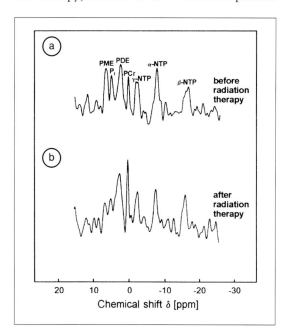

Fig. 11.13. *Localized ^{31}P NMR spectra of an intracranial lymphoma obtained before (a) and after (b) radiation therapy, where a dose of 24.2 Gy was applied. (Modified from Segebarth et al.[181])*

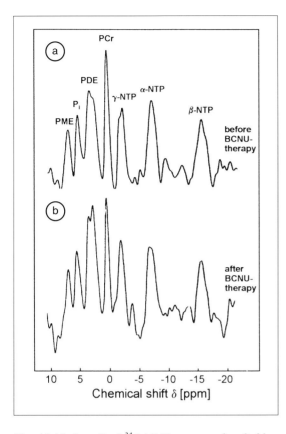

Fig. 11.15. Localized ^{31}P NMR spectra of a glioblastoma in a patient before (a) and after (b) superselective intra-arterial infusion of 1,3-bis-(2-chloroethyl)-1-nitrosourea (BCNU) over 3 hours. (Modified from Arnold et al.[100])

and 3 patients after intravenous 1,3-bis-(2-chloroethyl)-1-nitrosourea (BCNU) therapy. The most significant difference of the data obtained from the 2 groups of patients refers to the pH changes (ΔpH): intravenous administration of BCNU caused a transient acidosis (mean $\Delta pH \approx -0.15$) and intra-arterial administration caused an alkalosis (mean $\Delta pH \approx 0.15$) of the tumor tissue. This effect was seen before any changes could be detected by means of imaging modalities. The increase of pH was explained by membrane damage associated with the high local concentration of the drug in the intra-arterial route of administration, the acidosis by enzyme inactivation and accumulation of lactate. Alkalosis in soft tissue tumors is attributed to necrosis.[3,11,165,166,176]

Superselective catheter embolization of meningiomas with polyvinyl alcohol particles is a common presurgical treatment to minimize bleeding during subsequent neurosurgery. Localized ^{31}P MRS allows the detection of the ischemia caused by the embolization and was therefore applied to assess response to therapy and to classify successful obliteration. In this study,[97,98] localized *in vivo* ^{31}P MR spectra of the human brain were recorded by means of two-dimensional CSI with additional irradiation of the proton spin system to improve SNR via NOE of dipolar coupled ^{31}P and ^1H spins.[56,57,185]

Figure 11.16 shows localized ^{31}P MR spectra of meningioma and normal brain tissue (CSI voxel size $3 \times 3 \times 4$ cm^3) of the same patient obtained one day after embolization. In this case, PME:β-NTP, PDE:β-NTP, and PCr:β-NTP signal intensity ratios as evaluated by means of integration of Lorentzian line fits were 0.5, 4.0, and 1.0 before, and 1.1, 5.9, and 0.9 after embolization.[97] A decrease of high-energy phosphate levels and tissue pH, which reflect the ischemic process, were observed in meningiomas with high degree of embolization ($> 60\%$).[98] A complete depletion of the NTP pool was never observed. Preliminary data indicate that *in vivo* ^{31}P spectral parameters can be used for non-invasive estimation of individual success of embolization. The corresponding ^1H MRS study of Jüngling et al[186] is discussed in the next section.

Monitoring therapy by means of *in vivo* ^{31}P MRS has been performed in a variety of human tumors.[27] The data show that the method allows the detection of metabolic and physiologic

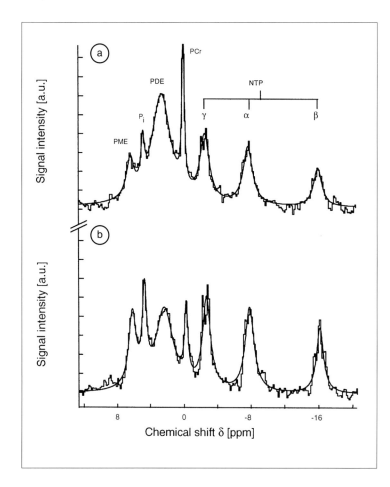

Fig. 11.16. ^{31}P chemical shift imaging (two-dimensional CSI) of the brain of a patient with meningioma after presurgical superselective catheter embolisation. (Modified from Knopp et al.[97]) Localized spectra from a voxel placed in the contralateral unaffected hemisphere (a) and in the meningioma (b). For quantitative evaluation Lorentzian lines are fitted to the measured data. Spectra differ in relative signal intensities of PME, P_i, PDE, and PCr, e.g. PDE : β − NTP = 6.7 in (a) and 3.1 in (b). The pulse sequence allows to sample 8 × 8 voxel of 3 × 3 × 4 cm^3 in a measurement time of 19 minutes with good signal to noise ratio.

effects in the neoplastic tissue in the initial phase after beginning of therapy and also the monitoring of tumor response in long-term controls. However, the results are not as convincing as are expected from experimental studies in animals, which clearly show that ^{31}P MRS can monitor effects of treatment in the tumor tissue immediately after onset of therapy. Possible reasons – besides the inherent physical limitations of the method (low sensitivity, large voxel size, overlapping resonances, broad signals) – are problems that arise from biological properties of human tumors, in particular their strong heterogeneity.

Moreover, many clinical MRS studies include only a small number of patients. The comparison of the results of different studies is often difficult, because of variations of applied treatment modalities and protocols and the lack of standardized examination methods (pulse sequence, measurement parameters, localization technique) and of techniques of quantitative analysis of *in vivo* MR spectra (baseline correction, peak integration, absolute quantification). There are attempts to overcome these problems in the near future.

In recent studies, Dewhirst et al,[163] Prescott et al,[179] and Sostman et al[166] applied *in vivo* ^{31}P MRS for monitoring response to therapy of soft tissue sarcomas in patients treated with a combination of radiotherapy and hyperthermia. In addition, the histology of the tumors was obtained after surgery. A strong correlation was found of pH and treatment outcome in patients with soft tissue sarcomas. This study showed that in osteosarcomas, 90% of necrosis after therapy is associated with disease-free survival of over 90%, whereas under 90% necrosis implies only 14% disease-free survival. The elevated pH of 7.30 ± 0.14 correlated to 98% necrosis, while pH = 7.17 ± 0.13 correlated to 44% necrosis ($n = 10$). Furthermore, a relation was found of treatment outcome and changes of PCr:PDE, PME:P_i, and NTP:PME signal intensity ratios. A reduction in NTP:PME ratio predicted a greater degree of necrosis ($\geq 95\%$ necrosis at surgery). No relationship of thermal parameters and change in pH was observed. Prescott et al[179] have verified that early changes in metabolic status during a course of hyperthermia and radiotherapy are predictive of treatment.

It appears that *in vivo* ^{31}P MRS is of significant benefit as an early monitor of response to hyperthermia and radiation therapy as well as for prognosis prior to therapy.

Proton MRS

Follow-up studies by means of *in vivo* ^{1}H MRS are still rare and mainly applied for the therapy of brain tumors. Heesters et al[187] observed reduced NAA and increased tCho and Lac signals in ^{1}H MR spectra of high-grade and unspecified gliomas before radiation therapy ($n = 9$). After radiation therapy (54–60 Gy), the tCho signal decreased, the Lac signal was absent (in the spectra of 3 out of 9 patients), and no recovery of the NAA intensity was seen.

Our own follow-up study showed an unchanged NAA signal intensity for a period of 41 days during fractionated radiotherapy (2 Gy per fraction, 5 fractions per week) in localized ^{1}H MR spectra of 2 patients with astrocytoma of grade II, whereas the tCr and tCho signals increased. This effect was observed after the first fraction and was significant after the fifth treatment (10 Gy) (Fig. 11.17).

In contrast to these immediate effects upon radiotherapy, Heesters et al[187] observed delayed ^{1}H spectral effects of the radiation when patients were examined 2–3 weeks and 2–3 months after treatment.

Jüngling et al[186] obtained localized ^{1}H MR spectra of meningioma before and after presurgical embolization therapy in 23 patients. Before treatment, the spectra showed high tCho and tCr signal intensity and no NAA and Lac resonances. The alanine signal, which was resolved in 7 of 16 spectra, was claimed to be specific for meningioma tissue. Pronounced variations of spectral data within the same individual were observed. After embolization, a strong Lac resonance was detected, which appeared immediately after the interruption of the blood supply and vanished towards the fourth day after embolization. While the Lac resonance disappeared, aliphatic signals increased within 4 days. These spectral changes are assumed to reflect the transition from vital tumor to necrotic tissue with a high concentration of mobile lipids.

Bongers et al[167] performed follow-up ^{1}H MRS studies in patients with Ewing's sarcoma, fibrous histiocytoma, and multiple myeloma. They observed changes of the signal intensities of tCho, tCr, and lipids. The changes of the tCho resonances (signal reduction) were more pronounced than that of the lipid signals. A distinct increase of the lipid signal was seen during

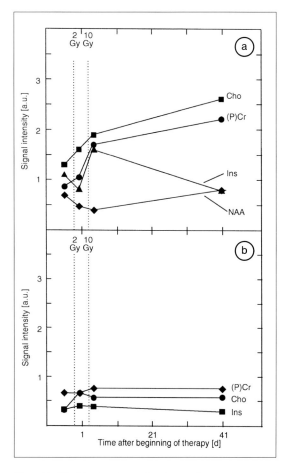

Fig. 11.17. Monitoring fractionated stereotactical radiation therapy (2 Gy per fraction, 5 times a week) by means of localized in vivo stimulated echo ^{1}H MRS. Examination of a 37-year-old patient with an oligodendroglioma of grade II shows spectral changes of the tumor (a) and of the non-radiated healthy brain of the contralateral hemisphere (b).

chemotherapy only in the spectra of the patient with Ewing's sarcoma. The effect is attributed to fatty acid degeneration upon treatment. Van Zijl et al[16] report similar findings in a patient with a squamous cell carcinoma.

Monitoring of diagnosis and therapy of bone marrow involvement in systemic neoplastic diseases, e.g. leukemia and malignant lymphoma, has been performed by means of proton CSI.[170–172] These studies showed that the relative lipid and water fractions derived from *in vivo* ^{1}H CSI data depend on the infiltration of the bone marrow. In serial examinations, an increase in lipid fractions was found, in agreement with the response to therapy proven by biopsies. Proton CSI proved to be a very useful method for spectroscopic examinations and monitoring of therapy of systemic bone marrow disorders.

DRUG MONITORING

Fluorine MRS

The antineoplastic drug, 5-fluorouracil (5-FU), is used in the treatment of head, neck, breast, and colorectal tumors. Its pharmacokinetics in the human body can be observed noninvasively by two techniques: positron emission tomography (PET) using the ^{18}F-labeled drug, and ^{19}F MRS. PET detects very low concentrations of metabolites with good spatial resolution; however, different compounds cannot be distinguished. A differentiation of the various metabolites is possible with MRS, because of its high spectral resolution.

^{19}F MRS allows to monitor the uptake of 5-FU in the tissue and its conversion into metabolic intermediates without interfering background signal. Experimental studies in animals detected *in vivo* ^{19}F MR resonances of 5-FU nucleosides and nucleotides (5-FUranuc), 5-fluoro-5,6-dihydrouracil (DHFU) and α-fluoro-β-alanine (FBAL) after administration of 5-FU. These signals can be resolved even at 1.5 T as a consequence of the large chemical shift range of fluorine.

In clinical ^{19}F MRS studies in which mainly patients with liver metastases of colorectal

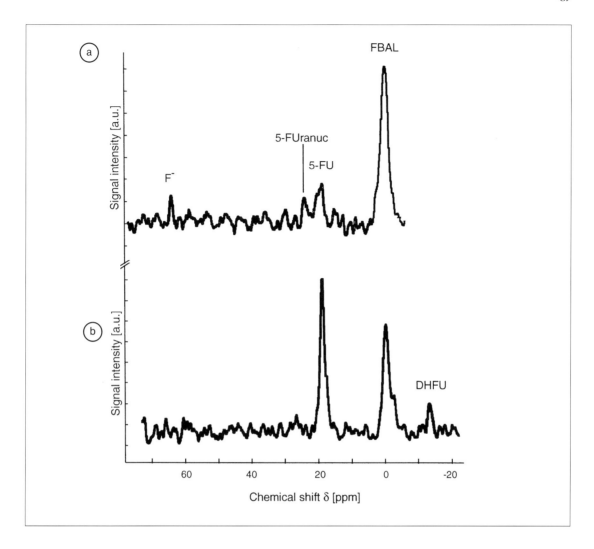

Fig. 11.18. In vivo ^{19}F MRS in tumor patients. (Modified from Schlemmer et al.[75]) (a) Spectrum obtained after intravenous administration of 1000 mg 5-FU (650 mg/(m^2 bs) in a 44-year-old male patient with liver metastases of a coecum carcinoma (sum spectrum obtained during 0–103 minutes after beginning of 5-FU administration). The signal of 5-FU nucleosides and nucleotides (5-FUranuc) is present. The assignment of free fluoride anions (F^-) is tentative. (b) Spectrum obtained after intra-arterial administration of 2000 mg 5-FU (1000 mg/[m^2 bs]) in a 56-year-old male patient with liver metastases of a rectum carcinoma (sum spectrum, obtained during 0–58 minutes after beginning of 5-FU administration). In this particular case of the study by Schlemmer et al,[75] FBAL was detected late and showed a comparable signal intensity in the sum spectrum with the 5-FU resonance. The presence of the 5-fluoro-5,6-dihydrouracil (DHFU) signal in this spectrum indicates a partial block of the catabolic pathway after the DHFU-producing step which results in an accumulation of 5-FU and slow production rate of FBAL. The enhanced 5-FU level correlates with response.

carcinoma have been examined, the kinetics of 5-FU and FBAL could be observed with good temporal resolution.[68,69,71,75] Only in a few cases, a weak unresolved resonance band was detectable; this was assigned to 5-FUranuc on the basis of its chemical shift[69,75] (see Fig. 11.3). The data show that even after administration of high doses of the drug by the intra-arterial route, the steady-state concentration of 5-FUranuc is below the ^{19}F MR-detectable level in most cases, in agreement with the low concentrations measured in biopsy material of metastases. The signal of the catabolic intermediate DHFU was observed in the spectra of only 1 of 30 patients examined; this resonates 13 ppm upfield from FBAL[75] (Fig. 11.18).

In a study of 18 patients with advanced colorectal adenocarcinoma, a correlation was observed for the mean drug level (5-FU bioavailability, measured by the normalized integral of the resonance in the sum of all spectra recorded during 50 minutes after the beginning of the drug infusion, I_{50}^{5-FU}) and the volume of liver metastases within the sensitive volume of the surface coil. No correlation was seen for FBAL, which indicates that the conversion of 5-FU to FBAL mainly takes place in the liver[188] and not in tumor tissue.

Responders showed significantly higher values of I_{50}^{5-FU} than patients classified as nonresponders ($P < 0.053$)[74,75] (Fig. 11.19). This suggests an accumulation and retention of 5-FU in tumor tissue, as well as a relation of 5-FU levels and tumor response. This is consistent with the concept by Wolf et al[70,189] of 5-FU tumor trapping, i.e. longer half-life of the cytostatic in tumor tissue than in peripheral blood in responsive patients. Spectral parameters and absolute concentrations of FBAL were not related to clinical response to treatment.

The results suggest the use of *in vivo* ^{19}F MRS to guide dose escalation schemes for the optimiza-

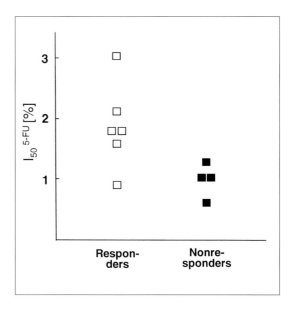

Fig. 11.19. In vivo ^{19}F MR signal intensities of 5-FU (I_{50}^{5-FU}) obtained during 0–50 minutes after beginning of intravenous 5-FU administration for responsive and non-responsive patients. (from Schlemmer et al [74]). Enhanced 5-FU levels in responders suggest an accumulation and retention of the cytostatic in tumor tissue of those patients.

tion of 5-FU bioavailability in each patient at the beginning of treatment, its use in assessing the modulation of 5-FU metabolism by other drugs in combined chemotherapy. As shown by Port et al,[72,73] a complete pharmacokinetic model (nonlinear three-compartment model,[72] population model with a non-linear two-compartment model as structural part[73]) can be applied to *in vivo* ^{19}F MRS data measured in patients during their 5-FU treatment. Analyses of this kind should greatly facilitate quantitative comparisons between patients or treatments.

REFERENCES

1 Cohen JS, Lyon RC, Chen C, Faustino PJ, Batist G, Shoemaker M, Rubalcaba E, Cowan KH. Differences in phosphate metabolic levels in drug-sensitive and -resistant human breast cancer cell lines determined by ^{31}P magnetic resonance spectroscopy. *Cancer Res* 1986; 46:4087–90.

2 Daly PF, Lyon RC, Faustino PJ, Cohen JS. Phospholipid metabolism in cancer cells monitored by ^{31}P NMR spectroscopy. *J Biol Chem* 1987; 262:14875–8.

3 Vaupel P, Kallinowski F, Okunieff P. Blood flow, oxygen and nutrient supply, and metabolic microenvironment of human tumors: a review. *Cancer Res* 1989; 49:6449–65.

4 Ng TC, Evanochko WT, Hiramoto RN, et al. ^{31}P NMR spectroscopy of *in vivo* tumors. *J Magn Reson* 1982; 49:271–86.

5 Evanochko WT, Ng TC, Lilly MB, et al. *In vivo* ^{31}P NMR study of the metabolism of murine mammary 16/C adenocarcinoma and its response to chemotherapy, x-radiation, and hyperthermia. *Proc Natl Acad Sci U S A* 1983; 80:334–8.

6 Naruse S, Hirakawa K, Horikawa Y, Tanaka C, Higuchi T, Ueda S, Nishikawa H, Watari H. Measurements of *in vivo* ^{31}P nuclear magnetic resonance spectra in neuroectodermal tumors for the evaluation of the effects of chemotherapy. *Cancer Res* 1985; 45:2429–33.

7 Naruse S, Horikawa Y, Tanaka C, et al. Observations of energy metabolism in neuroectodermal tumors using *in vivo* ^{31}P-NMR. *Magn Reson Imaging* 1985; 3:117–23.

8 Shine N, Palladino MA, Patton JS, Deisseroth A, Karczmar GS, Weiner MW. Early metabolic response to TNF in mouse sarcoma: a ^{31}P nuclear magnetic resonance study. *Cancer Res* 1989; 49:2123–7.

9 Smith TAD, Ojugo ASE, Leach MO, McCready VR. Choline and lactate levels in growing and regressing transplanted rat mammary tumours. In: *Proceedings, 2nd Annual Meeting, Society of Magnetic Resonance, San Francisco* 1994; 3:1328.

10 Semmler W, Gademann G, Bachert-Baumann P, Zabel HJ, Lorenz WJ, van Kaick G. Monitoring human tumor response to therapy by means of P-31 MR spectroscopy. *Radiology* 1988; 166:533–9.

11 Semmler W, Gademann G, Schlag P, et al. Impact of hyperthermic regional perfusion therapy on cell metabolism of malignant melanoma monitored by ^{31}P MR spectroscopy. *Magn Reson Imaging* 1988; 6:335–40.

12 Gademann G, Semmler W, Bachert-Baumann P, et al. ^{31}P MR spectroscopy and ^{1}H MR tomography of cervical lymph node metastases under chemotherapy in short-term follow-up studies. In: *Proceedings, 7th Annual Meeting, Society of Magnetic Resonance in Medicine, San Francisco* 1988; 2:715.

13 Karczmar GS, Meyerhoff DJ, Boska MD, et al. Response of superficial human tumors to therapy studied by ^{31}P MRS. In: *Proceedings, 8th Annual Meeting, Society of Magnetic Resonance in Medicine, Amsterdam* 1989; 1:432.

14 Arnold DL, Emrich J, Shoubridge EA, Villemure, JG, Feindel W. Characterization of astrocytomas, meningiomas, and pituitary adenomas by phosphorus magnetic resonance. *J Neurosurg* 1991; 74:447–53.

15 Bruhn H, Frahm J, Gyngell ML, Merboldt KD, Hänicke W, Sauter R, Hamburger C. Noninvasive differentiation of tumors using localized H-1 MR spectroscopy *in vivo*: initial experience in patients with cerebral tumors. *Radiology* 1989; **172**:541–8.

16 van Zijl PCM, Moonen CTW, Gillen J, et al. Proton magnetic resonance spectroscopy of small regions (1 ml) localized inside superficial tumors. A clinical feasibility study. *NMR Biomed* 1990; 3:227–32.

17 Karczmar GS, Meyerhoff DJ, Boska MD, et al. P-31 MRS study of response of superficial human tumors to therapy. *Radiology* 1991; **179**:149–53.

18 Meyerhoff DJ, Karczmar GS, Valone F, Venook A, Matson GB, Weiner MW. Hepatic cancers and their response to chemoembolization therapy. Quantitative image-guided ^{31}P magnetic resonance spectroscopy. *Invest Radiol* 1992; 27:456–64.

19 Redmond O, Stack JP, O'Connor NG, Codd MB, Ennis JT. *In vivo* phosphorus-31 magnetic resonance spectroscopy of normal and pathological breast tissue. *Br J Radiol* 1991; 64:210–16.

20 Redmond OM, Stack JP, O'Connor NG, et al. ^{31}P MRS as an early prognostic indicator of patient response to chemotherapy. *Magn Reson Med* 1992; 25:30–44.

21 Redmond OM, Bell E, Stack JP, et al. Tissue characterization and assessment of preoperative chemotherapeutic response in musculoskeletal tumors by *in vivo* ^{31}P magnetic resonance spectroscopy. *Magn Reson Med* 1992; 27:226–37.

22 Ross B, Helsper JT, Cox IJ, et al. Osteosarcoma and other neoplasms of bone: magnetic resonance spectroscopy to monitor therapy. *Arch Surg* 1987; **122**:1464–9.

23. Howe FA, Maxwell RJ, Saunders DE, Brown MM, Griffiths JR. Proton spectroscopy *in vivo*. *Magn Reson Q* 1993; 9:31–59.
24. Negendank W. Studies of human tumors by MRS: A review. *NMR Biomed* 1992; 5:303–24.
25. Radda GK, Rajagopalan B, Taylor DJ. Biochemistry *in vivo*: an appraisal of clinical magnetic-resonance spectroscopy. *Magn Reson Q* 1989; 5:122–51.
26. Bottomley PA. Human *in vivo* NMR spectroscopy in diagnostic medicine: Clinical tool or research probe? *Radiology* 1989; 170:1–15.
27. Steen RG. Response of solid tumors to chemotherapy monitored by *in vivo* ^{31}P nuclear magnetic resonance spectroscopy: a review. *Cancer Res* 1989; 49:4075–85.
28. Sostman HD, Charles HC, Rockwell S, et al. Soft-tissue sarcomas: Detection of metabolic heterogeneity with P-31 MR spectroscopy. *Radiology* 1990; 176:837–43.
29. Segebarth CM, Baleriaux DF, Luyten PR, den Hollander JA. Detection of metabolic heterogeneity of human intracranial tumors *in vivo* by H-1-NMR spectroscopic imaging. *Magn Reson Med* 1990; 13:62–76.
30. Ackerman JJH, Grove TH, Wong GC, Gadian DG, Radda GK. Mapping of metabolites in whole animals by ^{31}P NMR using surface coils. *Nature* 1980; 283:167–70.
31. Bendall MR, Foxall D, Nichols BG, Schmidt JR. Complete localization of *in vivo* NMR spectra using two concentric surface coils and rf methods only. *J Magn Reson* 1986; 70:181–6.
32. Blackledge MJ, Styles P, Radda GK. Rotating-frame depth selection and its application to the study of human organs. *J Magn Reson* 1987; 71:246–58.
33. Evelhoch JL, Crowley MG, Ackerman JJH. Signal-to-noise optimization and observed volume localization with circular surface coils. *J Magn Reson* 1984; 56:110–24.
34. Matson GB, Twieg DB, Karczmar GS, et al. Application of image-guided surface coil P-31 MR spectroscopy to human liver, heart, and kidney. *Radiology* 1988; 169:541–7.
35. Haase, A, Hänicke W, Frahm J. The influence of experimental parameters in surface coil NMR. *J Magn Reson* 1984; 56:401–12.
36. Bottomley PA, Foster TB, Darrow RD. Depth-resolved surface-coil spectroscopy (DRESS) for *in vivo* ^{1}H, ^{31}P, and ^{13}C NMR. *J Magn Reson* 1984; 59:338–42.
37. Sauter R, Müller S, Weber H. Localization in *in vivo* ^{31}P NMR spectroscopy by combining surface coils and slice-selective saturation. *J Magn Reson* 1987; 75:167–73.
38. Haase A. Localization of unaffected spins in NMR imaging and spectroscopy (LOCUS spectroscopy). *Magn Reson Med* 1986; 3:963–9.
39. Ordidge RJ, Connelly A, Lohman JAB. Image-selected *in vivo* spectroscopy (ISIS). A new technique for spatially selective NMR spectroscopy. *J Magn Reson* 1986; 66:283–94.
40. Merboldt KD, Chien D, Hänicke W, Gyngell ML, Bruhn H, Frahm J. Localized ^{31}P NMR spectroscopy of the adult human brain *in vivo* using stimulated echo (STEAM) sequences. *J Magn Reson* 1990; 89:343–61.
41. Frahm, J, Merboldt KD, Hänicke W. Localized proton spectroscopy using stimulated echos. *J Magn Reson* 1987; 72:502–8.
42. Frahm J, Bruhn H, Gyngell ML, Merboldt KD, Hänicke W, Sauter R. Localized high-resolution proton NMR spectroscopy using stimulated echoes: initial applications to human brain *in vivo*. *Magn Reson Med* 1989; 9:79–93.
43. Granot J. Selected volume excitation using stimulated echos (VEST). Applications to spatially localized spectroscopy and imaging. *J Magn Reson* 1986; 70:488–92.
44. Bottomley PA. Spatial localization in NMR spectroscopy *in vivo*. *Ann N Y Acad Sci* 1987; 508:333–48.
45. Brown TR, Kincaid BM, Ugurbil K. NMR chemical shift imaging in three dimensions. *Proc Natl Acad Sci U S A* 1982; 79:3523–6.
46. Maudsley AA, Hilal SK, Perman WH, Simon HE. Spatially resolved high-resolution spectroscopy by 'four-dimensional' NMR. *J Magn Reson* 1983; 51:147–52.
47. Maudsley AA, Hilal SK, Simon HE, Wittekoek S. *In vivo* MR spectroscopic imaging with P-31. *Radiology* 1984; 153:745–50.
48. Komoroski RA, Newton JEO, Walker E, et al. *In vivo* NMR spectroscopy of lithium-7 in humans. *Magn Reson Med* 1990; 15:347–56.
49. Bachert P, Bellemann ME, Layer G, Koch T, Semmler W, Lorenz WJ. *In vivo* ^{1}H, ^{31}P-{^{1}H} and ^{13}C-{^{1}H} magnetic resonance spectroscopy of malignant histiocytoma and skeletal muscle tissue in man. *NMR Biomed* 1992; 5:161–70.
50. Robitaille PM, Merkle H, Sako E, et al. Measurement of ATP synthesis rates by ^{31}P NMR spectroscopy in the intact myocardium *in vivo*. *Magn Reson Med* 1990; 15:8–24.

51. Hsieh PS, Balaban RS. ^{31}P imaging of the *in vivo* creatine kinase reaction rates. *J Magn Reson* 1987; 74:574–9.
52. Degani H, Alger JR, Shulman RG, Petroff OAC, Prichard JW. ^{31}P magnetization transfer studies of creatine kinase kinetics in living rat brain. *Magn Reson Med* 1987; 5:1–12.
53. Rees D, Smith MB, Harley J, Radda GK. *In vivo* functioning of creatine phosphokinase in human forearm muscle, studied by ^{31}P NMR saturation transfer. *Magn Reson Med* 1989; 9:39–52.
54. Mora BN, Narasimhan PT, Ross BD. ^{31}P magnetization transfer studies in the monkey brain. *Magn Reson Med* 1992; 26:100–15.
55. Luyten PR, Bruntink G, Sloff FM, Vermeulen JWAH, et al. Broadband proton decoupling in human ^{31}P NMR spectroscopy. *NMR Biomed* 1989; 1:177–83.
56. Bachert-Baumann P, Ermark F, Zabel HJ, Sauter R, Semmler W, Lorenz WJ. In vivo nuclear Overhauser effect in ^{31}P-{^{1}H} double-resonance experiments in a 1.5-T whole-body MR system. *Magn Reson Med* 1990; 15:165–72.
57. Bachert P, Bellemann ME. Kinetics of the *in vivo* ^{31}P-^{1}H nuclear Overhauser effect of the human calf muscle phosphocreatine resonance. *J Magn Reson* 1992; 100:146–56.
58. Evanochko W, Ng TC, Glickson JD. Application of *in vivo* NMR spectroscopy to cancer. *Magn Reson Med* 1984; 1:508–34.
59. Bárány M, Glonek T. Identification of diseased states by phosphorus-31 NMR. In: Gorenstein DG (ed). *Phosphorus-31 NMR*, Academic Press, San Diego, 1984:511–45.
60. Steen RG. Characterization of tumor hypoxia by ^{31}P MR: a review. *Am J Roentgenol* 1991; 157:243–8.
61. Oberhänsli RD, Hilton-Jones D, Bore PJ, Hands LJ, Rampling RP, Radda GK. Biochemical investigation of human tumors *in vivo* with phosphorus-31 magnetic resonance spectroscopy. *Lancet* 1986; ii:8–11.
62. Ng TC, Majors AW, Vijayakumar S, Baldwin NJ, et al. Human neoplasm pH and response to radiation therapy: P-31 MR spectroscopy studies *in situ*. *Radiology* 1989; 170:875–8.
63. Bruhn H, Frahm J, Gyngell ML, Merboldt KD, Hänicke W, Sauter R. Localized proton NMR spectroscopy using stimulated echoes: applications to human skeletal muscle *in vivo*. *Magn Reson Med* 1991; 17:82–94.
64. Jensen KE, Jensen M, Grundtvig P, Thomsen C, Karle H, Henriksen O. Localized *in vivo* spectroscopy of the bone marrow in patients with leukemia. *Magn Reson Imag* 1990; 8:779–89.
65. Schick F, Duda S, Laniado M, Jung WI, Claussen CD, Lutz O. Volume-selective MRS in vertebral bodies. *Magn Reson Med* 1992; 26:207–17.
66. Frahm J, Bruhn ML, Gyngell ML, Merboldt KD, Hänicke W, Sauter R. Spezielle MR-Methoden bei primären Knochentumoren II: Volumenselektive ^{1}H Spektroskopie. *Fortschr Röntgenstr* 1993; 159:325–30.
67. Behar KL, Rothman DL, Spencer DD, Petroff OAC. Direct measurement of macromolecule resonances in short-TE localized ^{1}H NMR spectra of human brain *in vivo* at 2.1 Tesla. In: *Proceedings, 2nd Annual Meeting, Society of Magnetic Resonance, San Francisco* 1994; 1:310.
68. Wolf W, Albright MJ, Silver MS, Weber H, Reichardt U, Sauer R. Fluorine-19 NMR spectroscopic studies of the metabolism of 5-fluorouracil in the liver of patients undergoing chemotherapy. *Magn Reson Imaging* 1987; 5:165–9.
69. Semmler W, Bachert-Baumann P, Gückel F, et al. Real-time follow-up of 5-fluorouracil metabolism in the liver of tumor patients by means of F-19 MR spectroscopy. *Radiology* 1990; 174:141–5.
70. Presant CA, Wolf W, Albright MJ, et al. Human tumour fluorouracil trapping: clinical correlations of *in vivo* ^{19}F nuclear magnetic resonance spectroscopy pharmacokinetics. *J Clin Oncol* 1990; 8:1868–73.
71. Glaholm J, Leach MO, Collins D, et al. Comparison of 5-fluorouracil pharmacokinetics following intraperitoneal and intravenous administration using *in vivo* ^{19}F magnetic resonance spectroscopy. *Br J Radiology* 1990; 63:547–53.
72. Port RE, Bachert P, Semmler W. Kinetic modeling of *in vivo* nuclear magnetic resonance spectroscopy data: 5-Fluorouracil in liver and liver tumors. *Clin Pharmacol Ther* 1991; 49:497–505.
73. Port RE, Schlemmer HP, Bachert P. Pharmacokinetic analysis of sparse *in vivo* NMR spectroscopy data using relative parameters and the population approach. *Eur J Clin Pharmacol* 1994; 47:187–93.
74. Schlemmer HP, Bachert P, Semmler W, et al. Correlation of response to 5-FU treatment and individual 5-FU bioavailability detected by *in vivo* ^{19}F MR spectroscopy. In: *Proceedings, 2nd Annual Meeting, Society of Magnetic Resonance, San Francisco* 1994; 1:132.
75. Schlemmer HP, Bachert P, Semmler W, et al. Drug monitoring of 5-fluorouracil: *in vivo* ^{19}F NMR study during 5-FU chemotherapy in patients with metastases

of colorectal adenocarcinoma. *Magn Reson Imaging* 1994; **12**:497–511.

76. Bachert P, Krems B, Ende G, Zabel HJ, Lorenz WJ. ^{19}F-[^{1}H] nuclear Overhauser effect of 5-fluorouracil and α-fluoro-β-alanine. In: *Proceedings, 2nd Annual Meeting, Society of Magnetic Resonance, San Francisco* 1994; **3**:1356.

77. Krems B, Bachert P, Zabel HJ, Lorenz WJ. ^{19}F-[^{1}H] nuclear Overhauser effect and proton-decoupling of 5-fluorouracil and α-fluoro-β-alanine. *J Magn Reson* 1995 (in press).

78. Krems B, Bachert P, Schlemmer HP, Zabel HJ, van Kaick G, Lorenz WJ. 2D ^{19}F chemical shift imaging of the metastatic liver of patients with colorectal tumors undergoing 5-fluorouracil chemotherapy. In: *Proceedings, 2nd Annual Meeting, Society of Magnetic Resonance, San Francisco* 1994; **3**:1163.

79. Ende G, Bachert P. Dynamic ^{13}C-^{1}H nuclear polarization of lipid methylene resonances applied to broadband proton-decoupled *in vivo* ^{13}C MR spectroscopy of human breast and calf tissue. *Magn Reson Med* 1993; **30**:415–23.

80. Moonen CTW, Dimand RJ, Cox KL. Noninvasive determination of linoleic acid content of human adipose tissue by natural abundance carbon-13 nuclear magnetic resonance. *Magn Reson Med* 1988; **6**:140–57.

81. Jue T, Rothman DL, Tavitian BA, Shulman RG. Natural-abundance ^{13}C NMR study of glycogen repletion in human liver and muscle. *Proc Natl Acad Sci U S A* 1989; **86**:1439–42.

82. Jue T, Rothman DL, Shulman GI, Tavitian BA, DeFronzo RA, Shulman RG. Direct observation of glycogen synthesis in human muscle with ^{13}C NMR. *Proc. Natl Acad Sci U S A* 1989; **86**:4489–91.

83. Heerschap A, Luyten PR, van der Heyden JI, Oosterwaal LJMP, den Hollander JA. Broadband proton decoupled natural abundance ^{13}C NMR spectroscopy of humans at 1.5 T. *NMR Biomed* 1989; **2**:124–32.

84. Beckmann N, Seelig, J, Wick H. Analysis of glycogen storage disease by *in vivo* ^{13}C NMR: comparison of normal volunteers with a patient. *Magn Reson Med* 1990; **16**:150–60.

85. Shulman GI, Rothman DL, Jue T, Stein P, DeFronzo RA, Shulman RG. Quantitation of muscle glycogen synthesis in normal subjects and subjects with non-insulin-dependent diabetes by ^{13}C nuclear magnetic resonance spectroscopy. *N Engl J Med* 1990; **322**: 223–8.

86. Beckmann N, Turkalj I, Seelig J, Keller U. ^{13}C NMR for the assessment of human brain glucose metabolism *in vivo*. *Biochemistry* 1991; **30**:6362–6.

87. Halliday KR, Fenoglio-Preiser C, Sillerud LO. Differentiation of human tumors from nonmalignant tissue by natural-abundance ^{13}C NMR spectroscopy. *Magn Reson Med* 1988; **7**:384–411.

88. Knopp MV, Hess T, Bachert P, et al. Magnetresonanzspektroskopie des Mammakarzinoms. *Radiologe* 1993; **33**:300–7.

89. Bellemann ME, Bachert P, Semmler W, Lorenz WJ. In vivo 2-dimensional ^{13}C chemical shift imaging of normal and neoplastic tissue in the human brain. In: *Proceedings, 10th Annual Meeting, Society of Magnetic Resonance in Medicine, San Francisco* 1991; **1**:463.

90. Ott D, Ernst T, Hennig J. Clinical value of ^{1}H spectroscopy of brain tumors. In: *Proceedings, 9th Annual Meeting, Society of Magnetic Resonance in Medicine, New York* 1990; **1**:105.

91. Ott D, Hennig J, Ernst T. Human brain tumors: assessment with *in vivo* proton MR spectroscopy. *Radiology* 1993; **186**:745–52.

92. Zülch KJ, Christensen E. Pathologische Anatomie der raumfordernden Prozesse. Springer, Berlin, Heidelberg, 1956.

93. Zülch KJ. Histological Typing of Tumours of the Central Nervous System. World Health Organization, Geneva, 1979.

94. Kernohan JW, Mabon RF, Svien KJ, Adson AW. A simplified classification of the gliomas. *Mayo Clin Proc* 1949; **24**:54–6.

95. Heindel W, Bunke J, Glathe S, Steinbrich W, Mollevanger L. Combined ^{1}H MR imaging and localized ^{31}P spectroscopy of intracranial tumors in 43 patients. *J Comput Assist Tomogr* 1988; **12**:907–16.

96. Heiss WD, Heindel W, Herholz K, et al. Positron emission tomography of fluorine-18-deoxyglucose and image-guided phosphorus-31 magnetic resonance spectroscopy in brain tumors. *J Nuc Med* 1990; **31**:302–10.

97. Knopp MV, Bachert P, Ende G, et al. Nuclear Overhauser enhanced ^{31}P chemical shift imaging for monitoring embolisation therapy of meningiomas. In: *Proceedings, 11th Annual Meeting, Society of Magnetic Resonance in Medicine, Berlin, Works in Progress* 1992: 1954.

98. Blankenhorn M, Bachert P, Ende G, et al. Clinical ^{31}P CSI in preoperative embolization therapy of meningiomas. In: *Proceedings, 12th Annual Meeting,*

Society of Magnetic Resonance in Medicine, New York 1993; **2**:1026.
99 Hwang YC, Mantil J, Boska MD, et al. Comparison of energy metabolism in various brain lesions via ^{31}P MRS and FDG-PET. In: *Proceedings, 11th Annual Meeting, Society of Magnetic Resonance in Medicine*, Berlin 1992; **2**:3610.
100 Arnold DL, Shoubridge EA, Emrich J, Feindel W, Villemure JG. Early metabolic changes following chemotherapy of human gliomas *in vivo* demonstrated by phosphorus magnetic resonance spectroscopy. *Invest Radiol* 1989; **24**:958–61.
101 Hubesch B, Sappey-Marinier D, Roth K, Matson GB, Weiner MW. P-31 MR spectroscopy of normal human brain and brain tumors. *Radiology* 1990; **174**:401–9.
102 Cadoux-Hudson TAD, Blackledge MJ, Rajagopalan B, Taylor DJ, Radda GK. Human primary brain tumor metabolism *in vivo*: a phosphorus magnetic resonance spectroscopy study. *Br J Cancer* 1989; **60**:430–6.
103 Ende G, Bachert P, Lorenz WJ. Quantification of the effects of composite-pulse ^{1}H-decoupling on *in vivo* ^{31}P MR spectra of the human calf muscle. In: *Proceedings, 12th Annual Meeting, Society of Magnetic Resonance in Medicine*, New York 1993; **3**:1143.
104 Bruhn H, Michaelis T, Merbold KD, Frahm J. On the interpretation of proton NMR spectra from brain tumors *in vivo* and *in vitro*. *NMR Biomed* 1992; **5**:253–8.
105 Frahm J, Bruhn H, Hänicke W, Merboldt KD, Mursch K, Markakis E. Localized proton NMR spectroscopy of brain tumors using short-echo time STEAM sequences. *J Comp Assist Tomogr* 1991; **15**:915–22.
106 Kugel H, Heindel W, Ernestus RI, Bunke J, du Mensil R, Friedmann G. Human brain tumors: Spectral patterns detected with localized H-1 MR spectroscopy. *Radiology* 1992; **183**:701–9.
107 Graham GD, Blamire AM, Howseman AM, et al. Proton magnetic resonance spectroscopy of cerebral lactate and other metabolites in stroke patients. *Stroke* 1992; **23**:333–40.
108 De Stefano N, Francis G, Antel JP, Arnold DL. Reversible decreases of N-acetylaspartate in the brain of patients with relapsing remitting multiple sclerosis. In: *Proceedings, 12th Annual Meeting, Society of Magnetic Resonance in Medicine*, New York 1993; **1**:280.
109 Mader I, Hagberg G, Roser W, et al. Investigations of time course of metabolite changes in primary Gd-enhancing MS plaques by proton MRS. In: *Proceedings, 12th Annual Meeting, Society of Magnetic Resonance in Medicine*, New York 1993; **1**:277.
110 Djurcic B, Olsen LR, Assaf HM, Whittingham TS, Lust WD, Drewes LR. Formation of free choline in brain tissue during *in vitro* energy deprivation. *J Cereb Blood Flow Metab* 1991; **11**:308–13.
111 Sijens PE, van Dijk P, Oudkerk M. Correlation between choline level and Gd-DTPA enhancement in patients with brain metastases of mammary carcinoma. *Magn Res Med* 1994; **32**:549–55.
112 Bruhn H, Frahm J, Gyngell ML, Merboldt KD, Hänicke W, Sauter R. Cerebral metabolism in man after acute stroke: new observations using localized proton NMR spectroscopy. *Magn Reson Med* 1989; **9**:126–31.
113 Luyten PR, Marien AJH, Heindel W, et al. Metabolic imaging of patients with intracranial tumors: H-1 MR spectroscopic imaging and PET. *Radiology* 1990; **176**:791–9.
114 Arnold DL, Matthews PM, Villemure JG. Combined proton and phosphorus MRS of human brain tumors *in vivo*: preliminary observations on the effects of tissue heterogeneity on metabolite ratios and pH measurements. In: *Proceedings, 8th Annual Meeting, Society of Magnetic Resonance in Medicine*, Amsterdam 1989; **1**:75.
115 Arnold DL, Shoubridge EA, Villemure JG, Feindel W. Proton and phosphorus magnetic resonance spectroscopy of human astrocytomas *in vivo*. Preliminary observations on tumor grading, *NMR Biomed* 1990; **3**:184–9.
116 Henriksen O, Wieslander S, Gjerris F, Jensen KM. *In vivo* ^{1}H-spectroscopy of human intracranial tumors at 1.5 Tesla. *Acta Radiol* 1991; **32**:95–9.
117 Alger JR, Frank JA, Bizzi A, et al. Metabolism of human gliomas: assessment with H-1 MR spectroscopy and F-18 fluorodeoxyglucose PET. *Radiology* 1992; **177**:633–41.
118 Fulham MJ, Bizzi A, Dietz MJ, et al. Mapping of brain tumor metabolites with proton MR spectroscopic imaging: clinical relevance. *Radiology* 1992; **185**:675–86.
119 Herholz MD, Heindel W, Luyten PR, et al. *In vivo* imaging of glucose consumption and lactate concentration in human gliomas. *Ann Neurol* 1992; **31**:319–27.
120 Naruse S, Faruya S, Ide M, et al. Specific metabolites in brain tumors detected by ^{1}H chemical shift imaging. In: *Proceedings, 11th Annual Meeting, Society of Magnetic*

Resonance in Medicine, Berlin, Works in Progress 1992: 1952.

121 Ernst T, Kreis R, Ross BD. Absolute quantitation of water and metabolites in the human brain. I. Compartments and water. *J Magn Reson* 1993; **B102**:1–8.

122 Kreis R, Ernst T, Ross BD. Absolute quantitation of water and metabolites in the human brain. II. Metabolite concentrations. *J Magn Reson* 1993; **B102**:9–19.

123 Michaelis T, Merboldt KD, Bruhn H, Hänicke W, Frahm J. Absolute concentrations of metabolites in the adult human brain *in vivo*: Quantification of localized proton MR spectra. *Radiology* 1993; **187**:219–27.

124 Sotak CH, Freeman DM. A method for volume-localized lactate editing using zero-quantum coherence created in a stimulated-echo pulse sequence. *J Magn Reson* 1988; **77**:382–8.

125 Sotak CH, Freeman DM, Hurd RE. The unequivocal determination of *in vivo* lactic acid using two-dimensional double-quantum coherence-transfer spectroscopy. *J Magn Reson* 1988; **78**:355–61.

126 Bellemann ME, Bachert P, Lorenz WJ. Volume-selective *in vivo* 2D multiple-quantum ^{1}H MR spectroscopy in man. In: *Proceedings, 11th Annual Meeting, Society of Magnetic Resonance in Medicine*, Berlin 1992; **2**:2150.

127 Lazeyras F, Charles HC, Bokyo O, et al. New perspectives in tumor grading by combined short echo/long echo ^{1}H spectroscopic imaging. In: *Proceedings, 11th Annual Meeting, Society of Magnetic Resonance in Medicine*, Berlin 1992; **2**:3604.

128 Negendank W, Zimmermann R, Gotsis E, et al. A cooperative group study of ^{1}H MRS of primary brain tumors. In: *Proceedings, 12th Annual Meeting, Society of Magnetic Resonance in Medicine*, New York 1993; **3**:1521.

129 Negendank W, Sauter R, Brown T, et al. Intratumoral lipids and lactate in ^{1}H MR *in vivo* in astrocytic tumors. In: *Proceedings, 2nd Annual Meeting, Society of Magnetic Resonance*, San Francisco 1994; **3**:1296.

130 Harris JR, Morrow M, Bonadonna G. Cancer of the breast. In: DeVita VT, Hellman S, Rosenberg SA (eds). *Cancer. Principles and Practice of Oncology*. Lippincott, Philadelphia, 1993: 1264–332.

131 Sijens PE, Wijrdeman H, Moerland MA, Bakker CJG, Vermeulen JWA, Luyten PR. Human breast cancer *in vivo* H-1 and P-31 MR spectroscopy at 1.5 T. *Radiology* 1988; **169**:615–20.

132 Merchant TE, Thelissen GR, de Graaf PW, Den Otter W, Glonek T. Clinical magnetic resonance spectroscopy of human breast disease. *Invest Radiol* 1991; **26**:1053–9.

133 Smith TA, Glaholm J, Leach MO, et al. A comparison of *in vivo* and *in vitro* ^{31}P NMR spectra from human breast tumours: variations in phospholipid metabolism. *Br J Cancer* 1991; **63**:514–16.

134 Kalra R, Wade KE, Hands L, et al. Phosphomonoester is associated with proliferation in human breast cancer. A ^{31}P MRS study. *Br J Cancer* 1993; **67**:1145–53.

135 Maris JM, Evans AE, McLaughlin AC, et al. ^{31}P nuclear magnetic resonance spectroscopic investigation of human neuroblastoma *in situ*. *N Engl J Med* 1985; **312**:1500–5.

136 Francis IR, Chenevert TL, Collomb LG, Gubin B, Glazer GM. Clinical P-31 MR spectroscopy of malignant hepatic tumors using 1D CSI technique. In: *Proceedings, 8th Annual Meeting, Society of Magnetic Resonance in Medicine*, Amsterdam 1989; **1**:70.

137 Meyerhoff DJ, Karczmar GS, Weiner MW. Abnormalities of the liver evaluated by ^{31}P MRS. *Invest Radiol* 1989; **24**:980–4.

138 Brinkmann G, Melchert UH. A study of T_1-weighted ^{31}Phosphorus MR spectroscopy from patients with focal and diffuse liver disease. *Magn Reson Imaging* 1992; **10**:949–56.

139 Meyer KL, Ballon D, Kemeny N, Koutcher JA. *In vivo* P-31 NMR 3D-CSI of colorectal hepatic metastases. In: *Proceedings, 12th Annual Meeting, Society of Magnetic Resonance in Medicine*, New York 1993; **2**:1025.

140 Cox, IJ, Sargentini J, Calam J, Bryant DJ, Iles RA. Four-dimensional phosphorus-31 chemical shift imaging of carcinoid metastases in the liver. *NMR Biomed* 1988; **1**:56–60.

141 Cox IJ, Bell JD, Peden CJ, et al. *In vivo* and *in vitro* ^{31}P magnetic resonance spectroscopy of focal hepatic malignancies. *NMR Biomed* 1992; **5**:114–20.

142 Dixon RM, Angus PW, Rajagopalan B, Radda GK. Abnormal phosphomonoester signals in ^{31}P MR spectra from patients with hepatic lymphoma. A possible marker of liver infiltration and response to chemotherapy. *Br J Cancer* 1991; **63**:953–8.

143 Meyerhoff DJ, Boska MD, Thomas AM, Weiner MW. Alcoholic liver disease: quantitative image-guided P-31 MR spectroscopy. *Radiology* 1989; **173**:393–400.

144 Kurhanewicz J, Thomas A, Jajodia P, et al. ^{31}P spectroscopy of the human prostate gland *in vivo* using a transrectal probe. *Magn Reson Med* 1991; **22**:404–13.

145 Narayan P, Jajodia P, Kurhanewicz J, et al. Characterization of prostate cancer, benign prostatic hyperplasia and normal prostates using transrectal ^{31}Phosphorus magnetic resonance spectroscopy: a preliminary report. *J Urol* 1991; **146**:66–74.

146 Franklin RB, Costello LC. Glutamate dehydrogenase and a proposed glutamate aspartate pathway for citrate synthesis in rat ventral prostate. *J Urol* 1984; **132**:1239–43.

147 Cooper JF, Farid EL. The role of citric acid in the physiology of the prostate: III. Lactate/citrate ratios in benign and malignant prostatic homogenates as an index of prostatic malignancy. *J Urol* 1964; **92**:533–6.

148 Pretlow TG, Harris BE, Bradley E, Bueschen AJ, Lloyd KL, Pretlow TP. Enzyme activities in prostatic carcinoma related to Gleason grades. *Cancer Res* 1985; **45**:442–6.

149 Costello LC, Littleton RR, Franklin RB. Regulation of citrate related metabolism in normal and neoplastic prostate. In: Sharma RK, Criss WE (eds). *Progress in Cancer Research and Therapy, Vol. 8: Endocrine Control in Neoplasia*. Raven, New York, 1978: 303–14.

150 Fowler AH, Pappas AA, Holder JC, et al. Differentiation of human prostate cancer from benign hypertrophy by *in vitro* ^{1}H NMR. *Magn Reson Med* 1992; **25**:140–7.

151 Thomas MA, Narayan P, Kurhanewicz J, Jajodia P, Weiner MW. ^{1}H MR spectroscopy of normal and malignant human prostate *in vivo*. *J Magn Res Com* 1990; **87**:610–19.

152 Tomoi M, Yoshida M, Shioura H, et al. Differentiation of human prostate cancer from benign hyperplasia by localized *in vivo* ^{1}H MR spectroscopy. In: *Proceedings, 2nd Annual Meeting, Society of Magnetic Resonance, San Francisco* 1994; **3**:1301.

153 Heerschap A, Jager G, Barentsz J, et al. Functional ^{1}H magnetic resonance in the identification and characterization of localized prostate cancer. In: *Proceedings, 2nd Annual Meeting, Society of Magnetic Resonance, San Francisco* 1994; **1**:274.

154 Narayan P, Kurhanewicz J. Magnetic resonance spectroscopy in prostate disease: diagnostic possibilities and future developments. *Prostate* 1990; **20(S)**:43–50.

155 Riedy G, Chan L. *In vivo* proton MR spectroscopy of the human kidney. In: *Proceedings, 2nd Annual Meeting, Society of Magnetic Resonance, San Francisco* 1994; **1**:233.

156 Hricak H. Phosphorus-31 MRS of the kidney. *Invest Radiol* 1989; **24**:993–6.

157 Balaban RS. MRS of the kidney. *Invest Radiol* 1989; **24**:988–92.

158 Chew WM, Hricak H. Phosphorus-31 MRS of human testicular function and viability. *Invest Radiol* 1989; **24**:997–1000.

159 Chew WM, Hricak H, McClure RD, Wendland MF. *In vivo* human testicular function assessed with P-31 MR spectroscopy. *Radiology* 1990; **177**:743–7.

160 Hendrix RA, Lenkinski RE, Vogele K, Bloch P, McKenna WG. ^{31}P localized magnetic resonance spectroscopy of head and neck tumors – preliminary findings. *Otolaryngol Head Neck Surg* 1990; **103**:775–83.

161 McKenna WG, Lenkinski RE, Hendrix RA, Vogele K, Bloch P. The use of magnetic resonance imaging and spectroscopy in the assessment of patients with head and neck tumors and other superficial human malignancies. *Cancer* 1989; **64**:2069–75.

162 Negendank WG, Crowley MG, Ryan JR, Keller NA, Evelhoch JL. Bone and soft tissue lesions: diagnosis with combined H-1 MR-imaging and P-31 MR spectroscopy. *Radiology* 1989; **173**:181–9.

163 Dewhirst MW, Sostman HD, Leopold KA, et al. Soft-tissue sarcomas: MR-imaging and MR-spectroscopy for prognosis and therapy monitoring. *Radiology* 1990; **174**:847–53.

164 Zlatkin MB, Lenkinski RE, Shinkwin M, et al. Combined MR imaging and spectroscopy of bone and soft tissue tumors. *J Comp Ass Tomogr* 1990; **14**:1–10.

165 Sostman D, Dewhirst M, Charles C, et al. Prognostic evaluation and therapy monitoring in human soft tissue sarcomas with ^{31}P MRS. In: *Proceedings, 9th Annual Meeting, Society of Magnetic Resonance in Medicine, New York* 1990; **1**:319.

166 Sostman D, Prescott DM, Dewhirst M, et al. MR imaging and spectroscopy for prognostic soft tissue sarcomas. *Radiology* 1994; **190**:269–75.

167 Bongers H, Schick F, Skalej M, Hess CF, Jung WI. Localized *in vivo* ^{1}H spectroscopy of human bone and soft tissue tumours. *Eur J Radiol* 1992; **2**:459–64.

168 Bongers H, Schick F, Skalej M, Jung WI, Einsele H. Localized *in vivo* ^{1}H spectroscopy and chemical shift imaging of the bone marrow in leukemic patients. *Eur Radiol* 1992; **2**:350–6.

169 Irving MG, Brooks NWM, Brereton IM, et al. Use of high resolution *in vivo* volume selected ^{1}H-magnetic resonance spectroscopy to investigate leukemia in humans. *Cancer Res* 1987; **47**:3901–6.

170 Gückel F, Semmler W, Brix G, et al. Diagnosis and therapy monitoring of bone marrow involvement in

patients with malignant lymphoma: MR tomography, chemical shift imaging and MR spectroscopy. *Magn Reson Imaging* 1989; 7(S):56.

171 Gückel F, Brix G, Semmler W, et al. Systemic bone marrow disorders: characterization with proton chemical shift imaging. *J Comp Assist Tomogr* 1990; 14:633–42.

172 Gückel F, Brix G, Semmler W, et al. Proton chemical shift imaging of bone marrow for monitoring therapy in leukemia. *J Comp Assist Tomogr* 1990; 14:954–9.

173 Roth K, Hubesch B, Meyerhoff DJ, et al. Noninvasive quantitation of phosphorus metabolites in human tissue by NMR spectroscopy. *J Magn Reson* 1989; 81:299–311.

174 Husted CA, Duijn JH, Matson GB, Maudsley AA, Weiner MW. Molar quantitation of *in vivo* proton metabolites in human brain with 3D magnetic resonance spectroscopic imaging. *Mag Reson Imaging* 1994; 12:661–7.

175 Naruse S, Horikawa Y, Tanaka C, et al. Evaluation of the effects of photoradiation therapy on brain tumors with *in vivo* P-31 MR spectroscopy. *Radiology* 1986; 160:827–30.

176 Ng TC, Vijayakumar S, Majors AW, Thomas FJ, Meaney TF, Baldwin NJ. Response of a non-Hodgkin lymphoma to ^{60}Co therapy monitored by ^{31}P MRS *in situ*. *Int J Radiat Oncol Biol Phys* 1987; 13:1545–51.

177 Semmler W, Gademann G, van Kaick G, Zabel H-J, Lorenz WJ. *In vivo* P-31 spectroscopy of human tumors and their response to therapy using a 1.5 Tesla whole body scanner. In: *Proceedings, 5th Annual Meeting, Society of Magnetic Resonance in Medicine, Montreal* 1986; 1:39.

178 Redmond O, Stack J, Scully M, Dervan P, Camey D, Ennis JT. Tissue characterization and spectral analysis of malignant tumors following chemotherapy. In: *Proceedings, 7th Annual Meeting, Society of Magnetic Resonance in Medicine, San Francisco* 1988; 1:432.

179 Prescott DM, Charles HC, Sostman HD, et al. Therapy monitoring in human and canine soft tissue sarcomas using magnetic resonance imaging and spectroscopy. *Int J Radiat Oncol Biol Phys* 1993; 28:415–23.

180 Negendank W, Murphy-Boesch J, Padavic-Shaller K, et al. Phosphomonoester changes and treatment response in non-Hodgkin lymphomas. In: *Proceedings, 2nd Annual Meeting, Society of Magnetic Resonance, San Francisco* 1994; 1:131.

181 Segebarth CM, Balériaux DF, Arnold DL, Luyten PR, den Hollander JA. MR image-guided P-31 MR spectroscopy in the evaluation of brain tumor treatment. *Radiology* 1987; 165:215–19.

182 Arnold DL, Shoubridge EA, Emrich J, Feindel W, Villemure JG. Metabolic changes in cerebral gliomas within hours of treatment with intra-arterial BCNU demonstrated by phosphorus magnetic resonance spectroscopy. *Can J Neurol Sci* 1987; 14:570–5.

183 Arnold DL, Shoubridge EA, Villemure JG, Feindel W. Phosphorus magnetic resonance spectroscopy of cerebral gliomas following treatment with intravenous BCNU. In: *Proceedings, 7th Annual Meeting, Society of Magnetic Resonance in Medicine, San Francisco* 1988; 1:333.

184 Ross B, Trop J, Derby KA, et al. Metabolic response of glioblastoma to adoptive immunotherapy: detection by phosphorus MR spectroscopy. *J Comp Assist Tomogr* 1989; 13:189–93.

185 Kolem H, Miyazaki T, Schneider M, Wicklow K, Sauter R. Nuclear Overhauser enhanced localized *in vivo* phosphorus spectroscopy using the CSI technique. In: *Proceedings, 9th Annual Meeting, Society of Magnetic Resonance in Medicine, New York, Works in Progress* 1990: 1331.

186 Jüngling FD, Wakhloo AK, Hennig J. *In vivo* proton spectroscopy of meningioma after preoperative embolization. *Magn Reson Med* 1993; 30:155–60.

187 Heesters MAAM, Kamman RL, Mooyaart EL, Go KG. Localized proton spectroscopy of inoperable brain gliomas. Response to radiation therapy. *J Neurooncology* 1993; 17:27–35.

188 Cabanac S, Malet-Martino MC, Bon M, Martino R, Nedelec JF, Dimicoli JL. Direct ^{19}F NMR spectroscopic observation of 5-fluorouracil metabolism in the isolated perfused mouse liver model. *NMR Biomed* 1988; 1:113–20.

189 Wolf W, Presant CA, Servis KL, et al. Tumor trapping of 5-fluorouracil: *In vivo* ^{19}F NMR spectroscopic pharmacokinetics in tumor-bearing humans and rabbits. *Proc Natl Acad Sci U S A* 1990; 87:492–6.

190 den Hollander JA, Luyten PA, Marien JH. Localized ^{1}H NMR, spectroscopic imaging of the human brain. In: Diel P, Fluck E, Günther H, Kosfeld R, Seelig J (eds). *NMR 27: In vivo magnetic resonance spectroscopy II: Localization and Spectral Editing*. Springer, Berlin, 1992: 151–75.

Index

In general the fully spelled out versions of terms are used as headings. A reference list of abbreviations and their full equivalents is on pages xii–xiii.

acetaminophen poisoning 125
adenocarcinoma
 prostatic 225
 therapy monitoring 231, 238
adenoma, pituitary 218–19
adenosine triphosphate (ATP)
 brain diseases 7, 8
 heart 75, 76, 78, 81–2
  correcting distorted signals 78–80
  diseases 82–3, 85–7
  transplants 84, 198–200
 liver 108, 109–13
  children 113
  diseases 121, 123, 124, 125–7, 132–4
  effect of metabolic manipulations 114–16
  transplants 192–6
 muscle 59, 62, 68
  recovery after exercise 68–9
  resting 65
 spleen 135
 transplant organs 179
  heart 84, 198–200
  kidney 186–8
  liver 192–6
  pancreas 201
 tumours 214, 226
adiabatic fast passage pulse (AFP) 44
AIDS dementia 147
alanine infusion, hepatic metabolic response 115–16
Alzheimer's disease
 ^{1}H MRS 12–13, 145
 ^{31}P MRS 12
astrocytomas 13–14, 215, 219, 221
 therapy monitoring 231, 232, 235

birth asphyxia 93–4, 95, 145
bone marrow
 neoplasms 227–8, 236
 transplants 202–3
brain, MRS
 amyotrophic lateral sclerosis 146–7
 clinical application 1–25, 139–59
 dementias 12–13, 145–7
 epilepsy 8–12, 101–2, 150–1
 ^{1}H MRS 141–3, 161–73, 214–15
  Alzheimer's disease 12–13, 145
  complicating factors *in vivo* 161
  epilepsy 8–12
  localization techniques 143, 162–3
  localized imaging 169–71
  magnetic field homogeneity 163–4
  metabolism studies 94–102, 145, 151–2
  multiple sclerosis 15, 16, 17, 147–8

 paediatric applications 96–102, 145
 phase encoding and MRS imaging 165
 practical aspects 161–73
 single volume 166–9
 stroke 3–6, 145, 146
 tumours 13–14, 149, 220–2, 235–6
 water suppression 164–5
localization 139–40
 ^{1}H MRS 143, 162–3
 ^{31}P MRS 141
metabolic disorders 97–100, 151–2
multiple sclerosis 15, 147–8
non–structural lesions 143
normal brain 139–43
^{31}P MRS 140–1
 dementias 12, 147
 epilepsy 8, 9–10, 12, 151
 multiple sclerosis 147
 paediatric applications 93–4, 145
 stroke 7, 8, 145
 tumours 13, 14, 149–50, 218–20, 230, 231, 232
paediatric applications 93–104
 hypoxic–ischaemic injury in neonates 93–4, 95, 145
 metabolism studies 93–102
pH measurement 140, 144
phosphorus spectra 140–1
proton spectra 141–3
serial study of chemical changes 144
stroke 3–8, 144–5, 146
time required 2
tumours 13–14, 144, 149–50
 diagnosis 217–22
 therapy monitoring 230, 231, 232, 235
breast tumours
 diagnosis 222–3, 224
 therapy monitoring 231

^{13}C MRS
 brain 15–20, 144
 corneal transplants 202, 203
 heart 87–8
 liver 105, 116–19
  metabolic disorders 134–5
  transplants 194
 muscle 59, 60, 61
  dystrophic 59, 60
  special characteristics 17–18
 tumours 212–13, 217, 224
^{133}Cs NMR 8
Canavan's disease 100, 152
cardiac *see* heart
cardiomyopathy 82–4
 dilated (DCM) 75, 82, 83, 86, 87

 hypertrophic (HCM) 82, 83
centrally ordered phase encode (COPE) 49
chemical shift 28
chemical shift dispersion 33
chemical shift imaging (CSI) 38, 39, 42, 48
 baseline errors 49–50
 bone marrow neoplasms 227–8, 236
 excitation errors 45
 heart 78
 liver 11, 108–10
  disease 121, 122–3, 125–7
 motion artifacts 48–9
  reduction techniques 49
 muscle 61
 small sampling matrices, errors arising 50–1
 spleen 135
 tumours 212
chemical shift selective imaging (CHESS) 34
 presaturation method 43
children, MRS applications 93–104
 birth asphyxia 93–4, 95, 145
 brain metabolism studies 93–104, 145
 liver 113
choline, brain 142, 166, 167–70
 dementias 147
 epilepsy 101–2
 metabolic disorders 151–2
 metabolism studies 96–7, 100
 multiple sclerosis 15, 16, 17, 147
 tumours 14, 149, 150, 221, 235
cirrhosis, liver 120–2, 124–5
citrate, prostatic tumours 225
convolution difference 51
corneal transplants 202, 203
coronary artery disease (CAD) 75 80, 86–7
creatine (Cr)
 brain 142, 166, 167–70
  epilepsy 101–2
  metabolism studies 96–9, 100
  tumours 14, 220–1, 235
 cardiomyopathy/heart failure 88
creatine kinase equilibrium 140–1
creatine kinase reaction 214
cyclosporin A (CSA) and kidney transplants 188–90

data reconstruction and analysis 38, 40
dementia, MRS 145–7
 ^{1}H MRS 12–13, 145–7
 ^{31}P MRS 12
deoxymyoglobin, heart 87, 88
depth resolved surface coil spectroscopy (DRESS) 47
 heart 76–7

247

Index

(DRESS) (cont'd)
   hepatic tumours 128–31
   muscle 61
diffusion effects 47
diffusion–weighted imaging (DWI) 1
   epilepsy 11
   stroke 6, 8, 11
Duchenne muscular dystrophy 59, 60, 69–70

echoes, stimulated 46, 162
echoplanar imaging (EPI) 1
eddy current suppression 34, 35
ependymomas 219
epilepsy, MRS 8–12, 150–1
   animal studies 8–10
   combined ^{1}H–^{31}P MRS 9–10
   ^{1}H MRS 8–9, 10–11, 151
   ^{31}P MRS 8, 10, 151
   temporal lobe epilepsy 10, 101–2, 151, 152
examination, image–guided spectroscopy 34–8
   localization techniques 36–8
   time scale 35
exercise studies
   muscle 62–5
      aerobic exercise 68
      disease 69–71
      electrical stimulation 64–5
      fatigue 67
      ischaemic exercise 68
      quadriceps 63–4
      recovery after exercise 68–9
   myocardial ischaemia 86

^{19}F MRS 7
   tumours 212, 215–7
      drug monitoring 236–8
fast rotating gradient spectroscopy (FROGS) 114
fish oils, effect on liver metabolism 116
5-fluoro-5,6-dihydrouracil (DHFU) 215, 236–8
α-fluoro-β-alanine (FBAL) 215–16, 236–8
5-fluorouracil (5–FU) 215–17, 236–8
free induction decay (FID) 28, 29
fructose metabolism
   disorders of 123–5
   liver 113–14

galactosaemia 125
gamma–aminobutyric acid (GABA), brain 167–8
   epilepsy 11
   tumours 222
Gibb's ringing 50
glioblastomas 149, 150, 218–19, 222
   therapy monitoring 232–3
gliomas 149, 218–19, 221, 235
glucose metabolism, liver 116, 118
   alanine infusion and 115–16
   diabetics 134–5
glutamine/glutamate, brain 166, 167–8
   metabolic disorders 98–9, 152
glyceroethanolamine (GPE), liver 106, 107
glycerophosphocholine (GPC)
   hepatic tumours 127
   liver 106, 107

glycerophosphoethanolamine (GPE), hepatic tumours 127
glycogen metabolism, liver 116–19
   diabetics 134–5
graft-versus-host disease (GVHD) 202–3

^{1}H MRS 30–4
   bone marrow transplants 202–3
   brain *see under* brain
   heart 88
      transplants 198
   liver 106–7, 120
      transplants 194, 200
   muscle 59, 60, 61
      dystrophic 59, 60
   transplant organs 189–90, 194, 198, 200
   tumours 212–13, 214–15
      diagnosis 212, 220–2, 223, 224–8
      therapy monitoring 235–6
   water suppression 164–5, 166
haemangiomas, cavernous, liver 126, 130
heart, MRS 75–91
   ^{13}C MRS 87–8
   cardiomyopathy 82–4
      dilated 75, 82–3, 86, 87
      hypertrophic 82, 83
   coronary artery disease 75, 80, 86–7
   ^{1}H MRS 88
   ischaemic disease 75, 86–7
   left ventricular hypertrophy 75, 82, 83
   methods 76–82
      distortion correction 78–80
      localization 76–8
      surface coils and optimizing sensitivity 76
   normal heart 75–6, 81–2
   ^{31}P MRS 75–87
   stress–testing 86–7
   transplants 84–5, 197–200
hepatitis
   alcoholic 121–3
   viral 120–1, 122, 123
histiocytoma 213, 235
Hodgkin's disease 126–7
Huntington's disease 147
hyperfine splitting 51–2

image selective *in vivo* spectroscopy (ISIS) 37, 44–5, 47
   heart 77
   kidney transplants 186
   liver 11, 108, 109, 110
      disease 121, 122, 127, 134
   muscle 61
infertility 226
intestine, small, transplantation 203–4
inversion recovery techniques, errors 44

J-coupling 51–2

kidney
   ^{31}P MRS 185–6
   transplants 175, 179–90
      clinical studies 184–6, 187–8
      dysfunction 186–90
      high-resolution urine studies 188–90
      organ preservation 176, 183–4
      pre-operative assessment 181–3, 184–5

lactate
   brain 142, 166–7, 168–9, 214–15
      epilepsy 8–9, 11, 18
      Huntington's disease 147
      metabolic disorders 97–8, 145
      multiple sclerosis 15, 16, 17
      pH and 7, 9
      POCE MRS 18–19
      stroke 3–6, 18–19, 145
      tumours 13, 14, 149, 221–2, 235
   heart transplants 198
lactic acidosis, congenital, neurological disorder in 97
Larmor frequency 27, 28, 29
Larmor relationship 33
left ventricular hypertrophy (LVH) 75, 82, 83
liver, MRS 105–135
   ^{13}C MRS 105, 116–19, 134–5
   cirrhosis 121–3, 124–5
   ^{1}H MRS 106–7, 120
   hepatitis 120–3
   metabolism
      disorders 123–5, 134–5
      fructose 113–14, 123–5
      glucogen 115–16, 118
      glycogen 116–19
      in liver disease 121
      phospholipids, effects of fish oils 116
   normal liver 106–19
      extracts of 106–7
   ^{31}P MRS 105–6
      children 113
      liver disease 120–34, 223
      localization methods 111–12
      metabolic manipulations 113–16, 123–5
      normal liver 106–16
      quantitation of metabolite levels 109–13
      relaxation studies 107–9
      spectral resolution 109
      transplants 192–7
   poisoning 125
   steatosis 120
   transplants 190–7
      preservation 192–4
      rejection 197
   tumours 125–32, 223
      metastatic 127–32, 215–16, 236–8
      response to therapy 132–4
localization techniques 36–8
localization techniques (continued)
   brain 139–40
   ^{1}H MRS 143, 162–3
   ^{31}P MRS 141
   heart 76–8
   liver 111–12
   muscle 60–2
   tumours 212
   see also names of specific techniques and categories of techniques
lymphomas 126–7, 133, 135, 221
   therapy monitoring 230, 231, 232

McArdle's disease 69, 70
magnetic field(s) 27–30
   field gradients 28, 30, 34, 35
   homogeneity 31–3
   ^{1}H MRS of brain 163–4

NMR scanners 30–4
    radiofrequency 27–8, 33–4
    static 27, 31–3
magnetic resonance angiography (MRA) 1
magnetic resonance imaging (MRI) 1
magnetic resonance physics 27–30
magnetic resonance spectroscopy (MRS)
    data reconstruction and analysis 38, 40
    image-guided
        examination 34–8
        physics of 27–30
    oncology 211–46
    paediatric applications 93–104
    problems, *in vivo* studies 41–53
        acquisition and processing errors 50–2
        excitation errors 45–7
        preparation errors 43–5
        spatial encoding artifacts 48–50
    quantification 41–2
    relationship with MRI 1
    scanners 30–4
    terminology 1
    transplantation 175–210
    whole body examination 27–40
    see also individual types and sites of MRS
magnetization
    longitudinal 27, 28
    transverse 28, 29
magnetization transfer effects 47
meningiomas 149, 218–19, 221
    therapy monitoring 233, 234, 235
metabolic diseases/disorders
    brain 97–100, 151–2
    liver
        ^{13}C MRS 134–5
        ^{31}P MRS 123–5
mitochondrial encephalopathics 97, 151
monoethylglycinexylidide (MEGX) test,
    liver transplants 191–2
motion artifact suppression technique
    (MAST) 49
motion of patient
    effects on CSI 48–9
    effects on excitation 45–6
    problems caused 42
multiple sclerosis 15, 16, 17, 147–8
multiple-echo selective acquisition (MESA)
    46
muscle, skeletal 55
    bioenergetics 58–59
    buffering capacity 66–7
    electrical stimulation 64–5, 67
    fatigue 67
    function 56–8
    metabolism 56–8
    MRS 58–73
        applications 65–9
        ^{13}C MRS 59, 60, 61
        exercise studies 62–5, 66–71
        ^{1}H MRS 59, 60, 61
        muscle disease 69–71
        ^{31}P MRS 2–3, 59, 60, 61, 65
        techniques 36–7, 58–65
    pennation angle 56
    pH
        measurements 65–6
        recovery after exercise 69
        resting 65
    structure 56–8
musculoskeletal system, tumours 226–7

therapy monitoring 230, 235
myeloma 235
myocardial infarction 85
myocardial ischaemia 75, 86–7
myoglobin, heart 87, 88
*myo*inositol, brain 166, 167, 169

N-acetyl (NA) groups, brain
    Alzheimer's disease 12–13
    epilepsy 11, 12
    multiple sclerosis 15, 16, 17
    stroke 3, 4
N-acetylaspartate (NAA), brain 142, 166,
    167–71, 214–15
    biochemical role 168
    dementias 12, 145–7
    epilepsy 11, 101–2, 151, 152
    loss in children 100–2
    metabolic disorders 151–2
    metabolism studies 95, 96–7, 100, 145
    multiple sclerosis 147–8
    stroke 3, 4, 145, 146
    tumours 14, 149, 220–1, 235
^{23}Na MRS 193, 198
neck, tumours 226
neonates
    brain metabolism studies 93–104, 145
    hypoxic-ischaemic encephalopathy 93–4,
        95, 145
neurinomas 221
neuroblastoma 132–3
neuroimaging, functional 1
nuclear magnetic resonance (NMR), types of
    measurement 1, 2
nuclear Overhauser enhancement (nOe) 45,
    214
nucleotide diphosphate (NDP), liver 106,
    107
nucleotide triphosphates (NTP)
    liver 106, 107
    tumours 128
    tumours 214
        brain 219
        breast 223
        liver 128
        therapy monitoring 228–35
        urogenital tract 224–6

^{17}O NMR 7
oncology *see* tumours
ornithine carbamoyl transferase (OCT)
    deficiency 98–100
osteosarcomas 235
oxymyoglobin, heart 87, 88

^{31}P MRS
    brain *see under* brain
    heart 75–6, 81–2
        disease 80, 82–4, 85–7
        methods 76–82
        stress-testing 86–7
        transplants 84–5, 198
    in vivo, problems 51
    kidney 185–6
        transplants 178–9, 181–8
        tumours 226
    liver 105–16
        diseases 120–34
        transplants 192–7
    muscle 2–3, 55, 59, 60

dystrophic 59, 60
resting 65
pancreatic transplants 200–2
small intestinal transplantation 203–4
transplant organs
    clinical studies 184–6
    dysfunction 186–8, 197, 201
    preservation 183–4, 192–4, 198
    viability assessment 178–9, 181–3,
        192–7, 200–1, 203–4
tumours 212–13, 214
    diagnosis 212, 218–20, 222–3, 224–6
    therapy monitoring 228–35
pancreatic transplants 200–2
    islet cells 201–2
    rejection 201
pH
    brain 140, 144
    muscle 65–6, 69
    tumours 226, 227
        therapy monitoring 229, 231, 232, 233,
            235
phosphate, inorganic (Pi)
    brain 7, 8, 10, 12, 93–4, 140, 151
    tumours 218–19
    heart 78, 81–2
        diseases 75, 85
        transplants 84–5, 199–200
    liver 108, 109–13
        diseases 121, 123–5, 127–8, 133–4
        effect of metabolic manipulations
            114–16
        transplants 194–6
    muscle 59, 65
        diseases 70–1
        recovery after exercise 69
        resting 65–6
    spleen 135
    transplant organs 179
        heart 84–5, 199–200
        kidney 181–2, 184–5, 186–8
        liver 194–6
    tumours 214
        brain 218–19
        breast 222
        therapy monitoring 228–31
phosphocholine (PC)
    breast tumours 223
    liver 106, 107
    tumours 127, 133
phosphocreatine (PCr)
    brain 166, 167–70
        disorders 10, 93–4
        tumours 218–19
    heart 75, 76, 78, 81–2
        correcting distorted signals 78
        diseases 75, 82–3, 85–7
        transplants 84–5, 198–200
    muscle 59
        diseases 70–1
        recovery after exercise 69
        resting 65
    transplant organs 179
        heart 84–5, 198–200
    tumours
        diagnosis 218–19, 223, 224–6
        therapy monitoring 229, 230, 232,
            233–4, 235
phosphodiesters (PDE)
    liver 106, 107, 108, 109–13

249

Index

phosphodiesters (PDE) (cont'd)
  children 113
  diseases 121, 123–34, 223
  effect of metabolic manipulations 114–16
  transplants 194–7
  transplant organs
    heart 200
    kidney 187–8
    liver 194–7
    pancreas 201–2
  tumours 214
    diagnosis 218–19, 222–4, 226
    therapy monitoring 228, 230, 233–4, 235
phosphoethanolamine (PE)
  breast tumours 223
  liver 106, 107
    children 113
    tumours 133
phosphofructokinase (PFK) deficiency 66, 70
phosphomonoesters (PME)
  brain
    disorders 10, 12, 13
    neonatal 93, 94
  liver 106, 107, 108, 109–13
    children 113
    diseases 121–34, 223
    effect of metabolic manipulations 114–16
    transplants 194–7
  muscle disorders 70
  spleen 135
  transplant organs 179
    kidney 181–2, 183, 185, 187–8
    liver 194–7
    pancreas 201–2
  tumours
    diagnosis 218–19, 224–6
    therapy monitoring 228–35
physics, magnetic resonance 27–30
pixel resolved spectroscopy (PRESS) 34, 37, 163
  brain 163, 164
  muscle 61
preinversion 44
presaturation 43
  errors 43
PROBE software, data reconstruction and analysis 40
prolactinoma 231
prostate tumours 224–6
proton MRS see ^{1}H MRS
proton observe-carbon edited (POCE) MRS 18–20

Rasmussen's syndrome 10–11
relaxation time errors 51–2
renal see kidney
resolution enhancement 51

respiratory gating, CSI 49
respiratory ordered phase encode (ROPE) 49
rotating frame zeugmatography (RFZ) 78

sarcomas 231, 235–6
scanners 30–4
seminoma 226
shimming 33, 34
  brain MRS 164
single voxel spectroscopy (SVS) 37–8
  brain 141
  heart 77–8
  muscle 61
  tumours 212
  see also names of specific techniques
sodium NMR 8
  see also ^{23}Na MRS
spin-echo techniques 162–3
spleen, MRS studies 135
status epilepticus (SE)
  DWI 11
  ^{1}H MRS 8–9
  ^{31}P MRS 8
stimulated echo acquisition method (STEAM) 34, 37, 39, 46
  muscle 61
stimulated echoes 46, 162
stroke
  blood flow measurement 7
  brain oxygen consumption 7
  choice of NMR measurements 2
  DWI 6, 8
  lactate evaluation 3–6, 145
  MRS 3–8, 144–5, 150–1
    ^{1}H MRS 3–6, 145, 146
    ^{31}P MRS 7, 8, 145
    POCE 19–20
surface coil localization 36, 37
  heart 76
  muscle 36, 37, 60
  tumours 212

testicular tumours 226
topical magnetic resonance (TMR) 60
transplantation 175–210
  bone marrow 202–3
  cornea 202, 203
  heart 84–5, 197–200
  kidney 179–90
    clinical studies 184–6, 187–8
    dysfunction 186–990
    high–resolution urine studies 188–90
    organ preservation 176, 183–4
    viability assessment 181–3, 184–5
  liver
    organ preservation 192–4
    rejection 197
    viability assessment 190–7
  organ viability assessment 178, 181–3, 184–5, 200–1
  ^{13}C MRS 194, 202

cornea 202, 203
^{1}H MRS 194, 202
kidney 181–3, 184–5
liver 190–7
^{31}P MRS 178–9, 181–3, 192–7, 200–1
pancreas 200–1
pancreas 200–1
preservation of organs 176
  fluids for 177–8, 183, 184, 193–4, 199
  heart 197–9
  kidney 176, 183–4
  liver 192–4
  ^{31}P MRS 183–4, 192–4
  tissue damage 176–7
small intestine 203–4
tumours 211–46
  bone marrow 227–8, 236
  brain 13–14
    diagnosis 149–50, 217–22
    therapy monitoring 144, 149–50, 230, 231, 232, 235
  breast, diagnosis 222–3, 224
  ^{13}C MRS 212–13, 217
  diagnosis 211–12, 217–28
  drug monitoring 236–8
  ^{19}F MRS 215–17
    drug monitoring 236–8
  ^{1}H MRS 212–13, 214–15
    diagnosis 212, 220–2, 223, 224–8
    therapy monitoring 235–6
  kidney 226
  liver 125–32, 223
    metastatic tumours 127–32, 215–16, 236–8
    response to therapy 132–4
  localization techniques 212
  musculoskeletal system 226–7, 230, 235
  neck region 226
  nuclei for in vivo MRS 212–17
  ^{31}P MRS 212–13, 214
    diagnosis 212, 218–20, 222–3
    liver 125–34, 223
    therapy monitoring 132–4, 228–35
  prostate 224–6
  spleen 135
  testis 226
  therapy monitoring 228–36, 231

urogenital tract tumours 224–6

voxel localization 36–8
  brain 141
  heart 77–8
  muscle 61
  tumours 212
  see also names of specific techniques

water suppression pulses, chemically selective 46